# BIGFOOT RESEARCH
# The Russian Vision

Dmitri Bayanov

Compiled by Christopher L. Murphy
Edited by Roger Knights

ISBN 978-0-88839-706-5


Second printing 2011

**Cataloging in Publication Data**

Bayanov, Dmitri
Bigfoot research : the Russian vision / Dmitri Bayanov ; compiled by Christopher L. Murphy ; edited by Roger Knights.

Includes bibliographical references.
Issued also in electronic format.
ISBN 978-0-88839-706-5

1. Sasquatch. I. Murphy, Christopher L. (Christopher Leo), 1941– II. Knights, Roger III. Title.

QL89.2.S2B39 2011 001.944 C2011-901042-9



Printed in China — Fpoint International Development Ltd.

*We acknowledge the financial support of the Government of Canada through the Canada Book Fund for our publishing activities.*

Published simultaneously in Canada and the United States by

**HANCOCK HOUSE PUBLISHERS LTD.**
19313 Zero Avenue, Surrey, BC Canada V3S 9R9
(604) 538-1114 Fax (604) 538-2262

**HANCOCK HOUSE PUBLISHERS**
1431 Harrison Avenue, Blaine, WA, USA 98230-5005
(604) 538-1114 Fax (604) 538-2262

***Website:* www.hancockhouse.com**
***Email:* sales@hancockhouse.com**

***Bigfoot*** is a familiar word today around the world. Since the middle of last century it indicates a mysterious giant primate of North America, who is also called Sasquatch. The name has since been borrowed and applied to similar beings in other parts of the globe. Bigfoot research is the business of a scientific discipline – ***hominology***, born in Russia half a century ago. The Russian vision of this research is distinguished by at least three peculiarities: it is based on the combined evidence regarding these primates obtained in different parts of the world, such as Russia, North America, China, and Australia in past centuries and modern time; it regards these humanlike beings as relict hominids (hominins by latest primate classification), i.e., the closest relatives of modern man, *Homo sapiens* (amazingly, the bottom line is that bigfoots may even be humans); and it firmly takes the existence of these still enigmatic bipeds for a biological fact, not a popular myth or a scientific hypothesis.

The purpose of the book is to substantiate these views and claims. The main philosophic question posed by it: What is it to be human?

The author, Dmitri Bayanov, is science director at the International Center of Hominology, Moscow, Russia. His previous books on this subject in English are *In the Footsteps of the Russian Snowman*, 1996, and *America's Bigfoot: Fact, Not Fiction*, 1997, published by *Crypto-Logos,* Moscow.

*Dedicated to the memory of*
*my professor of hominology*
*Boris Fedorovich Porshnev*
*(1905 – 1972)*

But if any one of them is real then as scientists we have a lot to explain. Among other things we shall have to re-write the story of human evolution. We shall have to accept that *Homo sapiens* is not the one and only living product of the hominid line, and we shall have to admit that there are still major mysteries to be solved in a world we thought we knew so well.

— John Napier, *Bigfoot*

# Contents

# Introduction

This book is a collection of my writings in hominology – the study of living non-*sapiens* hominids whose existence is denied by orthodox scientists. As for billions of laymen on earth, the subject is simply ignored. Only a tiny minority pay attention to it, merely out of curiosity. Not that there's anything wrong with curiosity. As my late American friend George Haas said, "A mark of simian is curiosity and we are simian folk too... Curiosity leads to information, and information may lead to understanding."

As for me, I am one of the few people in the world who pursue the subject for scientific and philosophic reasons. What has philosophy to do with living non-*sapiens* hominids? Philosophy's prime subject is Reason. Reason in the Cosmos is still a matter of conjecture. Reason on earth is embodied in man, *Homo sapiens*. So, from antiquity to modern times, philosophers have sought to understand how reason fares in man's body, how mind interacts with matter. Hence the never-ending interest of philosophers in self-knowledge and the nature of man.

We see this in ancient Greece in Chilon's dictum "Know thyself," favored by Socrates, and in the aphorism of Protagoras: "Man is the measure of all things." In the 18th century, the motto "Know thyself" was invoked in Latin (*Nosce te ipsum*) by Linnaeus in his *Systema Naturae*, wherein he dared place *Homo sapiens* side by side with *Homo troglodydes*, and both together with apes and monkeys in the single Order of primates. Earlier, in the Renaissance, human qualities were aptly described by the greatest playwright of the epoch:

> What a piece of work is a man! How noble in reason!
> how infinite in faculty!
> In form, in moving, how express and admirable! in action
> how like an angel!
> In apprehension how like a god! the beauty of the world!
> the paragon of animals!
> And yet, to me what is this quintessence of dust? man delights
> not me; no, nor woman neither.
> (*Hamlet*, II.ii)

Why "delights not me"? Because:

> ... man, proud man,
> Drest in a little brief authority,
> Most ignorant of what he's most assured,
> His glassy essence, like an angry ape,
> Plays such fantastic tricks before high heaven
> As make the angels weep.
>
> (*Measure for Measure*, II.ii)

Can man be "the measure of all things" if his nature is so contradictory? Shall we measure all things by a man like Adolf Hitler, or like Albert Schweitzer? And why is man's nature so contradictory and polarized? Some insight is provided by Julian Huxley (1887-1975), biologist, humanist philosopher, and director-general of UNESCO (1946-48). He was the grandson of Thomas Huxley, who was Darwin's staunchest supporter and follower. Said he: "And do not let us forget that man is extremely young. Not only on the evolutionary time scale, but also in the sense that he is not yet fully human. We have to plan for the full humanization of man. As a new dominant type, he is still very imperfect, and so are all his cultures and achievements, including our own." (*The Human Crisis*, 1963.)

What does it mean that man is "not yet fully human"? What exactly is the difference between "human" and "non-human"? Can we plan for the full humanization of man without sufficient knowledge in this respect? Obviously not, and that's where knowledge of living non-*sapiens* hominids should come in handy.

Philosopher Boris Porshnev, the founder of hominology, referring to the existence of these still enigmatic primates, came up forty years ago with a thought-provoking article, "Is a scientific revolution possible in primatology today?" He said therein: "Some say that modern man does not carry in his make-up any vestiges of his origin. But what if he does? The posing of this question alone sheds new light on the perspective of the possible scientific revolution." (*Questions of Philosophy*, 1966, No.3, in Russian.)

Such are, very briefly, some of the philosophic and scientific aspects of hominology. What's the problem then? The problem is that these priceless hominids stay stubbornly beyond the physical embrace of science, in spite of the fact that their presence is reported on all continents except Antarctica. This elusiveness is used by the scientific establishment to deny their very existence. These hominids manage to keep just out of the focus of science. How can this be? The answer is similar somewhat to the reason why the most notorious terrorists manage to keep for years at large despite all efforts to bring them to justice – they are hard to find.

The difference, however, is that millions of dollars are spent on the pursuit of terrorists, while hominologists must fund their investigations themselves. Their means are simply grossly inadequate for the practical solution of the problem. Pyotr Smolin, chief curator of the Darwin Museum in Moscow, emphasized the difficulty with a Russian proverb, "The elbow is close at hand but you can't bite it." And here's John Green's opinion of the matter: "... of the billions of research dollars and millions of man and woman hours of scientific talent, hardly a dollar or an hour is devoted to this quest. Why that should be so is, to me, the most intriguing mystery of all." ("Historical Overview and Basic Facts Involved in the Sasquatch or Bigfoot Phenomenon," *Journal of Scientific Exploration*, Vol.18, No.1, 2004.)

An eloquent solution of the mystery was offered by John Darnton in his science-fiction novel *Neanderthal* (1996), which was inspired by the saga of real hominoid searches in the past decades. In the novel, a skeptical scientist asks his colleague, who is engaged in the quest of extant Neanderthals: "If it's all there, as you say, how come no one but you and a few other nuts have heard of it?"

The colleague answers:

> Because it's disreputable. It's crackpot. It goes against the grain. Do you have any idea how vicious the scientific establishment – guys like you–can be when something threatening comes along? It's like any bureaucracy with a vested interest in the status quo, only worse. If a new theory surfaces that contradicts accepted wisdom, it's shot down–bang! – as soon as it's picked up on the radar. God forbid it should penetrate and get through to the masses!
>
> If it's only mildly threatening, it's subjected to ridicule. Journals weigh in, academics scoff, the popular press writes funny stories. But if it's something revolutionary like this, they play hardball and it gets the full treatment. Careers are ruined, people are run out of town, nothing appears in print. No one wants to look foolish. (...) The point of the story is the way the establishment reacted–the way it always reacts. It prefers to blot out something for which it has no ready explanation. It's an old story, older than Galileo. Science will turn to superstition and torture to defend its right to be wrong. (p. 57).

John Darnton knows what's what when writing about scientists with a vested interest in the *status quo*, for he was (and maybe still is) a reporter and editor for *The New York Times*. And if Julian Huxley is right, then mainstream scientists, like the rest of humanity, are also "not fully human."

But the explanation is not quite simple as that. Science is "a never-ending search for truth" (Ann Druyan), but it is not a straight and steady process. Zigzags, slowdowns, and standstills are also common. As is shown in this book, hairy "wild men" have been known throughout history, but from the middle of the 18th century to the middle of the 20th, all references to them were regarded as legendary by European scientists. The reason for this correlates with the general status of evolutionary biology, primatology, and anthropology. Modern civilization owes its existence and technological feats not to these disciplines but to the queen of sciences – physics. When using a cell phone, working at a computer, or flying aboard airliner, I feel great pride and gratitude for the human intellect and human hands that make such marvels possible. But then my thoughts move to the price we pay for technological progress, and the glad feelings promptly evaporate.

A wit once said that "progress is the development of more machines to provide more people more time to be bored in." Ancient civilizations flourished on the labor of slaves and collapsed because of slavery. Modern civilization thrives on machines and is bound to collapse because of injustice and violence, not so much towards people, as towards Nature. More and more people realize that unchecked mechanical progress is suicidal and leads to catastrophe.

That is why Julian Huxley titled his work "The Human Crisis" and called for man to be "a partner with Nature, not the lord or the conqueror of Nature." He advised to change the set of our minds, to re-think and re-formulate our values. He wrote that nationalism is no longer ethical, consumerism is not ethical. He said it's more important to explore the inner world of the mind than the world of outer space. He said education is very low on any list of priorities today. He held that "the most important sciences today are [i.e., should be–D.B.] not physics and chemistry and their applications in technology, but evolutionary biology and ecology and their applications in scientific conservation." To sum up, he believed "The humanist revolution is upon us."

What has all this to do with hominology? It shows the conditions under which the discipline was born and is developing. If primatology and anthropology, with derivatives, are not yet favored disciplines, how can hominology be? On the other hand, it is possible that no discipline but hominology will give impulse and impetus to the revolution envisaged by the eminent humanist.

Let us give thought to the fact that there were no unemployed in prehistory, that pre-*sapiens* hominids weathered the Ice Ages, and that they probably never suffered from boredom, the curse of modern man. I also muse sometimes that relict hominids have survived just to teach us to be not the conquerors but the partners of Nature.

I include these thoughts in the introduction to acquaint the reader with a relevant side of the author's philosophy. And I beg the reader's pardon for repetitions of my views and concepts in different works offered here – repetitions that came about because I wrote on different occasions and for different audiences, wishing to put across what I believed to be certain truths. I find some justification for this in the sayings, "Repeating is the mother of teaching" and "Truth does not pale from repetition." But if you dislike repetitions, just skip them when you come across them.

In conclusion, I express the deepest gratitude to Chris Murphy and Roger Knights for investing lots of time and effort in preparing this book for publication; to Christian Le Noël and Nikolai Startsev for adding some historic hominid images to the collection illustrating this book.

One final remark: The book itself is only an introduction and prelude to the subject, while the live show is yet to come.

*Moscow, October 2006*

SECTION 1

# Furry Fellows in the Mirror of Fantasy

## Goblins and Brownies in the Flesh

### *Part I.*
### How I Peeped into the "Goblin Universe"

As a young man I devoted much thought to the plight of my countrymen and people in general. Along with others I thought that so-called human nature had a lot to do with the destiny of mankind. Human nature is the result of man's origin and evolution; hence I became keenly interested in anthropology and problems of anthropogenesis.

In 1964 I read an article by Professor Boris Porshnev, claiming the present-day existence of Neanderthals. The claim greatly surprised and intrigued me. So I met the professor, read his voluminous book on "relict hominoids" (1963), and found his arguments weighty and persuasive. What remained to be checked, I thought, was the veracity of purported eyewitnesses to whom Porshnev referred, or rather the very existence of such witnesses. Their alleged presence in the Caucasus was most incredible. It sounded no less fantastic than, say, relict hominids in California! Yet Porshnev's colleague, French-born Marie-Jeanne Koffmann (usually called simply Zhanna in Russia), was investigating the matter and collecting eyewitness accounts right in the central north Caucasus Soviet republic of Kabardino-Balkaria. So in the summer of 1964 I joined Koffmann's expedition in Kabarda, where the object of her search was called "almasty."

The expedition became an eye-opening event in my life. It turned out that people claiming sightings and repeated close encounters with the almasty did really exist, and not only among the natives of the Caucasus, but also among newcomers and strangers to local customs and traditions. With great amazement I became suddenly conscious of a tremendous "knowledge gap" in science: marvelous and priceless information was in the possession of lay people and absent among scientists. And it was a time of "legend come to life" for me, because many things which I had taken for mythology turned out to be stark reality.

To my surprise, the local people saw nothing unusual or surprising in the existence of almasty; on the contrary, they were puzzled by our ignorance and surprise. Their explanations of the phenomenon ranged from purely natural to purely supernatural. One local man said, "There are wild goats, wild sheep, wild pigs; why not wild people?" On the other hand, an old Kabardian Moslem gave us this explanation: "All around us are invisible ghosts and spirits. When an evil spirit grows old it becomes visible and turns into a shaitan [devil]. That's how almasty comes about." "A shaitan-pensioner," wisecracked one of our group. "That's right," said the old man without a hint of a smile.

I was surprised even more upon hearing that the notion of "devil" was applied to the almasty by ethnic Russian witnesses, unfamiliar with the local folklore and mythology. A Russian man told me that when he was making hay and went to a nearby lake for water, he came across a tent of reeds from which hairy legs stuck out. Then a hairy wild man appeared from the tent and walked away. The witness had never heard the word "almasty" and used the Russian word "chort" (devil) to indicate what he saw.

Another ethnic Russian, livestock specialist Nadezhda Serikova, related a dramatic episode of her encounter with almasty in the winter of 1956. Having arrived in the Caucasus from central Russia, she rented lodgings in the Kabardian village of Zalukokoazhe and one night was frightened nearly to death by an almasty who quietly entered her room. When she noticed him the creature was squatting beside her bed and seemed ready to jump at her. The young woman became paralyzed with horror but managed to utter, "Whence you here?" At which the intruder dashed out of the room, leaving behind a choking stench.

The rest of the night Nadezhda could neither sleep nor move. In the morning, her Kabardian woman neighbor, surprised that Nadezhda did not go to work, came in and asked, "What's the matter?" "I saw the devil at night," answered the Russian woman. After hearing the explanation, the Kabardian woman said, "Don't you worry. It's not really a devil. It was an almasty. He won't hurt you. He is fed by a family in the neighborhood and stays in their lumber-room in the daytime."

Nadezhda did not know what "almasty" means and asked the Kabardian to explain. The latter tried to recall the exact Russian word for that kind of almasty and soon remembered: "domovoy," which literally means "domestic one," but is usually translated into English by dictionaries as "brownie."

The witness gave me the most detailed description of the creature, leaving no doubt in my mind that what she had seen was neither a hoaxer nor an ape that had run away from a circus. Nadezhda was not religious (which was not unusual in the atheistic Soviet Union), so I asked her why on earth she took an almasty for a devil. She answered, "I had never seen such a creature, but when I was a little girl my grandmother taught me to pray and used to say that devils will punish people who do not pray. So, seeing that creature, I immediately remembered grandmother and her words about devils."

That was my introduction to the reality of the "devils" and "brownies." The story of an almasty living in a lumber-room was confirmed to

Expedition leader M. J. Koffmann (left) and your author sporting Caucasian shepherds' cloaks, beside Koffmann's mini-car, the ever-ready prize for almasty. Village of Pervomaiskoye, July 1964.

me by other villagers besides Nadezhda Serikova, but my attempts to resolve the problem by asking the owner of the lumber-room himself came to nothing. He denied any knowledge of almasty. Neighbors can spread rumor and gossip a lot, but coming to grips with actual contactees is a different matter. Following ancient traditions and taboos they flatly deny any contacts with homins and even their very existence. If you see a group of Kabardians on a village street and start asking them questions about almasty, they first look at you in surprise and then women start leaving the group, as if you said something indecent.

There was also a rumor of an almasty fed by a family in another village, that of Pervomaiskoye. It was confided to Koffmann by her Kabardian friends, but the family in question was not Kabardian, but of a different nationality, namely Karachai. Relatives of Koffmann's friend Muhaddin (Misha for short) lived in Pervomaiskoye and informed him that their next-door Karachai neighbors were feeding a male almasty that stayed in the daytime in the loft of their house. The creature had been sighted as recently as a month before. The Karachai family consisted of four sisters whom Misha referred to simply as "the girls," although they were not young at all. They had fine names: Beecha, Jaga, Batyk and Mariam. The latter was a Communist Party member and Deputy of the local Soviet, a very prestigious distinction at the time.

When Zhanna Koffmann got wind of that she said to Muhaddin: "Misha, help me get that almasty. Tell 'the girls' the State will give them a lot of money for it. And I will give you my car if you pull off the deal." She had a Zaparozhets mini-car and it was then an attractive offer. So Muhaddin discussed the matter with his relatives, who were first appalled by the idea; but when he said to them, "Look, it's not a Kabardian but a Karachai almasty that I propose to sell," the relatives consented. Then Misha went to Koffmann and said, "All right, Zhanna, I will help you get a Karachai almasty. Let's go to Pervomaiskoye."

In high spirits and full of hope, Muhaddin, Koffmann, I, and two other members of our team speeded up to Pervomaiskoye. It seemed that for the first time ethnic frictions and rivalries prevailing in the Caucasus would serve a useful purpose. We stopped over at the house of Muhaddin's relatives and he went to negotiate behind closed doors with "the girls." He was met by the Soviet Deputy and Communist Party member Mariam and said to her, "My relatives and your next-door neighbors have told me that you've been having a hard time keeping that almasty of yours. Zhanna [her name was known to everyone in Kabarda] has come to Pervomaiskoye with me. She says you will get big money, enough for the four of you in old age, if you turn over your almasty to the State."

Zhanna, ever tinkering with the car engine, after rough rides on unpaved mountain roads, Kabarda hills in background. Expedition base at village of Sarmakovo, Kabarda, June 1964.

Mariam hesitated and went to ask her three sisters, who answered with a unanimous and unconditional "nyet."

Despite these setbacks, on my next, 1965, expedition to the Caucasus, I was still hopeful of a chance to present an almasty to science by dint of an obliging contactee. It was with this hope that I made the acquaintance of a tall and robust Kabardian named Anen Psonukhov, aged 52, a resident of Zalukokoazhe. Several villagers confided to me that Anen "cohabits with an almasty woman." To prove that it was not an empty rumor, they made the following points:

1) Anen is a bachelor; 2) He lives alone and never allows anybody to enter his home, the windows of which are always shuttered and curtained. Once, when some men tried to enter his home, Anen, a pitchfork in hand, chased them away; 3) His late mother was known to be in contact with almastys; 4) A female almasty was seen in and around his orchard, most recently that very summer.

On hearing that, I recalled the words of a young Kabardian named Pate, from the village of Sarmakovo, who gave me an account of his two quite realistic and credible sightings, and then added that a friend of his was cohabiting with an almasty.

"How come?" – "Yes, she visits him three times a year. He has four children by her."

"Where are they?" – "They stay with her in the wild."

"How is he dating her?" – "By means of little sticks." (?)

Koffmann (standing behind the cart) with Kabardians of Sarmakovo village, members of Ali Kardanov family who rented rooms for expedition base, Kabarda, May 1974. Old quadruped, standing in front, belonged to Koffmann and happened to be called Fox.

"So she doesn't speak?" – "She can say one word in Kabardian."

"Which?" – "Give!"

Quite a useful word for an almasty, I thought, although I doubted very much that part of Pata's story. In my diary it was marked "fantastic" and I thought it unworthy of investigation.

The Anen case gave me a second chance. Provided with his home address, I went down to see for myself. Unlike other houses, facing the village street, his old home was hidden in the middle of a large and densely overgrown orchard at the very end of the settlement, next to the cemetery and a pasture. For the next three days I would stubbornly come to that place to "lecture" the alleged contactee. And for three days, against all the strict rules of Caucasian hospitality, Anen would fail, not only to invite me inside his home, but even inside his garden. Every time I had to call him out and talk to him in the street. I must say that in all my subsequent travels I have never experienced such an inhospitable stance by a local man.

The gist of my talk was the importance of almasty for science and the great reward to anyone who would help with the discovery. The gist of his retort was that almasty was nonexistent, it was nothing but an invention by those who spread rumors and slander about him. Our conversation went as follows:

"Why don't you let people in and prove them wrong?"

Expedition members, in search of almasty bones, heading to a huge crack in rock face. Second in line, Bayanov; third Koffmann, July 1964.

"Because I live alone, my rooms are untidy. I don't want others to see that." (Yet the clothes on Anen were clean and tidy.).

"All right, you haven't seen an almasty, but people say they have seen one and in your very orchard. So let me stay there at night and maybe I'll happen to see it, too."

"No, that would be bad."

"Why? Please explain!"

Silence from Anen. By the way, other local people never refused to allow me to stay day and night in their gardens and orchards after a reported sighting there. After three days of hard "bargaining," it became clear to me that if Anen was really in contact with a female almasty, he would not avow it for any riches in the world, and not only because of the traditional taboo but because he was said to be cohabiting with her.

If victory proved as elusive as the almasty itself, then at least I managed to collect some interesting local beliefs regarding almasty-human contacts. A young woman confided to me what she heard from the old men: "Almasty man is less likely to become attracted to people. He is tetchy and if offended would leave the homestead. Almasty woman, on the contrary, is tender and attached to her human master, and if she becomes his wife, she does not stand another woman in the house. If her master dies, she weeps bitterly."

I was told that if an almasty regularly visits a homestead, the dogs get accustomed to it and don't bark. One man told me: "Almasty would never show up to a hero or a coward, only to one in between. Why? Because the hero will shoot it and the coward will get sick."

Almasty is believed to be capable of "mind reading" at any distance: "As soon as you leave Moscow for the Caucasus, he already knows it."

I met an 80-year-old man, Ibrahim Bombasir, a Persian, who was born in Mosul (north Iraq). He was taken prisoner by the Russians in World War I, got married in Kabarda and remained there for good. His Russian was very poor, so he told me through an interpreter that his family in Mosul kept a female almasty as a domestic servant. He was then a boy of ten. He said a needle was stuck into her breast and she could not get it out. While the needle remained there she would not leave people. She asked everybody to get it out. I inquired how she was asking that and was told: "In Arabic." He also added something that impressed him as a young boy but sounded incomprehensible to me, probably because of inadequate translation: "When she sleeps blood [is seen]." Later I was told by Kabardians that the almasty sleeping place on the floor is stained with blood. Ah, the menses, I guessed.

According to legend, Kabardians have a different ploy to keep the female almasty servant from leaving her job. It is necessary to tear out a hair from her head and hide it. She will remain with people until the hair is found. The rest of the tale is usually this: While the people were away the almasty remained with their child and made it show her where the hair was hidden. Having gotten it, she threw the child into the boiling water (or milk) and made away.

Another crime by almasty was revealed to me as a fact. A man told me that his grandfather had gone to bathe in a river and never returned. Almasty killed him. "How? Why do you think it was almasty?" - "People said almasty tickled him to death." I was skeptical, but later, studying published folklore, became reconciled to the idea of tickling-to-death homins. (See *In the Footsteps of the Russian Snowman*, pp.173-74.)

I was told that, having spotted people, almasty would stand still, sometimes for a long time, and would never immediately run on a surprise encounter but would first step quietly backward. And that viewed from behind an almasty seems to be without arms, because they are hanging in front. Nearly all locals, including chil-

Author holds a widely available item of almasty food.

dren, when asked to describe almasty, would first of all say, "Their eyes are different from ours," and put two fingers of both hands in a vertical, not horizontal, position to their eyes. The exact anatomical meaning of this gesture remains a riddle to this day.

There were almasty sightings by the locals in their gardens and orchards at the very time of my presence in Kabarda, and what impressed the eyewitnesses was the unbelievable speed of running by almasty, especially young ones, when they chose to run.

I should not create the impression that collecting relevant information was always simple and easy. Looking for eyewitnesses, you are referred to someone else, and that person would deny that the sighting occurred recently, as you were first told; it happened long ago and most details were forgotten. Once I was directed to a witness who said to me:

"Yes, I saved an almasty from the dogs on a pasture."

"Please describe it." - "I can't."

"Why?" - "I didn't look at it."

"Why?" - "Was not interested."

There is not enough space in this paper to describe all I learned on my first two expeditions to the Caucasus; for example, the story by an ethnic Russian who claimed to have seen in 1928 or 1929, when he was young, a hairy, stocky, and swarthy three-year-old boy, born by a woman allegedly kidnapped by a "bear." According to the man, the boy was taken from his mother by the authorities and sent to an institution in Moscow. True or not, it would be worthwhile to examine the records of corresponding institutions not only in Russia but across the world, with an eye to their possible hairy inmates.

As a result of those two expeditions I came to the following four conclusions: 1) Almasty is a reality; 2) Almasty has not matured for civilization; 3) Civilization has not matured for almasty; 4) Porshnev was correct in his theory that the so-called "popular demons," such as "devils," "goblins," "brownies," etc., etc., were a reflection of what he termed "relict hominoids" in the flesh. "Your snowman is nothing but a wood goblin" [i.e., a mere superstition] charged his critics. "Yes, a wood goblin," answered the professor, "only the other way around. The wood goblin is a snowman, i.e., a paleoanthropological relic."

Back in Moscow (and here I reiterate what I told Craig Heinselman in our 2001 interview), I plunged into reading literature on folklore, demonology, and the history of religion. I was fascinated by what opened to my eyes, my mind having already been opened by Porshnev's theory and what I had learned in the expedition. It became clear to me that folklore and demonology, or what anthropologist John Napier called the Goblin

Universe, is the richest source of hominology, very realistic but totally misunderstood and misinterpreted by academic specialists on folklore and mythology. Soon I came up with a work whose title could be translated into English as *In Defense of Devilry*. The work was never published in Soviet years and no folklorist ever agreed to cooperate with me, despite my friendly approaches.

When the country's political situation began to change, I enlarged my original work, changed the title to *Wood Goblin Dubbed Monkey: A Comparative Study in Demonology*, and after addressing in vain many publishers, at last succeeded in finding one who published it in 1991. I sorted out in it volumes of published folklore of the many peoples in the Soviet Union, focusing on the most realistic descriptions of the appearance, behavior, and habits of their "demons."

Folklore not only supports what we learn from eyewitnesses, but provides details and particulars that I never heard from them, because it contains observations and memories amassed and compressed over hundreds of years.

In a recent message to me, Keith Foster, a keen Bigfooter from Colorado, refers to Theodore Roosevelt's book, *Wilderness Hunter* (1893), to the effect that "Roosevelt's native companion did not want to go into a certain area for fear of the 'devils' there. Roosevelt called them 'forest hobgoblins.'" I am glad to see such similarities in the "demonology" of Russia and America.This is yet another feather in the hat of hominology.

## *Part II.*
## Wood Goblin Dubbed Monkey: A Comparative Study in Demonology

The following is a summary of my book in Russian on folklore and demonology as seen and analyzed from the viewpoint of hominology. Also discussed are some relevant points that came up after the book's publication in 1991. First some words about the title.

Folklore informants often compare wood goblins and other folk demons with apes and monkeys. In the Chuvash language, of the ethnic people in the Volga area, one and the same word, "arsuri," means both wood goblin and monkey, even though monkeys are native neither to the local woods nor to the whole of Russia. A Chuvash folklorist, the author of a scholarly work on the arsuri, finds this circumstance very strange and inexplicable, but for the hominologist the matter is no riddle at all. I think the word arsuri for the wood goblin is the ancient and original term, and, subsequently, when the Chuvash became aware of foreign arboreal animals resembling their goblins, they applied to those animals the name of their goblins. Another name for their wood demon is "upate" which translates as "half-man."

It is quite probable that the Greek word "pithekos" (ape, monkey) was also originally applied to the homin, because etymologically it is connected with the Latin "foedus" (abominable), the Lithuanian "baisus" (horrible) and the Russian "bes" (devil). H. W. Janson, in his book *Apes and Ape Lore in the Middle Ages and the Renaissance,* writes that, "the ape was to be viewed as the kin of demons and monsters, rather than as an 'ordinary' animal…" (p. 75). According to primatologist Eman Friedman, it was the presumed demonological connection of apes and monkeys that so delayed the development of primatology, which acquired a scientific basis only in the 19th century.

Here is another example from my book. An old hunter in the backwoods of Siberia told a scholar, "I don't know if monkeys exist or may be made up. But I have seen the leshy [Russian wood goblin] more than once with my own eyes."

Why did he mention monkeys in this case and doubt their existence? Because he had seen them only in pictures and they reminded him of the creature seen more than once in the flesh. The scholar, Professor K. Platonov, in his book, *Psychology of Religion* (1967), cites the old hunter's words as an example of "outright superstition."

Academic folklorists and demonologists refer to the "heroes" of their books, i.e., "devils," "goblins," "brownies," etc., by such names as "fabulous beings," "creatures of fantasy," "irreal characters," "mental constructions," etc. Accordingly, they focus attention on the fabulous and imaginary. In this respect, the hominologist's objective is the opposite of theirs. To get at goblin biology and ethology he has to amass and sort out as much folklore material as possible (from as many lands and regions as possible), taking into account first and foremost not what folklorists say, but what their folk informants related. That is why it has to be a "Comparative Study."

And the hominologist is not deterred or deceived by folk terminology, for he is well aware that folklore calmly applies "irreal names" to real beings. A telling example of this charming practice is offered by John McKinnon in his book *In Search of the Red Ape* (1974). In Borneo, in 1969, during an outing in the jungle, the author suddenly "stopped dead," amazed at what he saw. It was a footprint so like a man's yet so definitely not a man's that his skin crept and he felt a strong desire to head home. Back at camp McKinnon asked his Malay boatman what animal could make such tracks. "Without a moment's hesitation he replied 'Batutut,' but when I [McKinnon] asked him to describe the beast he said it was not an animal but a type of ghost." It follows that ghosts, spirits, and other "irreal characters" do not necessarily levitate, leaving no foot tracks on the ground.

So how does the hominologist tell reality from fantasy in folklore and demonology? In the same way an educated person tells real things from imaginary ones in fairy-tales and legends. The hominologist is helped in this task by his knowledge of homins who combine both human and simian traits, being however neither apes nor *Homo sapiens* humans. That is yet another reason for a comparative study.

It should also be mentioned that the names "snowman," "relict hominoid," or any eyewitness accounts collected by hominologists are not to be found in my book. It includes nothing but the published material by professional folklorists, ethnographers, and lexicographers. I did this on purpose in order to beat the opposition with their own weapon – by showing that folklore and demonology alone, seen through the eyes of hominology, graphically reveal the reality of homins. When such material is put side by side with the eyewitness accounts, footprints, and photographic evidence collected by hominologists, the positive conclusion becomes irresistible.

## The Biblical Connection

Among historical examples cited in the book, references to the Bible may be of special significance for readers in the West because of the Holy Book's ubiquity in Europe and America. The beings of interest to us are mentioned, for example, in Isaiah 13:21 and 34:14, i.e., in a prophecy against Babylon. The prophet says that Babylon shall be destroyed, turned into a waste land, and "wild animals of the desert" shall come to live there.

Along with such denizens of the desert as ostriches, jackals, and hyenas, the Bible in Russian mentions "the leshy" (wood goblin)! How come wood goblins in the desert? The question intrigued me and begged an answer. In search of it I discovered that the earliest edition of the Bible in Russia (in old Slavic) has "bes" (devil) instead of "leshy" in those verses in Isaiah. I then looked up the Authorized Version of Isaiah in English and discovered "satyrs" in the corresponding places. So I opened Encyclopedia Britannica, 1961, vol. 20, p. 11, and read, in part, the following:

Satyrs, in Greek mythology, spirits half-man, half-beast. [...] In Italy often identified with the fauni. In the Authorized Version of Isaiah xiii. 21, xxxiv. 14 the word "satyr" is used to render the Hebrew "se'irim" (hairy ones). A kind of demon or supernatural being known to Hebrew folklore as inhabiting waste places is meant; [...]. They correspond to the "shaggy demon of the mountain pass" (azabb al-'akaba) of old Arab superstition.

So what did the "hairy ones" alias the "shaggy demon of the mountain pass" alias wood goblins alias satyrs alias devils have to do with the ruins of Babylon? Various translations of the Bible answer as follows: they "will leap about," they "will dance," they "shall call to each other," they "shall cry out to one another." Well, Isaiah would have made a good hominologist, I thought. After all, it was not he who called the hairy ones by such names as goblins, satyrs and devils. He used the term derived from the creatures' biological characteristic.

I then happened to look up the New International Version of the Holy Bible, and what did I see? "Wild goats" instead of "satyrs!" "And there wild goats will leap about...", "and wild goats will bleat to each other." What a leap from reality!

## Folklore on the Origins of Demons

Hebrew folklore has it that God created the se'irim on the Sabbath eve, and therefore did not have time to make them fully human. But Russian peasants had a different opinion on the matter. When the peasant's son inquired, "Daddy, what is meant by the devil, the leshy, the domovoy? What is the difference between them?" The adult peasant answered, "There is really no difference. They say that when God created man, satan was eager to create, too, but no matter how hard he tried he could only make devils, not men. When God saw that satan had already produced several devils, He ordered Archangel Gabriel to dump satan and his goods from heaven. Gabriel did so. The devil that fell on a wood became the leshy (wood goblin), another, that fell on a field, became the polevoy (field

Jewish demon, a "hairy one" in the original Hebrew text of the Bible, and "devil," "satyr," "wild goat" and "he-goat" in various translated Bibles. (Illustration borrowed from the Universal Jewish Encyclopedia, article "Demons.")

goblin), and a third, that fell on a house, became the domovoy (domestic demon, brownie). That's how they came about and got different names. But actually all devils are alike."

In Bielorussia folklorists recorded the following legend:

> Adam and Eve had a dozen pairs of children. When God came to look at them, they showed Him six pairs, and hid the other six pairs under an oak. So, like we come from those six pairs shown to God, they (the demons) come from the other six pairs. Their number is the same as ours, only they are invisible because they are hidden from God.

A Moslem tale, "On the Origin of Almasty, Jinn and Div," relates that Adam and Eve had an argument. He said that children originate from him. She said they originate from her. To resolve the problem, they agreed to abstain from sex for a certain time, keeping their semen separately. From Adam's semen came living men, and Eve's fluids produced creatures that turned into Almastys, Jinns and Divs.

The "shaitan-pensioner" version, related by a Kabardian Moslem and cited in Part I of this work, relates to this topic as well, but it is not included in my book, which refers only to published material.

Less civilized people, living in the lap of nature, had a different and more realistic view on the subject. Thus the Mansi, living in the taiga of Siberia, say that in making people, gods used two materials: clay and larch timber. As soon as people made of larch were produced, they dashed into the forest. Those are "menkvs" (wood goblins). Slow moving beings, made of clay, became ordinary people. Their lifespan is short; arms made of clay, legs made of clay, what's the use of them? If man falls into water, he drowns; if the weather is hot, water comes out of him. If men were made of larch, they would be hardier and wouldn't drown in the water.

There are many other folklore versions of the theme, including the belief that demons arise from dead people who were not buried or were buried the wrong way. What is interesting and important for the hominologist in such tales and legends is people's wish to explain both great likeness and great difference between man and demon, and not the essence of the explanations, arising from fantasy and superstition.

## The Proverbial Connection

Folklorists define the proverb in this way: An apt and colorful expression summarizing people's observations and reflections regarding various sides of real life.

Citing this definition in my book, I note that the proverb has two meanings: one direct and literal, referring to "various sides of real life," and the other indirect and figurative, applied to various episodes and developments of social life. Thus, when people say "It never rains but it pours," or "A bird in the hand is worth two in the bush," or "One shouldn't look a gift horse in the mouth," they use literal, real life meanings in a figurative sense.

So I ask what is the real meaning of the numerous proverbs and sayings of all peoples of the world referring to the devil and other demons. The Russians say, "The devil is not so ugly [or fearsome] as he is painted." The English say, "The devil is not so black as he is painted" and "to paint the devil blacker than he is." The Russians say, "The devil is swarthy from birth, not from the sun." They also say "Brown devil, gray devil, still a devil." Does this not mean that the creators of these proverbs did know the look of the devil?

The Russian equivalent of the English, "Still waters run deep," is "Devils dwell in a quiet slough (pool)." For the hominologist the real meaning of the proverb is quite clear.

The famous 19th-century lexicographer Vladimir Dahl offers other proverbs and sayings reflecting the devil's aquatic preferences. "To be led to the devil, like the devil to the marsh, "Given a marsh, given the devils," "When devils dive nothing but bubbles arise," "A job [a work assignment] is not a devil, won't disappear into the water," "Worms in the earth, devils in the water, crooks in the court, where can a man go?"

Some more sayings from Vladimir Dahl's *Dictionary of the Russian Language*: "You are as big as the devil [or leshy] but still small in the mind," "You are clever and strong but can't beat the leshy," "Leshy is mute but vociferous," "To roar like a leshy", "Infected with the devil's fleas and lice," "The devil brushed himself and lost his brush."

An Arab proverb goes "Azrata min ghoul" (stinking like a ghoul); also quite a familiar sign. A synonym for "demon" in Russian is "unclean spirit." Demons collectively are referred to as "nechistaya sila" (unclean power).

When the Kabardians say "to catch the almasty by head hair," they mean to pull a thing off.

The advice and wish "Go to the devil!" and "The devil take you [him, her]" seem to be international. When a needed person appears at last after a long wait the Russians say, "Where has the devil been carrying you?" Enlightened by the Albert Ostman case, the hominologist knows that the latter saying is a reflection of real life as well.

## Morphology and Manners

Folklore on demons confirmed all I knew about the homin anatomy and behavior and added things I did not know. The demonic beings are hairy manlike bipeds, often bigger and always stronger than man. There are male and female demons, as well as their offspring. A shock of hair is sometimes mentioned on the heads of males, but bald-headed demons are on record as well. Females boast of long-hanging or flying head hair, sometimes disheveled, sometimes brushed.

The Komi people in the north of Russia say their wood goblins have hair-covered ears. One folklore item in Siberia mentions hair on female breasts. And an item on the Chuvash female wood goblin tells of a hair covered body "except the genitalia," which is a simian characteristic!

The hair color ranges from black to white, with lots of browns and reds, and is likened to the fur of animals native to the particular geographic area (reindeer, bear, camel, goat, buffalo).

The attribute of hairiness is present in the local names of demons, from the Hebrew "se'irim," to the medieval European "pilosus," to the Russian "volosatik" and "volosatka" (literally "hairy one" for male and female).

The color of the skin is swarthy, with a reddish, or yellowish, or grayish tinge.

The pointed, cone-shaped head is a usual feature, even reflected in the names of Russian devils and goblins: "shishko," "shishiga," from "shishka" (cone).

The eyes appear big at night when they shine "like stars." Facial features are not, to put it mildly, attractive, since folklore uses the word "muzzle" in reference to a demon's face. Lack of a neck is mentioned in one item from Siberia.

Folklore dwells a lot on the enormous size of a female demon's breasts, calling them "huge" and even "frightening." Mentioned is the size of "about one arshin," an old Russian measure which equals 71 centimeters, or 28 inches. It's noteworthy that hugely hypertrophied breasts, so-called "mamma pendula," hanging down to the middle of the thigh, are sometimes registered in human females, probably as an atavistic trait.

In many cases, from different geographic areas of Europe and Asia, the enormous demoniac mammae are said to be carried thrown over the shoulders. Accordingly, there is an observation in folklore that female creatures carry infants on the back and suckle them with breasts thrown over the shoulders. As for *Homo sapiens*, women with breasts thrown over the shoulders are reported among the aborigines of Australia and Africa.

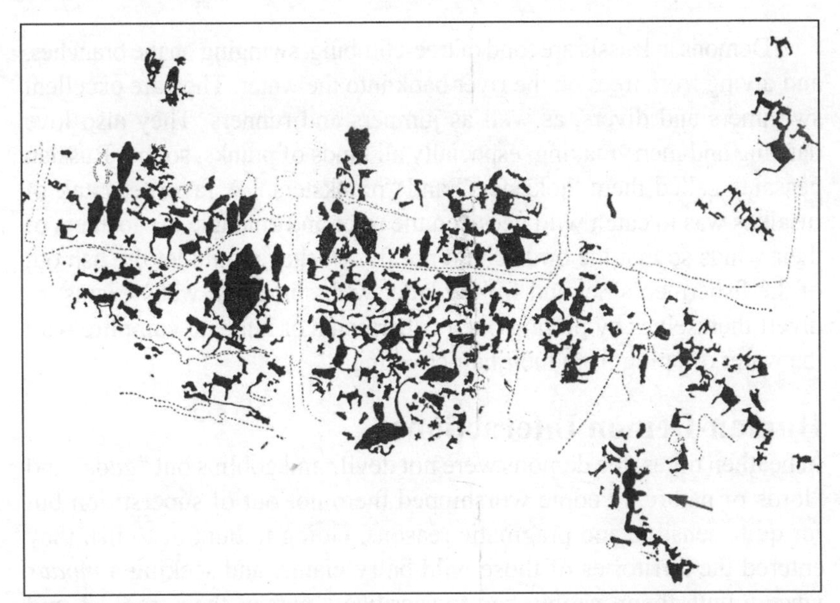

*"Devil's Footprints"* is the archaeologists' name for a set of petroglyphs (rock carvings) in Karelia (in the west of Russia, bordering on Finland). The scene is a marvelous relic of the heathen past, when the homin was worshipped as the lord of nature and offered bountiful sacrifices. He is shown in the lower right corner of the picture, with a number of true homin features, including the exaggerated big foot. The enormous phallus is not a reflection of anatomy, but rather a symbol of fertility and maybe of satyriasis. The ancient artist carved not only the figure of his god but also a set of his footprints coming across a mass of sacrificed animals, all of which represent the true local fauna. I call this picture of petroglyphs "Bigfoot of Karelia." The illustration is borrowed from the book in Russian *Risunki na skalakh* (Pictures on the Rocks) by Yuri Savvateyev, published in Petrozavodsk, capital of Karelia, in 1967.

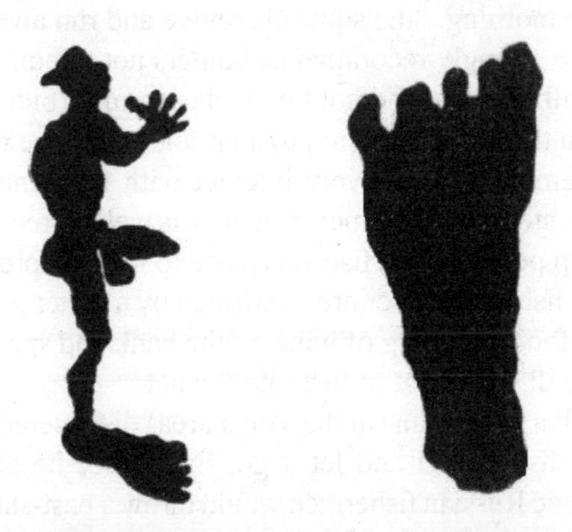

Bigfoot of Karelia at his best and his footprint that well resembles that of his North American counterpart.

Demons in Russia are fond of tree-climbing, swinging on the branches, and diving from trees on the river bank into the water. They are excellent swimmers and divers, as well as jumpers and runners. They also love dancing and merrymaking, especially all kinds of pranks, so that Russian peasants called them "jokesters" and "pranksters." A favorite prank of rusalkas was to catch wild geese on the river and entangle the feathers of their wings so that the birds could not fly. Or they would let the fish out of the fishermen's net and fill the latter with slime and water-plants, or divert themselves by putting out a fishermen's or hunters' campfire with the water dripping from their hair covering.

## Human-Demon Interactions

In heathen times, the demons were not devils and goblins but "gods" and "lords of nature." People worshipped them not out of superstition but for quite sensible and pragmatic reasons. Going to hunt or to fish they entered the territories of those wild hairy giants, and seeking a *modus vivendi* with them, people had to sacrifice a part of their trophies and catches to the homins. That is the origin of religious sacrifices, whose echoes are still reverberating in folklore.

One folklore item from the European part of Russia, cited in my book, says that in olden days hunters "had to prepare gifts for the 'lord of the forest' for allowing them to hunt on his property." In later times the relationship "progressed" and an item from Siberia says that hunters there engaged in barter trade with wood goblins: the latter supply squirrels and get generous gifts of vodka in exchange. It is most remarkable that squirrel bodies are delivered at night and if the hunters fail to skin them before morning, "the squirrels revive and run away."

Folklore strongly recommends hunters not to build their cabins on the forest path of the wood goblin. And custom forbids whistling in the forest and in the home so as not to alert and invite the goblin.

Folk demons also actively interact with fishermen. That homins partake (i.e., steal) of fishermen's catches is well on record, but that they can also help people catch fish was news to me. According to Georgian folklore, all fish in the river are controlled by a water goblin. If a fisherman leaves food and a jug of wine on the bank and speaks nicely of the demon, he will send a lot of fish into the net.

A Mordva fisherman (in the Volga area) discovered a crying goblin child in the fishing net and let it go. Ever since he always had good catches. Ethnic Russian fishermen would throw a bast-shoe into the water and yell: "Hey, devil, drive fish into our net!"

But the demons' greatest contractors were herdsmen. It is reported that in Russia they made secret "contracts" with wood goblins who helped pasture the herd, find lost cows, and protect them from wolves and bears. The service was paid for with food and animals from the herd. Such deals were popular with the peasants, but kept strictly secret because they were viewed as very sinful by the Orthodox Church. It is worth mentioning that in ancient Rome fauni were said to protect herds from wolves, and a celebration was held in their honor on the 15th of February, called Lupercalia.

Bald demon depicted on a Greek vase.

Another kind of interaction and category of homin whom I call "visiting demons" are those who approach human habitation for one reason or a combination of them. The most common is food, another clothes, a third the warmth of the hearth. An item from Tajikistan says that when the children asked their mother to give them more pancakes for supper, the mother answered, "If I give you more, what shall we leave for the adjina? She will come at night, and finding nothing may become angry."

There are stories in Tajikistan that when the cry of an infant is suddenly heard from a barn, it means that a demon has given birth. People give food to her, "she eats, takes the baby, and goes away."

In Georgia, the ancient clan of Naraani was said to have befriended a dev. They "fed him well," leaving food warm in the ashes of the hearth. When the family went to sleep, he would come and have his fill.

If food is not offered, the demons would steal it, all kinds of it, especially vegetables and fruits from gardens and orchards.

As a rule, demons are seen naked, but there are many exceptions, and clothing is the next item of interest for them to come into contact with humans. It is advised, when encountering a goblin in the wood, to offer it bread or a piece of clothing, even a torn-off sleeve if nothing else is available. On record are Ukrainian and Bielorussian songs telling how rusalkas beg human girls to give them shirts, no matter how old or tattered. No wonder, demons usually sport threadbare garments, often worn the wrong side out. As a result, when Russians saw a man in a shirt worn inside out, they used to say: "Look, he is (dressed) like a leshy!"

The leshy were said to approach campfires built by lumberjacks or hunters in order to warm themselves in cold weather, and it is said that

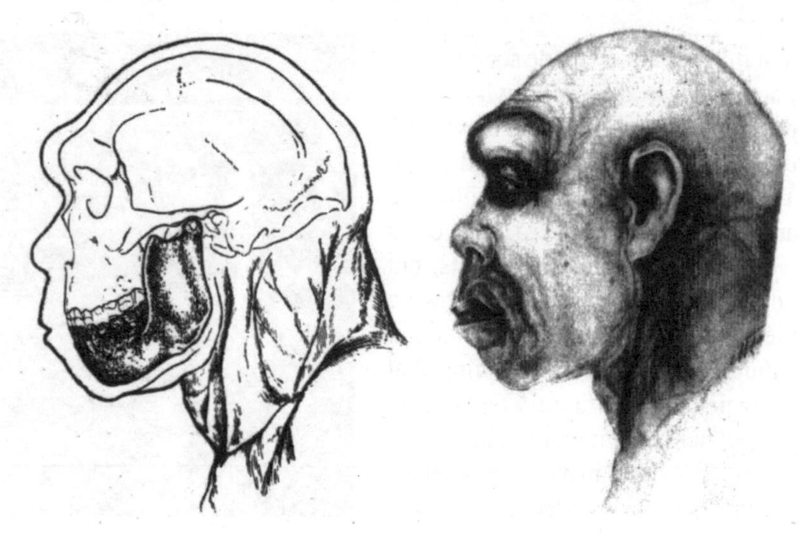

Reconstruction of the head of a paleoanthropus based on the fossil cranium, bearing resemblance to heads of demons.

they "turned away their muzzles," apparently because of the bright light. They also took care that flying sparks did not touch their hair.

Seeking warmth they also entered peasant bath huts and barns for crops stocked there. It is reported that a leshy, festooned with icicles, entered a barn and put out a fire with melting ice. In contrast, in the summer they would come up to a campfire not for warmth but to put it out.

Folklore is insistent that demons love human children. Hiding from adults, they often come in view of children and even play with them when adults are not around. They are also said to calm down crying babies and, inevitably, as a result of such fondness occasionally take human kids with them. In Bielorussia, a wood goblin was "charged" with stealing a cradle with a baby and hiding it in a birch tree. In the Novgorod province, a boy of 13 was kidnapped by a wood goblin. Four years later the boy returned naked and unable to speak.

As mentioned earlier, a Mordva fisherman released a young water demon caught in the net. And there is a Georgian story about an old man who came at night in the wood across a crying boy about 9 years old. He took the boy home and when he and his wife looked at the foundling in the light of the fire burning in the middle of the hut, they realized it was a "chinka" (wood goblin), because he was hairy all over and had red eyes. The old woman began to scold the old man for bringing in a goblin. They tied him up to a bench and showed him to many people in the morning. All agreed that it was a chinka. When untied, the young goblin escaped.

## Their Mortality

There was a time, which lasted far longer than the whole of written history, when humans were in the minority and homins in the majority. Then, thanks to agriculture and animal husbandry, human numbers substantially increased and the ratio began to reverse. It was then that kings and high priests, first in Persia, later in Israel and other lands, forbade the worship of "unclean half-man half-beasts," condemned them as "pagan gods," and introduced monotheistic religion instead of polytheistic. The hairy homin was turned into a "devil" and declared an enemy.

The echo of those events is heard most clearly in the Persian epic *Shah Namah*, based on popular legends and traditions. It describes fierce battles waged by the kings of the country against an enemy called alternatively "divs" and "devils." Though the latter displayed miracles of physical force, they battled with their bare hands against human warriors wielding swords and spears. Inevitably, pagan gods suffered heavy casualties, proving beyond doubt their mortality.

Geographer and traveler Pausanias (2nd century A. D.), in his *Description of Greece*, says that the silenus race must be mortal, since their graves are known. He also says that when satyrs grow old, they are called silenoi.

There are many examples of demon killings in folklore. According to one item from Siberia, a reduction in wood goblin numbers there was due to the appearance of hunting guns. Some tales relate that hunters, having killed a demon, cut off parts of its body, sometimes the head, as souvenirs and valuable trophies. Obviously, encounters with human beings wielding firearms boded no good for "mythical beings" and that is a reason for their legendary seclusion.

There are also plenty of beliefs that demon killers suffer inevitable retribution for the deed. Chuvash folklore intimates that in a village where "upate" (half-man) were killed, population no longer increased. Tatars had similar beliefs, and when they saw a little poor village, they used to say, "Shurale kargagan" (condemned by shurale, the latter word meaning wood goblin). An example from Azerbaijan mentions a hunter who fired pointblank at a "biaban-ghouli," who fell to the ground, then stood up and ran away, leaving behind a bloody trail. After that, the hunter sold his gun and never hunted again. Asked why, he answered, "After that all my children died." A parallel First Nations tale was published in 1929 in Canada by J. W. Burns and reprinted by John Green in *The Sasquatch File*, 1973, p. 11.

Cases of demons imprisoned by humans are also numerous in folklore. A creature, especially young, could get entangled, as already

Assyro–Babylonian demon Humbaba, lord of cedar forests in the mountains, who did not allow people to cut them. According to legend, King Gilgamesh, helped by his befriended wildman Enkidu, fiercely battled with Humbaba and killed him.

mentioned, in a fishing net. To catch migratory birds, the Russians used to hang a huge net on the trees of a forest vista. It happened sometimes that instead of wild ducks and geese, the hunters found a devil in the net. The technical term for that kind of net is "pereves." So there appeared a proverb, "popalsya kak bes v pereves," (caught like a devil in a net).

From Tatar folklore we learn that the inhabitants of a village, tired from the tricks by a shurale (wood goblin) that troubled their herd of horses every night, spread tar on the back of the best horse and by this ploy caught a she-demon who had tried to ride that horse.

But the surest and most ancient method of catching demons was by intoxicating them with wine, of course. In ancient Greece it was used by King Midas, who caught a silenus; in Italy by King Numa Pompilius, who caught a faun. Being so rare and impressive, these events were recorded by legend. The only modification in the method in Russia is that wine is replaced by vodka. A tale from Abkhazia had it that a wood goblin that meddled with hunters' traps was caught only after imbibing a bucketful of vodka.

## The Sexual Connection

Folklore and demonology present this as one of the most prominent factors in human-demon relations. To begin with the ancient world, according to legend the Babylonian King Gilgamesh habituated and befriended the half-man half-beast Enkidu with the help of the priestess of the goddess of love, Ishtar. Enkidu is said to have been shaggy with "hair that sprouted like grain," he ate with the gazelles and drank with the wild beasts at their waterholes. He protected wild animals from hunters, so a hunter went to King Gilgamesh with a request for help. The king recommended that the hunter take a priestess of Ishtar with him to the waterhole and instruct her to take off her clothes, thus enticing Enkidu away from his animal friends. The ruse succeeded and the wildman enjoyed the woman's favors for a week, being gradually persuaded to eat bread and drink wine with the shepherds. He became their friend and helped them by driving lions away from the flocks. Subsequently Enkidu found himself in the palace of Gilgamesh and became the king's best friend and aid in hunting. He also helped Gilgamesh in fighting the monstrous demon Humbaba, actually a wildman in the forested mountains of Lebanon.

Lustfulness was a distinguishing trait of satyrs in ancient Greece. Historian Diodorus Siculus wrote that, "this animal [!] shamelessly seeks cross-breeding." The situation is reflected and recorded in the medical terms "satyriasis" and "nymphomania."

But for readers in the West, most significant and impressive is one more reference to the Holy Bible. Among the commandments by the Lord that Moses gave to Israel was this: "And they shall no more sacrifice their victims to devils, with whom they have committed fornication. It shall be an ordinance for ever to them and their posterity." (Leviticus 17:7, *The Holy Bible*, Douay Version, reproduced from the first edition of *The Old Testament*, printed at Douay in 1609.)

Another translation in *The Holy Bible*, London, 1850: "And they shall no more offer their sacrifices unto devils, after whom they have gone a whoring. This shall be a statute for ever unto them throughout their generations."

A third version, published in *The New English Bible*, Oxford, 1970: "They shall no longer sacrifice their slaughtered beasts to the demons whom they wantonly follow."

Let us note that, according to the Hebrew text, Moses did not use the words "devils" or "demons" in this commandment by the Lord. Again the term "se'irim" (hairy ones) was used, which presented a sticking point for the translators. "Hairy ones," and moreover sacrifices to and fornication with them, called for an explanation; "wild goats" would not fit in this case. So "devils" and "demons" were found to be preferable terms, for who does not know that devils and demons are seducers and perverters of mankind?

Christianity also condemned "pagan gods" for lustfulness. Saint Augustine wrote that fauns and satyrs, "called at present incubuses," have intercourse with women. "This has been testified to by so many people and so positively that it would be insolent to deny this." In the Middle Ages in Europe, many victims of the Inquisition were tortured and condemned to burn for sex with demons.

In Asia, the 12th century Persian scholar Nizami al-Arudi wrote that "the Nasnas, a creature inhabiting the plains of Turkestan, of erect carriage and vertical stature, [...] is very curious about man. [...] And if it sees a lonely man it abducts him and is said to be able to conceive by him. This, after mankind, is the highest of animals…" Modern scholars say the Nasnas is an imaginary creature, a kind of faun.

Sexual relations with demons is a topic present in all works on folklore that I read and referred to in my book. In Tajik folklore, the female demon "pari" seeks the love of a hunter and pays him with wild goats that she sends him in gratitude.

In Chuvash folklore, the female arsuri (goblin dubbed monkey) would run in the wood in front of a man, laughing impudently, showing him her genitalia and beckoning to him. The name "arsuri" is applied by the Chuvash to a shameless woman.

In Circassian folklore it is said that the shaitan and his female partner jinne can be caught. However it is not advisable to catch a shaitan because he will offer strong resistance. Jinne is a different matter. If caught, she can be used as a woman. Sometimes she herself is seeking sex with humans, coming to herdsmen for the purpose.

In Bielorussian folklore there is a beautiful poetic incantation intended for young male peasants in case they are accosted by an enamoured rusalka. It is pointed out that the man should not look at her, but at the ground, and say the following (in my rather inadequate translation):

> Water dweller, wood denizen, wild, unruly and whimsical girl! Go away, get away, don't show up at my homestead! […] I kissed the golden cross and abide by the Christian faith, so can't mix with you. Go to the pine forest, to the forest lord. He has prepared a bed of moss and grass and is waiting for you. You are to sleep with him, not with a Christian like me. Amen.

Here a quote will be in place from *In the Footsteps of the Russian Snowman:*

> The most characteristic trait of rusalkas, known from folklore and poetry to all Russians, is their habit of accosting a young man bathing in a lake or river, and, with much delight and merriment, 'tickling him to death' (i.e., until he drowns). And that is exactly what the 'madwoman' in the case of Turgenev was prepared to do: 'touched his neck, his back and his legs with little cacklings of delight.' Had Turgenev not been 'a splendid swimmer,' Russian and world literature would have suffered a heavy loss from the hands of a rusalka.
>
> The business of tickling, as part of lovemaking, is ascribed not only to Russian rusalkas, but also to their counterparts in the folklore of Tatars, Bashkirs, Kazakhs, etc. (Let's recall a case in Kabarda that I

related in Part I of this work.) And what is most interesting and important, it is also typical of anthropoid apes, as observed by the famous English primatologist Jane Goodall. (…) Returning to rusalkas, the danger of drowning from their caresses was once so real for Russian countrymen that people invented a very simple but potent means of defense: just a pin or needle held at ready while bathing. Folklore has it that it is enough to give rusalka a pin-prick or show her a needle to make her flee. (Ibid., pp. 173, 174).

## Crossbreeding

The basic difference of demons from all real creatures, including apes and monkeys, is their desire of sexual relations with man. Clearly, this circumstance is responsible for their unprecedented and unique role in the history of mankind. A Russian specialist on oriental folklore and the Koran wrote in 1893 about the demons called "jinn": "The peculiarity of their nature is that they can have sexual intercourse with people."

A natural question then is: What comes as a result of such intercourse? Folklore is quite talkative on this score. An item from Siberia: "Sometimes a she-devil cohabits with hunters in the forest and becomes pregnant from them, but she tears the infant apart at its very birth." The Circassian jinne can also kill her crossbreed baby, in case her human husband reveals her presence to his neighbors.

A success story in crossbreeding is reported by Kazakh folklore, telling of a horse herdsman who encountered a female almasty in the steppe and thought, "Be it a shaitan or a human, it doesn't matter." He lived with her and "they had three children born to them."

Bashkir folklore explains the origin of the name of the Shaitan-Kudey clan by the fact that once a brave Bashkir caught and married a female shaitan and their posterity formed the said clan. Nogai folklore notes the rapid growth and unusual strength of the offspring of their legendary hunter Kutlukai and his almasty wife. Their son became a national hero and all Nogai nobility descend from him.

If we give credence in this respect to folklore, then hominology is faced with the question: What is the genetic status of "demons," i.e., homins, in relation to *Homo sapiens*?

"Good" species are not supposed to produce fertile crossbreeds. Still, division into species and subspecies of closely related organisms is often a matter of speculation and agreement. Primatologists are aware of fertile hybrids of different monkey species. Another case in point is the example of wolves and coyotes, considered to be different species. Yet they carry the same number of chromosomes and there exist no genetic barriers to their interbreeding. If not for behavioral differences, which keep them separate, one species would have long ago absorbed the other.

The homin-human situation appears to be similar; the barrier to crossbreeding is neurological and behavioral, not genetic. For these reasons it can be overcome in principle and in practice, but the process has been "invisible" and very protracted.

One more example in favor of this view is a quote from *Essays on Russian Mythology* (1916) by D. K. Zelenin: "People believe that if a rusalka is made to wear the cross, she will become a human being. Such cases are reported from the Vladimir Province, where two boys married baptized rusalkas."

As regards North America, Dr. Ed Fusch reports crossbreeds between Indians and the "Stick Indians" (Sasquatch, "Night People") in *S'cwen'yti and the Stick Indians of the Colvilles* (1992). (Posted by Bobbie Short on her Bigfoot Encounters site and supplied to me by the late Don Davis.)

## The Domovoy

Regarding contact with humans, demons can be divided into those: 1) who avoid any contacts (Russian folklore calls them "free" and "free roaming" beings); 2) who meet with people occasionally (I call them "visiting demons"). Reasons for contact, as already mentioned, can be food handouts or payment for certain services or "barter trade"; warmth of the hearth, sexual contacts; 3) who are staying permanently on a homestead.

The latter subdivision includes a) she-demons "married" to humans and, b) former "visiting demons," grown old and unable to provide for themselves on their own. For obvious reasons, category "3a" is shrouded in great secrecy and there is little information on it. Category "3b" is less cryptic and provides some interesting material. The general Russian term for a demon dwelling in a homestead is "domovoy" (literally, "domestic one"). There are male and female "domestic ones," but most information is about the male.

An 18th-century book, *Descriptions of Old Slavic Heathen Fables*, has this to say about the domovoys: "These imaginary half-gods were called 'geniuses' by the ancients, the Slavs called them 'protectors of places and homes,' while modern superstitious simpletons take them for 'domestic devils.'"

Domovoy is described as an old, sometimes very old, man, hairy all over, and with a pointed, cone-shaped head. He lives in the stable or the cattle-shed, and as noted by folklorists, "deals more with the livestock than with people"; "caring for the cattle is his most frequently mentioned function in folklore."

Russian domovoy. A drawing by artist Ivan Bilibin who studied and illustrated Russian folklore. The drawing was made in 1934 long before the birth of hominology. The domovoy is shown inside a peasant home. Note his coned head and long finger nails

The great 19th-century lexicographer Vladimir Dahl, in his work, *On the Beliefs, Superstitions and Prejudices of the Russian People*, wrote that:

> Generally speaking, domovoy is not a malicious character, but rather a mischievous one. If he likes a person or a person's home, he becomes an obedient servant; but if he dislikes someone, he can evict that person or even cause his death. His service can be like this: cleaning, sweeping and tidying up in the house at night; above all he likes horses, cleaning and brushing them, and plaiting their manes and tails; sometimes he mounts a steed and rides from one end of the village to the other. Sometimes domovoy plays pranks on women, especially if they are silly and loudmouthed, [...]. Results of his pranks can be seen in the morning, for example, all dishes can be found in the slops bucket, stools and benches broken or piled up in a corner of the room. It is remarkable that domovoy does not like mirrors; some think that mirrors can repel him from the room where he likes to play pranks [...]. In some regions nobody would pronounce the word 'domovoy' out of fear; that is why he has so many other names, including the honorary appellation 'grandfather.'

At night domovoy can stroke a sleeping person's face with his hand, which feels furry, and with long cold fingernails. In the winter one can see his tracks in the snow near the stable, but the maker of them is only seen on rare occasions. "Domovoy hates the curious," says Dahl. He describes a special religious procedure which, according to peasant beliefs, helps in seeing a domovoy. The latter is then observed in the light of a candle crouched and hiding in a corner of a cattle-shed or a stable. "Then you can talk to him." In his *Dictionary*, Vladimir Dahl says that, "Domovoy can be seen on Good Sunday's night in the cattle-shed; he is shaggy, but nothing else can be remembered because he knocks out memory."

Still, not everybody's memory is completely knocked out, for other folklorists add certain details to those supplied by Dahl. Such, for example, as "domovoy is about the size of a bear and covered with long soft fur of a dark brown color. But it is impossible to see his figure in detail because he quickly disappears." Another folklorist writes that, "He is like a man of medium height, stooping, wide-shouldered, stocky, covered with long hair (in color can be dark or white or skewbald)." An item from Siberia says that domovoy can be seen "in the image of a man resembling an ape depicted in pictures, but only much more ugly than a monstrous ape."

To avoid coming face-to-face with such a monster, Russian peasants had the custom of repeated coughing when going out into the yard at night.

As for domovoy's vocal ability, Dahl writes that the "grandfather" can tease horses by neighing like a horse. Another folklorist says domovoy can "frighten people at night by yelling or crying." Sounds made by domovoy were believed to be prophetic: if he laughs, it's a good omen, and if he makes "hoo, hoo, hoo!," it's a bad omen.

It was mentioned earlier that domestic demons were also called "protectors of places and homes." An item from Siberia, supporting this belief, goes like this:

> Domovoy does not like it when someone enters his territory without permission. For example, if a stranger comes to a house to spend the night, he should address domovoy with the words, 'Grandfather, let me spend the night here.' Otherwise domovoy can do bad things to the stranger at night; can press him or even throw out of the house. He especially dislikes tipsy strangers. Should a drunken man spend the night in somebody else's home without asking domovoy for permission, the latter would not fail to throw the man out. Such cases are known, and the victims were frozen to death in the yard, or got the feet and hands frozen if this happened in the winter.

But most of all the domovoy cares for cattle and one of his many appellations is, “cattle feeder.” “If domovoy loves a horse or a cow, or sheep, he steals hay for them from the homestead owner if the latter does not feed his animals well. If the owner has no hay at all and the animals remain hungry, domovoy goes to another homestead and steals hay there for his animals.” There are even stories about “domovoy fights over hay which they steal from each other. In some cases domovoys would appeal to humans for help. […] In the Kaluga Province peasants said that when domovoys are fighting, it is necessary to cry: ‘Hey, our own, beat that stranger!’”

Domovoy’s love of cattle was not entirely altruistic, of course. “A housewife, going into the yard or cattle-shed, would catch him milking a cow, but on seeing her he would immediately disappear.”

A salient feature of domovoys noted by folklore is their cryptic character: “Domovoy hates the curious.” By leading a nocturnal and secluded way of life they managed to escape observation not only by scholars and educated people, but even by most peasants amongst whom they were living, remaining always mysterious and legendary creatures.

There are many references in folklore to the leshy (wood goblin) as big as Bigfoot, but no mention of similar big domovoys. Apparently giant homins were too frightening and gluttonous for the role of a domestic demon. So there must have been selection of candidates of appropriate size for the domovoy niche.

Vladimir Dahl cites a proverb which is rhymed in Russian and translates: (domovoy) “has abandoned the devil but has not joined people either.” The word “devil” implies here a giant homin free roaming in the wild. And I see at least two reasons for domovoy’s failure to “join people.” Firstly, his advanced age, when it is too late to learn human ways; secondly, people’s overly emotional behavior in the presence of the domestic demon.

On the whole, folklore studies confirmed and clarified for me some things I learned on expeditions to the Caucasus, as described in Part I. Both folklore and field investigations have convinced me that the domovoy is still a reality.

## The Brownie

According to the Russian-English dictionary, the English for “domovoy” is “brownie.” Having learned that, I began to look for literature on the subject, and was delighted to discover in the best public library catalog a pointer to, *The Brownies: Their Book*, by Palmer Cox, published in New York in 1887. Yet great was my disappointment when I ordered and opened the book; it turned out to contain only verses which had nothing to do with brownies in the flesh. Later I read the following about this book in *Encyclopedia Britannica:*

> In 1883 the Brownies created by Palmer Cox, author and illustrator for *St. Nicholas Magazine*, were introduced to American children. Suggested by Scottish legends but modified to fit the contemporaneous scene, Cox's creatures of fantasy delighted children for 30 years. His series of drawings, "The Brownies," in *St. Nicholas Magazine* depicted the astonishing adventures of a race of benevolent little people.

The only bit of prose in Cox's book was this introduction:

> Brownies, like fairies and goblins, are imaginary little spirits, who are supposed to delight in harmless pranks and helpful deeds. They work and sport while weary households sleep, and never allow themselves to be seen by mortal eyes.

Having found no special work on the theme, I had nothing but dictionaries and encyclopedias to go by. Their entries on "brownie" have been written and "modified" by folklore scholars, but even so present certain points of interest for the hominologist. The following is what I read and copied after publication of my book on folklore in 1991:

*The Concise Oxford Dictionary of Current English,* first edition 1911:

> **brownie:** Benevolent shaggy goblin haunting house & doing household work secretly.

*The Oxford English Dictionary,* second edition, vol. 2, Oxford, 1989:

> **brownie.** Also browny, and with capital initial. (denominative of BROWN, with somewhat of diminutive force). A 'wee brown man' often appears in Scottish ballads and fairy tales. A benevolent spirit or goblin, of shaggy appearance, supposed to haunt old houses, esp. farmhouses, in Scotland, and sometimes to perform useful household work while the family were asleep.

*Webster's New International Dictionary,* 1947:

> **brownie:** (From its supposed tawny or swarthy color). A good-natured

goblin supposed to perform helpful services by night, such as threshing, churning, and sweeping.

*The Everyman's Encyclopaedia,* London, New York, 1913, Vol. 3:

**Brownie,** in the folklore of Scotland a goblin of the most obliging kind. He was never seen, but was only known by the good deeds which he did. He usually attached himself to some farmhouse in the country, and he was only noted by the voluntary labor which he performed during the night. He would churn, or thrash the corn, or clean all the dairy utensils, or perform some equally good-natured labour. His work was always done at night. The country people had great faith in the good works of the B. and believed in him implicitly. His reward was usually a dish of cream. The B. bears a strong resemblance to Robin Goodfellow in the Eng. and the Kobold of Ger. literature, whilst some comparison can be made between him and the household gods of the Roms. And of the domovoy. The Bs. were often the cause of the mysterious disappearance of things, and in this respect can be compared with the Jans, or Jennis, of the Arabs, and also to the pixies of South-western England. Practically every known folklore has its special fairy which can be compared to the brownie.

*The New Encyclopedia Britannica,* Vol. 2, Chicago, 1993:

**Brownie,** in English and Scottish folklore, a small industrious fairy or hobgoblin believed to inhabit houses and barns. Rarely seen, he was often heard at night, cleaning and doing housework; he also sometimes mischievously disarranged rooms. He would ride for the midwife, and in Cornwall he caused swarming bees to settle quickly. Cream or bread and milk might be left for him, but other gifts offended him. If one made him a suit of clothes, he would put it on and then vanish, never to return. The boggart of Yorkshire and the bogle of Scotland are hostile, mischievous brownies indistinguishable for poltergeists. See also puck.

*The Encyclopedia Americana,* International Edition, Vol. 4, New York, 1973:

**Brownie,** a household spirit or goblin of English and Scottish folklore. Usually pictured as a tiny old man wearing a brown hood and coat, the brownie would attach himself to families and help with the chores.

At night, while the family slept, the brownie would sweep rooms, clean pots and pans, and occasionally help with farm animals. Stories were told of a brownie riding horseback to fetch the midwife at child birth or helping his master to win at checkers. Mischievous as well as helpful, brownies were thought to take revenge, when criticized, by breaking dishes, spilling or souring the milk, or turning the animals loose. A helpful brownie would have a bowl of milk or cream and a special cake left for him, but any other kind of reward or wages would only anger him. If given a new suit of clothes, the brownie would put them on, chanting, "Gie brownie coat, gie brownie sork / Ye'll get nae mair o'brownie's work," and disappear, never to return. Similarities can be seen between the brownie and the boggart, or bogle, also in the folklore of Scotland, and such household spirits as the Kobold in Germany and the nisse in the Scandinavian countries.

## Their Gifts and Abilities

The history of man's relations with homins is full of ambivalence. The wild hairy bipeds were believed at one time or another, or at the same time, to be gods, semi-gods, devils, half-men and wildmen. Accordingly, views on their gifts and abilities have been varied and contradictory. One exception though is the unanimity of opinion regarding their physical endowment. All popular demons of both sexes are far better athletes than humans. Many folk tales relate of athletic competitions between man and demon, and every time man would resort to ruse and trickery to "win" the round.

On record is Pliny the Elder's phrase in *Natural History*: "the Satyrs have nothing of ordinary humanity about them except human shape." The hominologist, at his present stage of knowledge, tends to both agree and disagree with the ancient scholar. The beings in question are undoubtedly very different from ordinary humanity, and at the same time they are like human beings not only in shape but in many other respects as well.

The ancients believed satyrs to be gods and semi-gods, which did not prevent Hesiod from saying that these "brothers of mountain nymphs [were] an idle and worthless race." Idle...if this means that satyrs and their ilk do not earn a living by labor, it is correct. For all we know today, they lead an animal way of life. We also know today that some animals make and use tools that help them obtain nourishment. How about demons in this respect?

There is mention of clubs in the hands of wood goblins, but no mention of stone tools, just use of stones as projectiles. There are though references to tools taken from man. Rusalkas, for example, were seen with a pestle in the hand; they were often described combing their hair with combs, apparently taken from peasant bath huts which they visited; one item said the comb was made of a fish backbone.

A peasant once observed a rusalka standing in the water and looking into it as if into a mirror, smartening herself up. This shows, I wrote, how immensely close the rusalka intellect is to human. Folklore avers that rusalkas make wreaths of flowers, sedge, and tree branches, and put them on their heads. Let us also note that satyrs, nymphs, fauns, etc., are often depicted adorned with wreaths.

Pan, the great god of flocks and shepherds, when tired of striking panic into man, would start playing on a flute. There are also pictures of satyrs on Greek vases doing the same. Pan is even credited with inventing the shepherd's flute, the syrinx. Satyrs, nymphs, oriental paris, and Russian rusalkas love dancing and merrymaking, which is credible enough, but I always doubted that demons not only dance but also play music and invented a musical instrument. So I wondered why the Greeks credited them with such gifts. Recently I happened to read Dr. Henner Fahrenbach's report on sasquatches imitating "even short phrases on a flute." This prompted me to think that when a Greek shepherd played on a flute, Pan and company, well hidden in the wood, simply imitated the sounds, and hence the origin of the legend.

Demons can wear clothes, given by humans or stolen from them. The clothes are usually old, tattered, and worn inside-out. There is mention of wood goblins tearing off bast from trees and trying to make bast shoes (maybe in imitation of similar work by peasants). One item tells of a rusalka that made a cradle for her baby out of a birch-tree bark. In this connection let us recall Albert Ostman's words about sasquatches: "...they had some kind of blankets woven of narrow strips of cedar bark, packed with dry moss. They looked very practical and warm – with no need of washing."

I mentioned already various activities of demons helping humans - in hunting, fishing, pasturing, as well as in household work. Such activities are viewed very positively in folklore, with only a few exceptions. For example, regarding the Georgian dev mentioned earlier, it is said that when people were making hay on a hill, during the night the dev carried all the haystacks to the hilltop, while hay was needed in the valley below. "The people thought to themselves: 'Why wouldn't he carry the stacks down instead of uphill?' The next night the dev brought all the hay down."

The work of household she-demons is highly praised, but it is noted that they can't bake bread because they burn their hands. In regard of fire, it is clear that demons are not afraid of it: they approach campfires and hearths to warm themselves and they are able to put out fire, but are never said to be able to make it.

Demons can laugh; in sorrow their women and children would weep. They can sing, whistle, and imitate cries of various animals and voices of people (males, females, and babies).

As for the crucial question of speech, the answer in folklore is generally negative. Several examples to this effect are cited in my book, including the method, recommended by the Jewish Talmud, of telling a demon in the dark. If you happen to run into someone in the dark and can't tell who it is in front of you, the Talmud recommends saying "Shalom!" (Hello). If the greeting is not returned, chances are you are facing a demon. The same device is mentioned in Georgian folklore, using Georgian "Gamarjoba!" instead of "Shalom."

Folklore mentions demons resorting to gestures and fingers when communicating with humans. Vladimir Dahl writes that demons "sing without words," that their mumbling heard from a distance can be taken for speech, and peasants would interpret it in a jocular way (as if meaning "Walked, found, lost" or "Worse off every year"), but when coming face-to-face with a demon it would become clear that he is speechless.

Discussing communication abilities of demons-alias-homins, we have to take into consideration the long-held views in favor of their so-called extrasensory perception. At the 1978 Sasquatch conference in Vancouver, Dr. James R. Butler contributed a paper entitled "The Theoretical Importance of Higher Sensory Perceptions in the Sasquatch Phenomenon." He wrote:

> The term 'telepathy' (feeling from a distance) was coined by Frederick W. Myers as early as 1882 to denote 'the communication of impressions of any kind from one mind to another, independent of the recognized channels of sense.' [...] Our present civilization has perhaps overemphasized the value of the cerebral cortex because of our reliance on language and our insistence upon rationality. Also, the absence of selective environmental pressures has not genetically favored the continued development of HSP. [...] There is increasing documentation supporting the existence of unknown sensory channels in both human beings and other animals. If we accept this documentation, we would have to theoretically accept the probability of its occurrence in a Sasquatch. [...] It is hopeful that the Sasquatch investigation will soon shift from the pages of mythology into the physical and behavioral sciences. When increased emphasis is placed upon direct behavioral observations of Sasquatches, collecting data from field

> observations may not be as easy as it sounds unless we are prepared to apply new methodologies aligned with the problems imposed by HSPs. (In *The Sasquatch and Other Unknown Hominoids*, ed. by V. Markotic and G. Krantz, Calgary, 1984, pp. 207, 213, 215.)

It is also of relevance here what Arnold J. Toynbee, a well-known British historian, had to say on the subject of language and telepathy:

> Man shares with some other species of living creature the powers of communicating with his fellows telepathically. In human society, however, this faculty has been pushed into the margin of intercourse by language. This more effective medium of communication is possessed by all human beings in all societies and is one of the distinctive marks of being human. In spite of our command of language, telepathy is still an indispensable means of communication for human beings too. [...] However, language is a more copious means of communication than telepathy is. Like telepathy, language can communicate feelings and impulses, but it can also communicate thoughts, which telepathy can convey, if at all, only when they have a strong emotional charge—and emotion is the enemy of intellectual clarity and objectivity. (*Change and Habit* by Arnold J. Toynbee, Oxford University Press, London, 1966, p. 16.)

We have learned recently from Boris Porshnev's archive that he accepted the possibility of telepathic abilities in relict hominids. There are certain hints and signs in folklore and demonology, including the material of this work, suggesting the presence of homins' hypnotic and telepathic powers. Similar signs are also present in accounts of witnesses collected by investigators. Such evidence, along with sporadically reported indications of homins' speech faculty, is the subject of both on-going and future studies.

* * *

Folklore is the richest source of information for the hominologist and, at the same time, an obstacle that has to be overcome on the way to the truth. *Wood Goblin Dubbed Monkey serves this double aim. The book ends up with the question: "Will goblins help the world of science to open its eyes on what was clear to Boris Porshnev over twenty years ago?"*

# A Note on Folklore in Hominology

## Abstract

Mythology and reality can be closely interconnected, as in the case of hominology (the study of unknown living hominoids/hominids). While folklorists tend to dismiss real hominoids, the existence of mythological hominoids is a necessary, though not sufficient, condition of the existence of real hominoids. The factual origin for some hominoid myths should be given consideration.

The relationship between "realists" and "folklorists" in hominology has not been easy or productive, and this has induced me to re-examine its background and to try to lay down some basic rules.

There are philosophers who insist that "reality" exists only in the mind of the beholder. I know of no logical argument to counter this assumption, which can be regarded as an extreme case of "folklorism." Presumably, such a philosopher, if kidnapped by a sasquatch like in the case of Albert Ostman, would be consoled by the thought that the drama is only taking place in his head.

On the other hand, we know that an archaeologist, Heinrich Schliemann, who, proceeding from the ornate imagery of the ancient Greeks, confronted the world with the reality of Troy. Schliemann was a realist, and there can be little doubt that if he and other archaeologists had asked and followed the advice of "folklorists" on the reality of Troy, its precious relics would still be lying underground.

This example shows that there can be totally different entities bearing the same name, and our failure to recognize and differentiate such entities leads to a lot of confusion and useless arguments. The name Troy applies, on the one hand, to a figment of an ancient poet's imagination, studied by specialists in literature and mythology, and, on the other hand, to a real historical city, whose study is the business of archaeologists and historians.

Of course, the two entities are interconnected in some way; one was the cause of the other, and both can have some overlapping characteristics, but, on the whole, their natures are so different that it would be most unwise to judge the one, say the historical city of Troy, by our knowledge of the other, the mythological Troy.

I believe the same analogy applies in hominology. There are *real* hominoids (that is, creatures of biology—we know this from several categories of evidence combined), and there are *imaginary* ones (those of mythology). Our opponents say that one kind is quite enough (those of mythology), which dispenses with the necessity for real ones. But I say nay—the existence of mythological hominoids is a necessary, though not sufficient, condition of the existence of real hominoids. The argument was set forth by us in 1976 as follows:

> Folklore and mythology in general are an important source of information for science. But hominologists look for myths about these creatures not only to find a real basis for the myths and to supplement their knowledge of the problem. They also need the myths as such, for they are yet another "litmus test" confirming the historical reality of hominoids. If, in the course of history, people had encounters with "troglodytes," then these most impressive beings could not have escaped the attention of the creators of myths and legends. Of course, the reality of relic hominoids cannot be supported by recourse to folklore alone, but neither can it be refuted by such references, as our opponents have attempted to do. Is the abundant folklore, say, about the wolf or the bear not a consequence of the existence of these animals and man's knowledge of them? Therefore, we say that, if relic hominoids were not reflected in folklore and mythology, then their reality could be called into question. Fortunately, this channel of information is so wide and deep that much work can be done in this sphere: it is necessary to re-examine and re-think a good many anthropomorphic images playing important roles in folklore and demonology [Bayanov and Bourtsev 1976].

The last sentence above seems to find support in the words of Wayne Suttles:

> If there is a real animal, shouldn't there be better descriptions in the ethnographic literature? Not necessarily. Anthropologists do not consciously suppress information, but they sometimes do not know what to do with it. There are ethnographies of peoples whom I know to have traditions of Sasquatch-like beings that make no mention of such traditions; I suspect that these omissions occur not because the writers had never heard of the traditions but because they did not know how to categorize them [Suttles 1972].

I wish ethnographers in the U.S.S.R. would make such a scientifically fruitful admission. Why is it difficult for ethnographers to categorize such material? Probably because they have no idea what is real and what is imaginary in it. And the fact that the informants do not know either cannot be of much help to the scientist, who should always attempt to draw a line between fact and fiction.

Hence, ideally, "realists" and "folklorists" in hominology should sit down together and, without violating each other's territory, sort out the mountain of folklore on hominoids. When Suttles says that "a large non-human primate would not really steal women" (Suttles 1972), I am afraid he trespasses on the turf of other kinds of experts. When a nineteenth-century Russian ethnographer said that the large breasts of a female wood-goblin ("forest woman") had been made-up by ignorant peasants to symbolize heavy precipitation, he simply ascribed his own ignorance and fantasy to his informants. What about the image of a "tree-striker" that has the habit of "knocking down dead trees" (Suttles 1972)? Well, if it's a hominoid's way of feeding on larvae, the image has a basis in reality.

In the abstract of a paper (Suttles 1980) presented at the 1978 Manlike Monsters Conference at the University of British Columbia, Suttles asked: "If the Sasquatch is a real animal, why should there be several Indian images, some rather different from the usual non-Indian image?" I hink it is the folklorist who is to ponder this question. To ask it of the realist would be like asking Schliemann to account for every flight of Homer's fancy.

## References cited


Bayanov, Dmitri, and Igor Bourtsev. 1976. The mysterious biped (in Russian). *Science and Religion* no. 6, 39.

Suttles, Wayne. 1972. On the cultural track of the Sasquatch. *Northwest Anthropological Research Notes* 6(1):82.

———. 1980. Sasquatch: The testimony of tradition. *In Manlike monsters on trial: Early records and modern evidence,* ed. Marjorie M. Halpin and Michael M. Ames. Vancouver: University of British Columbia Press.

# The Yahoo Quote

Jonathan Swift,
*Gulliver's Travels,* 1726;
Part IV. A Voyage to the Country of
the Houyhnhnms;
Chapter VIII.

TRAVELS
INTO SEVERAL
Remote NATIONS
OF THE
WORLD.
In FOUR PARTS.
By LEMUEL GULLIVER,
Vol. I.
LONDON:

*Lemuel Gulliver*

The Author of *Gulliver's Travels* relates several particulars of the Yahoos (homins) that I present here.

They would approach as near as they durst, and imitate my Actions after the Manner of Monkeys.... They are prodigiously nimble from their Infancy.... I observed the young Animal's Flesh to smell very rank, and the Stink was somewhat between a Weasel and a Fox, but much more disagreeable.... They are strong and hardy, but of a cowardly Spirit.... They dig up Roots, eat several Kinds of Herbs, and search about for Carrion, or sometimes catch Weasels and Luhimuhs (a Sort of wild Rat) which they greedily devour.

Nature hath taught them to dig deep Holes with their Nails on the Sides of a rising Ground, wherein they lie by themselves; only the Kennels of the Females are larger, sufficient to hold two or three Cubs.

They swim from their Infancy like Frogs, and are able to continue long under Water, where they often take Fish which the Females carry home to their Young. And upon this Occasion I hope the Reader will pardon my relating an odd Adventure.

Being one Day abroad with my Protector the Sorrel Nag, and the Weather exceeding hot, I entreated him to let me bathe in a River that was near. He consented, and I immediately stripped myself stark naked, and went down softly into the Stream. It happened that a young Female Yahoo standing behind a Bank, saw the whole Proceeding; and inflamed by Desire, as the Nag and I conjectured, came running with all Speed, and leaped into the Water within five Yards of the

Place where I bathed. I was never in my Life so terribly frightened; the Nag was grazing at some Distance, not suspecting any Harm; she embraced me after a most fulsome Manner; I roared as loud as I could, and the Nag came galloping towards me, whereupon she quitted her Grasp, with the utmost Reluctancy, and leaped upon the opposite Bank, where she stood gazing and howling all the time I was putting on my Cloaths.

This was Matter of Diversion to my Master and his Family, as well as Mortification to my self. For now I could no longer deny that I was a real Yahoo, in every Limb and Feature, since the Females had a natural Propensity to me as one of their own Species. Neither was the Hair of this Brute of a Red Colour (which might have been some Excuse for an Appetite a little irregular) but black as a Sloe, and her Countenance did not make an Appearance altogether so hideous as the rest of the Kind; for I think, she could not be above Eleven Years old."

— (Jonathan Swift. [1726] 1941. *Gulliver's Travels*.Oxford: Basil Blackwell, 250–51).

The incident in the river is described so realistically (minus the Nag, of course) that something similar might have happened to Swift himself or he must have heard about such incidents from witnesses in Ireland. In Russia, female Yahoos were called Rusalkas, and they were known to accost male peasants bathing in rivers and lakes. See Dmitri Bayanov, *In the Footsteps of the Russian Snowman,* pp. 167–76.

"I was never in my Life so terribly frightened," said Gulliver. "I have never felt such fear in my life," said the eminent Russian writer Turgenev, who had had a narrow escape from a lovesick rusalka when he took a swim in a secluded nook of a woodland river. Swift presents Yahoos as degraded humans. Turgenev thought that his "monster" ("like a female gorilla") was "a madwoman." Both were wrong.

As human populations increased and homins decreased, a disproportion must have appeared between homin male and female numbers: males, as more adventurous and mobile, probably perished sooner than females. Hence incidents as described above.

The name Yahoo is applied to homins in some countries of the former British Empire, and it is not clear whether it was coined by Swift and borrowed from his book or Swift himself borrowed the name from travelers and used it in the book.

(First cited in *Bigfoot Co-op*, June/August 2004, p. 9)

# SECTION 2
# Samples of Sighting Reports in Russia

## Snowmen Around Lake Baikal

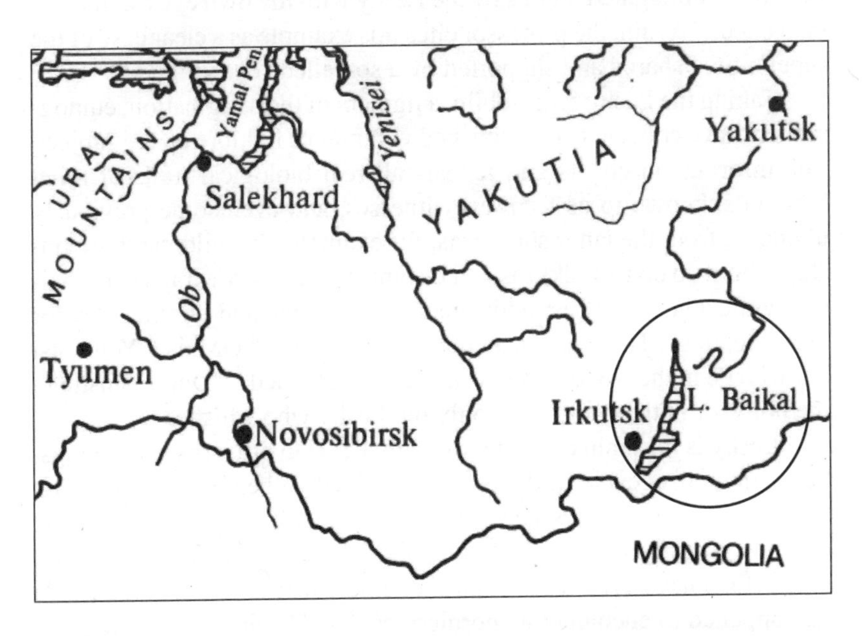


A Russian translation of Charles Stoner's book *The Sherpa and the Snowman* was published in Moscow in 1958—the peak of general interest in the problem, spurred by the Himalayan expeditions in search of the Yeti. In a foreword to the book, Dr. G. D. Debets, a noted Soviet anthropologist, asked the question "Can the stories told by the Sherpas be taken as indisputable evidence of the Yeti's existence?" and answered:

> No, they cannot. This author happened to visit the upper reaches of the Ilim River in the Irkutsk Region [adjoining Lake Baikal], in 1929. Today the area is crossed by a railroad but at the time it could not boast a single school or a literate person among the local population. The author was often told stories about the Leshy, and with details not in the least inferior to the Sherpa stories about the Yeti.

Leshy (from the Russian word "les"—wood, forest) used to be, and still remains, a sign of popular "ignorance" and "superstition" in the eyes of educated Russians. Russian-English dictionaries translate "Leshy" as "wood-goblin," "goblin of the woods," with the mark "folklore." By comparing Yeti to Leshy, Dr. Debets implied that the belief in both was the result of a superstition.

Professor K.K. Platonov, in his book *The Psychology of Religion* (Moscow 1967), has this to say: "I had occasion to converse with an old Transbaikal [i.e., east of Baikal] hunter who said: 'I don't know if apes exist or just imagined, but I saw the Leshy with my own eyes, and more than once.'" Again, the professor cites this example as a clear case of the superstition, shared and supported by a so-called "eyewitness."

Taking the Leshy for a goblin, a figment of the imagination, ethnographers have collected and published volumes of folklore on the subject. The information covers and reveals all real biological traits of relict hominids, known to us from eyewitnesses, and even some previously unknown from the latter source, as, for example, the wildman's friendship with wild dogs. Folklorists find some of the Leshy's traits paradoxical, while a hominologist finds them quite normal and familiar. As, for example, this: "The Leshy is dumb but vociferous." Or this: "You must not whistle in the wood—the Leshy can be offended." "Don't whistle in the home or in the wood—it's only the Leshy who whistles."

Leshy is the main character of many Russian proverbs and sayings, being interchangeable in them with the devil. When a person appears after a long absence, Russians sometimes say "Where has the leshy (the devil) been carrying you?" (Remember Albert Ostman!) In my Baikal region file I have a reference to an eyewitness who said that in the 1920s he happened to encounter a "hornless devil." This helps to explain the proverb "The devil is not as ugly (terrible, scary) as he is painted." In Russian icons the devil is usually painted with horns. It turns out that in reality the horns are absent.

Another item in the file says that the leshy can strike a friendship with a hunter and bring him squirrels in exchange for vodka refreshments. It sounds like a joke and fable, but folklore (and not only folklore!) about “joint ventures” of hunters and wildmen, from the Gilgamesh epic onward, is so abundant that we should not reject it out of hand.

“According to Transbaikal Cossack stories, water devils come out of the water to play with bathing children.” Wildman’s love of human children is a ubiquitous feature of folklore. One unfortunate result of this love is that children happen to be kidnapped by wildman.

One more item from Transbaikal Cossacks: “Sometimes a she-devil falls in love with hunters in the wood and becomes pregnant therefrom, but the child conceived by man is torn apart by her at the very moment of birth.”

But enough of folklore. One of my correspondents in the region, Victor Lushnikov, wrote that his father, Innokenty Lushnikov (1871–1920), was Editor of the newspaper *Baikal*. As a boy, Victor used to hear eyewitness accounts about “forest people” told to his father by local hunters. According to his great-grandmother, when she was young, her girlfriend was kidnapped by a Tungu (the Even name for wildman).

Another correspondent, Victor Cherepakhin, a resident of the city of Chita (east of Baikal), an artist by profession, wrote me of what he witnessed in June 1968 in the northern part of the Chita Region, where he was painting landscapes. At a certain moment, during his work in the mountains, he felt “a kind of uneasiness.”

> It felt like I was being watched. I turned cautiously and began to scrutinize the surroundings. At first I did not notice anything, but then, continuing to peer, I suddenly saw a manlike creature standing by a tree. It was dark, darker than the pine trunks, rather tall, 190–195 cm, maybe even taller. It looked bulky, despite its height. Trying to see better, I was standing still. The creature also stood motionless. Then it turned to one side, bent down, as if picking up something from the ground, and disappeared in the wood.

The artist hastened to leave the place.

> Later, recollecting what I saw, I concluded that the way the creature moved indicated it was devoid of clothes. The movements were neither slow nor hurried, very natural and resembling those of man. And at the same time they were somewhat different from human. I had then, and retain now, after so many years, the feeling that I happened to touch an innermost secret of Nature.

Now one of the last letters from the Chita Region, received in

May 1996 from Yuri Lagunov, 46. It begins with the story about Yuri's friend, "Uncle Vasia," who flatly declined Yuri's plan to go hunting in a certain place rife with hares. The reason was divulged only after they emptied a bottle of vodka. It turned out that in 1969, Uncle Vasia had an encounter there with a huge, wild, "hairy muzhik." In the evening, the latter entered half of his body into the hunter's hut and eyed the stunned owner of the hut for a minute, and then quietly departed. (Did he want to strike a "deal" with the hunter?) Uncle Vasia snatched his gun and remained seated with it through the night, unable even to shut the door. That's why he declined Yuri's proposal to go hunting in that place. The letter goes on as follows:

> At 46 I have understood: a man opens up only in a moment of drunken sincerity. Hunters may see and shoot and even kill a snowman but keep silent. If you talk, people will laugh and mock, say it was a bear, you just imagined it, etc. I didn't believe that story either, until my own incident.
>
> This happened in 1975, when I was 25. I worked as a film projectionist in the settlement and was at the time on leave. It was haymaking time, the 16th of August. My father and I, we had mowed some grass and waited for it to dry. Everybody who has been hunting or fishing know what a great appetite you get there, and haymaking whets your appetite even more. So father says to me 'Why don't you go hunting on the salines? Maybe you'll kill something.' So off I go to the salines, about two kilometers from the settlement. I had only four buck-shot cartridges. I was seated on a small platform, raised two meters from the ground. That way mosquitoes and midges are less annoying. Hell, I had forgotten to bring anti-mosquito ointment. So I was smoking to get the mosquitoes away. The gun was charged, the cock raised.
>
> The sun had set down behind the hill. Just the right time for gurans (wild goats) to show up. Suddenly I hear a guran cry some 100–150 meters from the salines, and then run away. I smoked again to check the wind direction. It was all right.
>
> What could have frightened the goat? So I sit and wait; the dusk is getting thicker, no sound is heard. Now I begin to feel sort of scared. I'd been hunting goats on the salines many times without a problem. Why this fear now? So I am trying to reassure myself, saying inwardly there's nothing to fear, I've got a gun with me, the settlement is not far, the bears don't attack in the summer.
>
> Suddenly I hear a growling on the hill. Must be wild dogs. The growling stops and my fear subsides a bit. A full moon is rising from

behind the hill. Visibility is good now. Not a whiff of air, not a leaf moving on the shrubs and trees. Then I feel being watched. I turn the head—Oh, terror! There he is, the snowman, standing 5–7 meters away, half-turned to me. Broad face, wide shoulders, the chest like that of a boxer, the height about two and a half meters. Mine is 180 centimeters and I reached the platform from the ground with a fully outstretched hand. He was taller by half a meter or more than the platform.

When I saw him, fear gripped me in earnest. I thought: am I asleep and dreaming? But the cigarette burned my hand and I dropped it. Slowly I moved the gun and pointed at him. If he attacks, I fire. But what's the use? It's like shooting at an elephant with buck-shot. At 25 you want to live. I was in love and hoped for a better life. Today I would shoot without a second thought, just to prove the snowman exists. But at that time, aiming at him and trembling with terror, I began to pray to all gods that they make him disappear.

Mind you, I am not a believer, but, as the saying goes, a muzhik would remember God only when it thunders. Having crossed myself and the snowman, I whisper 'Jesus Christ, and the Son*, and the Holy Spirit, I entreat you, do make him disappear!' Nothing happens. He is there. I pray 'Great and Holy Allah, I entreat you, do make him disappear!' No go. He is there. I pray 'Great and Holy Buddha, I entreat you, do make him disappear!' Nothing happens. He is there.

Now the moon is high in the sky and sheds enough light on him. His eyes are burning, I see his hairy face, his developed torso, the chest, the muscles on the arm. How long we eyed one another, I don't know, I had no watch. At last I saw him raise the hand, move a branch slowly, and disappear in the brush.

Some minutes later I heard a cry I never heard before. And a similar cry far away. I thought it was an echo. But no, some time later I hear a far cry again and a cry in response quite near. I realize there are two of them, and pray to all gods again that daybreak come sooner.

As day was breaking I see a guran approach. I don't shoot, just watch it moving. I figure, if the goat is startled and cries, it means the snowman is nearby. The goat passed by without fear. It means the way out is free for me. I lighted a cigarette and headed towards the road, carrying the gun at the ready and bypassing dense clusters of shrubs.

---

* The witness's belief that Jesus Christ and the Son are not the same shows him to be truly an unbeliever (when it does not "thunder").

When I reached the road I heard the snowman growl on the other side of the gully. The growling was uninterrupted. A dog growls with breaks, when it inhales the air; a bear roars abruptly, letting the steam out in one blow. This thing was growling without a letup while I was running away along the road to the settlement. In the morning sounds are heard afar. When I reached home mother had already milked the cow and, seeing my eyes almost out of their orbits, asked what's the matter? I then stammered out what had happened.

Two remarks in conclusion. Firstly, I am sorry that my translation has failed to reproduce the folksy flavor of this "confession." Secondly, I am very glad that Yuri Lagunov did not resort to the gun in his confrontation with the Leshy. Every gun shot at our hairy woodland cousin only postpones, not brings closer, the hour of their recognition by science. It is only by establishing friendly contact with them that researchers will be able to study them in earnest and take protection and preservation measures.

(First published in *Bigfoot Co-op*, June 1999, pp. 7-11)

# Latest from Siderov

Sighting in the Far North: Back in Tomsk after his expedition, Georgy Sidorov learned of a homin sighting in the far north of Siberia, in July 1999. Following his field report, he informed me of the event by letter. A party of geologists, three men and two women, engaged in prospecting, pitched camp at the foot of a low hill, near a nameless lake in the tundra, 45 km from the Byrrang Mountains in the Taimyr Peninsula, which is the northernmost projection of Siberia. Their two tents stood on dry ground on the bank of a brook flowing into the lake.

That day, four expedition members had gone off prospecting, while one woman (who insists on remaining incognito) stayed at camp. Sitting by the brook she was sorting out and classifying the collected rock samples. As she told Sidorov, she was feeling warm and fine. It was about noon, July 10th. Suddenly she felt she was being watched. She turned and became stupefied. From the northern side of the hill, which was still covered with ice, a huge hairy man was marching towards her. For some time she was unable to collect her wits. Then she realized who it was but was unable to budge. Her carbine was by her side but she forgot about it. So she stayed put when the "snowman" came up to her.

The witness said his eyes looked human despite his overall beastly appearance. In her estimate, he stood about three meters in height and was covered with thick brown hair, almost black on the chest and back. His neck seemed non-existent, his arms and legs very strong. She also remembered a flat nose with straight nostrils on a gorilla-like face, with a powerful lower jaw. But in spite of his frightening looks his eyes were not malicious.

He looked at her and chuckled. Then he circled her but never touched. "Had he touched me, I'd have died right away," she said. Then he squatted and started "talking" to her. She insists she heard sounds resembling human speech. "The voice was low and even pleasant." But mostly he "was signaling" her not vocally but "with lips" (!?).

Having finished his short "speech", the snowman stood up, looked in the tents, touched the hanging linen, and went off on his way. The witness was unable to stand up from her collapsible stool until the arrival of her colleagues. Her legs simply failed her. Stammering, she told them of the event and, although the tracks were clearly in view around the tents, she was immediately ridiculed.

"That's all there is to this regrettable story," concludes Georgy. "The hominoid himself attempted contact but the human failed to respond and show him hospitality ... If we'd ever had such a fluke!"

The story gives the hominologist a feeling of déjà vu. As if programmed, the humans behaved in the usual way: the witness first stupefied, then ridiculed; others distrustful, ignoring the footprints ....

The homin's character and behavior are also familiar: self-confident, inquisitive, not devoid of a sense of humor (judging by his chuckling), non-aggressive when unprovoked. Of great interest and importance is additional information (to already existing) on the subject of vocalization resembling human speech. It touches on the crucial aspect of hominology and the origin of man. His "signaling with lips" seems apelike (has to be checked with primatologists).

What, to my knowledge, is unprecedented, is the region of the encounter. The Taimyr Peninsula is as far north as the middle part of Greenland, it's more northerly than the Kola Peninsula, the Chukchi Peninsula, and the whole of Alaska. That our hairy cousins survive at such inhospitable and chilly latitudes is reassuring enough. Stories like that will be repeated over and over again, until hominology turns from a Cinderella of science into a princess of primatology.

(First published in *Bigfoot Co-op,* June 2000, pp. 3–4)

# A Story that Rang the Bell and a Mystery

As economic conditions in Russia are slowly improving, old members of our seminar begin again to attend its sessions, Yuri Krashnikov and Yakov Polyakov among them. At a recent session they recalled an episode of their expedition to Tajikistan in July-August 1984. It happened not far from Lake Pairon, visited by me in 1982 (see *In the Footsteps of the Russian Snowman*, p. 116). At the end, Yuri Krashnikov mentioned something that sat me straight up and brought to mind my contributions to the *Bigfoot Co-op* two decades ago.

In 1981, Warren Thompson of The Bay Area Research Group (California) sent me Archie Buckley's "Report on Sasquatch Field Findings," from which I quoted as follows:

> They (sasquatches) are excellent mimics. In order to communicate and conceal their presence, they will at times employ the mimicking of other animals, birds, and natural sounds. We have heard them bark like dogs and coyotes, whistle, and pound rocks. I have even had them repeat my voice. Another sound they occasionally use when your presence is near—is one which has the phonetics of a small brass bell; as if one were to tinkle a bell softly four or five times, then stop for several seconds and repeat (*Bigfoot Co-op*, October 1981).

We had ample confirmation of the homins' mimicking ability in Eurasia, both by folklore and witnesses, but I had never heard of their making bell sounds. So when in the same year (1981) I came across mention of bell sounds while reading *A Note on the 'Wild People' of Tibet* by Johan van Manen, I hastened to share that point with colleagues through the *Co-op*. The relevant text is offered in the above mentioned work as an appendix entitled "Strange Phenomena in the Himalayas," one passage of which reads as follows:

> But even stranger than the musical sounds made by eddies and gusts of winds whistling up glens and ravines, is the beautiful clear tinkle of a bell heard sometimes on lonely nights. The sound has been heard in Switzerland, and in the Andes, and in the Himalayas by travelers and mountaineers. A missionary once heard it in Western China and his letter, I think I am right in saying, produces one or two [testimonies] from people who said they had heard it in remote parts of Scotland. Naturally some will say that the bell must have been round the neck

Software developer Yury Krashnikov, Smolin Seminar member, who heard mysterious sounds in the mountains of Tadjikistan in August 1984

of some concealed cow or sheep, but if that explanation were acceptable to the men who have described the phenomenon there would be no mystery at all. The whole point is that the tinkle of the bell has been heard in places where, according to those who tell the tale, there was no possibility of cattle or sheep being present. The sound, too, is described as indescribably beautiful and as persisting only for a few seconds. It is heard and is gone (*Bigfoot Co-op*, April 1982).

Now I return to Yuri Krashnikov's story. During a hike he and Yakov Polyakov, overtaken by darkness, had to spend the night on a mountain slope, away from camp. The moonless night was pitch-dark. They went to sleep and then woke to the sounds of someone walking nearby. They heard stones rolling and the crackling of twigs in rhythm with the steps. They were approaching. Yuri Kraskhnikov:

> When the distance between us and the source of the sounds equaled several meters, the sounds stopped. There was silence. We also kept silent and did not budge. Suddenly, from the place where the movement stopped, came sounds resembling the tinkle of a small bell. [Hearing

> that I almost jumped from my chair!—D.B.] The sounds repeated periodically, once or twice a second. And their source began to move around us, counterclockwise, at a distance of several meters. No other sounds, those of steps, for example, were heard. Taking into account that we were far away from the stream and silence was complete, we must have heard all sounds that were produced. But we heard nothing but the tinkling of a bell. What's more the tinkling seemed to come from the air above us. Having circled us and returned to the starting point, it stopped. Then we heard the sounds of retreating steps.
>
> In the morning we detected indistinct marks of big tracks on the ground strewn with fine-grained rock.... The 'bell sounds' we heard were very peculiar. The description is only approximate, referring to something familiar. They were TOO clear, so clear they seemed almost unreal. If I heard them now, and not 17 years ago, I'd sooner describe them as 'computer-made' or 'synthesized.' Yakov and I compared our impressions and they coincided in minutest detail.

Both witnesses had never heard of Archie Buckley's observations, nor of the mysterious bell sounds in the Himalayas and other places, and they were much fascinated to hear all that from me.

(First published in *Bigfoot Co-op*, October 2001, pp. 3–4.)

# Evidence From The Carpathians

The Carpathians is a mountain chain in Central Europe, embracing parts of Poland, Slovakia, Hungary, Rumania, and Ukraine. It's a watershed for the Baltic and the Black Seas, well wooded, and in some areas still rich in wildlife.

Hominological evidence from the region (namely, two cases of homins male and female, captured in the 18th century) is cited in my book *In the Footsteps of the Russian Snowman*, pp. 162–166. Evidence from the 19th century is absent in my files, but there is some from the 20th, provided by a pen-friend named Vyacheslav Zinov. In 1981, he and his friend hiked in the mountains of the Transcarpathian Region of Ukraine (the Region's central city is Uzhgorod) and quite by chance touched on the subject of "snowman" in a chat with an old local man, named Vasil Dobron, aged 70.

Vyacheslav Zinov sent me his sketch depicting the theme of a tapestry he saw in a museum. The creature in the upper corner is locally known as "woodman".

When Dobron was 20 (so it must have been in the 1930s), he pastured sheep in the mountains and had heard from the old men that "devils" and "goblins" were sometimes running in the forest, that they were big (two meters tall), covered with black hair, but otherwise man-like. They sometimes stole sheep, but were afraid of dogs.

It was toward the end of summer, the weather was dry, so Dobron took a vessel and went down to fetch water from the river. He was armed with a hunting gun. When he reached the steep slope from which a path led down to the river he saw below "a man in a fur coat and leather trousers sitting on the bank and drinking water." Vasil Dobron:

> I made the gun ready and ran down the path. At first he did not see me, but when there were about 50 meters between us he stood up. As I was approaching he turned toward me and looked at me. I stopped. There were 25-30 meters between us. I aimed the gun at him and asked: "Who are you? Aren't you a bandit? A socialist run away from prison?" [At the time, that part of Ukraine belonged to Czechoslovakia, i.e., it was not under Soviet power—D.B.]

The stranger was silent. Then he began mumbling and howling. Only then did I realize he was not a human being. It was his black hair covering his body that I'd taken for a fur coat. He was about 190 centimeters tall, arms as thick as a girl's legs, the face also covered with hair, except the eyes and the nose.

I then cried: "Get away, devil!" and fired at him out of fear. He tore along the river bank faster than a horse, went up a rock like a goat and vanished in the forest. Ever since I've always prayed in the morning and in the evening, and thought that if I had not worn a cross I wouldn't have prevailed over that devil.

A second case was after the war, perhaps in 1955. I went bear hunting with a carbine. Went up high in the mountains and suddenly heard a bear roaring. I approached cautiously through the brush, listening all the time. The roar came from one place. I figured the bear killed an elk or something and was gorging. Well, I thought, I'll get him. I went further and saw a she-bear trying in vain to climb up a thin tree, clawing the bark and roaring angrily. And up in the tree was another creature, probably a young bear. Coming closer, carbine at ready, I saw it was not a bear but a human or a devil in the tree. He was also making sounds, sort of hissing and grumbling. Now it looked just like the devil I'd seen at the river, only smaller, one meter and a half tall, a young devil that is. Also all in black hair. He stood up full height on a branch ready to jump to a bigger tree while the she-bear was almost ready to break the thin tree. The young devil than jumped but failed to reach the other tree and fell to the ground, somersaulted and tore off on two legs so that even lumps of earth flew up from under his feet, and the she-bear darted after him. I crossed myself and ran in the opposite direction.

Such is an eyewitness's story from the Carpathians of the 20th century, with its familiar motif of "devilry," gun shooting, and homin/ursus antagonism.

(First printed in *Bigfoot Co-op*, October 2001, pp. 5–6.)

# Homins between Moscow and St. Petersburg

In 1996, when *In the Footsteps of the Russian Snowman*, was published, I happened to read a booklet which made me fear my book missed the most important story of all. The booklet had been supplied by a colleague who bought it in a provincial town. It was written by an Oleg Ivanov, titled *Avdoshki*, published at the author's expense, in 1996, in Malaya Vishera. The latter is a town on the railroad Moscow-St. Petersburg, not far from the city of Novgorod.

The booklet bewildered me a great deal. Its impact was like that of Albert Ostman's story. I mean the claimed encounters were about as close as those of Ostman, or rather those of Jane Goodall with chimpanzees or Dian Fossey with gorillas.

Could I believe that? With new, liberal times in Russia all sorts of sensations had flooded the media, the kind we used to see in your tabloids. And yet some bells in the booklet seemed to be ringing true. I had never heard of Oleg Ivanov. A decade and a half earlier Moscow hominologists were hoodwinked by another Ivanov, living in a village some 200 km from Moscow. He also claimed close encounters with snowmen in the woods around his village. It took us months to conclude the man was not mentally balanced. Was Ivanov number two a repeat of number one? I hastened to invite him to our seminar in Moscow.

Now more about the booklet itself. "Avdoshki" is the plural of "avdoshka," a local endearing or jocular name for the Russian "leshy" (wood goblin). The introduction is written by the author's cousin, Vera Senoyedova, who mentions two sightings by her father, and relates a local legend about a beautiful village girl kidnapped by a "man of the forest." When the girl's fiancé found her at last in the woods, she refused to go home, saying: "I live with the woodman. He has bewitched me." The young man killed both the girl and the wildman and brought their corpses to the village. The priest of the village said to the girl's father: "Your daughter had sold herself to the devil. Take their bodies back to the forest. They can't be buried at the cemetery." So, according to legend, the girl and the wildman are buried side by side in the forest.

The booklet's first chapter is titled The First Encounter. It happened in 1960, when Oleg Ivanov was 12 years old. He and his friend Tolik went fishing and came across huge manlike footprints, which they followed and came in view of four hairy creatures in the forest, two big and two small. The boys took fright and, while fleeing, ran into a bog and started sinking. They cried desperately for help, and suddenly it did come .... No, not from the wildmen, but from an unfamiliar man. He cut a pole, threw it to the boys, and pulled them out onto firm ground. Having saved them, the stranger showed the boys the way home. Oleg ran home barefoot because his rubber high boots remained in the quagmire. This earned him a good thrashing from his father.

A subsequent chapter is devoted to the stranger who saved the author's life. He is called Alexander Komlev, aged 40 at the time, and nicknamed "Professor" by the boys. This man is not only the main character of the booklet but, if Ivanov is to be believed, the unchallenged hero of world hominology. "Professor" is claimed to have spent 20 years observing the "forest people" in the manner Jane Goodall observed wild chimps. He saw not only how they lived but also how and where they died. He followed them for hundreds of miles on migration routes. He was a successful "match-maker" among the avdoshkas, helping them to meet and "marry," in order to counteract the adverse effect of their falling numbers.

Sounds like science fiction, doesn't it? If not, why hadn't we heard of Alexander Komlev before? Ivanov gives a weighty reason. "Professor" had worked at a scientific research institute, and in 1953, during Stalinist repressions, was exiled to the Arkhangelsk Region, in the north, where he slaved cutting timber. He escaped from the prison camp, and roaming through the forest encountered "forest people." His fascination with them was so great that he switched direction of research for the rest of his life. For this purpose, but also under the pressure of circumstances, he adopted their way of life, which greatly facilitated his contact with them.

In a secretive way the fugitive joined civilization, having probably changed his name. So Alexander Komlev is not his real name. The real name is not known. Understandably, "Professor" was not talkative about his past, and information on it was gleaned by Ivanov from sundry bits of conversation when he and "Professor" became friends. During winters "Professor" worked at various jobs, mostly on construction sites, and devoted all his summers to life with the avdoshkas. He and Oleg met from time to time in the forest, being in pursuit of their common interest. And sometimes Oleg received letters from Komlev, when the latter traveled far and wide, following his hairy wards.

Quote: "In the summer of 1976, I went off on a business trip to the city of Perm. When I came back Pavel Nikiforov told me that "Professor" had died. The hospital buried him as a person without relatives.... Later on, the cemetery was enlarged, new graves appeared in that place, and the grave of "Professor" no longer exists. I was told that before he departed he came and sat on a bench by my house, wishing to hand to me his old black bag with his notes. Having learned that I was absent he went to the forest. Pavel Nikiforov wanted to take the bag and pass it to me, but Komlev refused."

What a plot for a novel, eh? Looks like perfect fiction, but a "Professor" is also mentioned by Vera Senoyedova in an Introduction, and

Oleg Ivanov signing copies of his book in a public library. He claims his white hair is the result of a face to face encounter with a wildman.

Ivanov refers in the booklet to other real people who claim to have met Alexander Komlev. So you can imagine how eagerly I looked forward to Ivanov coming to Moscow and talking at our seminar. I asked him to bring along his friend and co-witness Tolik, but he wrote back that Tolik was no longer alive, having been poisoned to death by sub-standard vodka (a common cause of death over here; thank goodness Oleg is not a drunkard). I asked him to bring and show us a letter by "Professor," but he replied that "Professor" had asked him to destroy his letters, and Oleg did. Damn it!

Upon coming to Moscow, Ivanov turned out to be a big, strong man, with snow-white hair. He says it turned white when he was 15 and met a male avdoshka face to face for the first time. The shock was such that it made him ill for a long time and his hair became white as a result.

Some people at the seminar took Ivanov's story to be a piece of artistic prose based on a copious reading of the "snowman" literature. I was inclined to take it for a mixture of truth and fiction, seeing the latter clearly in some places and doubting in others. Ivanov did admit when we met that he had added some fiction to a true story, saying he was advised to do so to increase sales of the booklet. He seems to be quite indifferent to the damage this does to the value of his story for hominology. And I never got a clear answer from him why he had never tried to contact us, hominologists, in Moscow. (Why Komlev did not advertise his discoveries is clear enough.)

All in all, Ivanov's visit to Moscow did not dispel some of my colleagues' skepticism and the uncertainty of others. So this summer I asked a member of the seminar, Marina Smolyaninova, to go to Malaya Vishera to try and find other witnesses, besides Oleg Ivanov. I also asked her to search for some documentation of Alexander Komlev's existence (a record in the hospital for example.)

Marina was fairly successful on the first point, having interviewed not only a number of presumed local eyewitnesses but also a very important outsider. He is Dmitri Panov, 62, a retired engineer, residing in St. Petersburg, who had bought a house in a village not far from Malaya Vishera. He claims a good avdoshka sighting while gathering mushrooms in a swampy spot in the forest in July, 1998. He also saw what he took to be the creature's tracks.

Marina's report at the last seminar has favorably swayed the members' opinion. All agree now the Malaya Vishera area calls for serious fieldwork. The question is where to get the staff and the money, with us poor veterans growing poorer and older, and young blood not forthcoming under the country's harsh economic conditions.

I called for further efforts to substantiate with documents Komlev's existence (Marina was not successful on this score the first time) and to neatly separate truth from fiction in the booklet. If only a half of what Ivanov says is true (and he insists almost all is true, with very little exception), it's fantastic. Alexander Komlev's still unrecognized feat should become an inspiration for all dedicated workers in hominology.

(First published in *Bigfoot Co-op*, December 1999,pp.3-7)

SECTION 3

# The Chinese Scene

## On the Trail of the Yeren

**(From the Journal *Asia and Africa Today*)**

Scientific investigations into the problem of relict hominoids began in China in the late 1950s under the impact of the Yeti expeditions in Nepal and the analogous research going on in the Soviet Union, where in 1958 the USSR Academy of Sciences set up a Commission to study the "snowman" question. The Commission gathered a lot of relevant information, which was then published in four collections of *Information Materials* and in the monograph of the late Professor Boris Porshnev, entitled *The Present State of the Question of Relict Hominoids* (1963). The work contains twenty pages of terse text devoted to information about hominoids in China. Professor Porshnev's sources included such eyewitnesses as Bai Xin, director of the PLA Film Studio, Professor Ho Wailu, a well-known historian, and Professor Pei Wenchung of the Sinjiang branch of the Chinese Academy of Sciences.

Professor Pei Wenchung informed Professor Porshnev, in particular, about the Himalayan expedition organized by the All-China Federation of Sports. The expedition members included scientific workers of the PRC Academy of Sciences and biology professors of Peking University. The investigation was conducted from May to July of 1959 and was stopped as a result of the flare-up of the national movement in Tibet. The expedition interviewed the local population on the subject of "snowman" and found a hair of an unknown animal. A microscopic examination of the hair showed that it was different in structure from that of the bear, the yak and the other known mammals of the Tibetan fauna. These and other results of the Chinese scientists' research in the Himalayas were reported by zoologist Xiu Wen at a paleoanthropological conference in December 1959. The same year the newspapers *Renmin ribao* and *Guanmin ribao* carried an interview with Professor Porshnev and published his article on the riddle of the "snowman," written in co-authorship with the zoologist Dr. S. Kleinenberg.

Researches on this subject in China, and contacts with their Soviet counterparts, were interrupted by the "cultural revolution." However, in 1977 the Chinese Academy of Sciences returned to the problem of Yeren, after having mounted an expedition to the mountainous regions of central China in Hubei, Shanxi, and Sichuan provinces. The venture had been prompted by a telegram received from six workers of the Forest Zone Party Committee, who had reported sighting a "wildman" in the area of the Shen-nongjia mountain range in Hubei Province. Taking part in the expedition were research workers of the Institute of Paleoanthropology and Vertebrate Paleontology and the Peking Natural History Museum. The results of the expedition and the subsequent investigations were published in the Chinese magazines *Hua Shi (Fossils), Kasue Shiyan (Scientific Experiment), China Reconstructs,* in the newspaper *China Daily* and in other publications, as well as in the international scientific journal *Cryptozoology*.

## Earliest Historical Records and Folklore

In their efforts to solve the riddle of Yeren, Chinese scientists turned, in particular, to historical data and folklore in order to compare ancient reports with modern. According to anthropologist Zhou Guoxing of the Peking Natural History Museum, Chinese historical documents (among them written chronicles), contain numerous references to a "wildman," which is also given such names as "man-bear," "hairy man," "mountain monster," and others.[1]

In the period of the Warring States (475-221 B.C.) the poet Qu Yuan wrote a poem about "shangui" (a mountain monster), implying, according to some scholars, Yeren; and during the Ming Dynasty (1368-1644), the great pharmacologist Li Shizhen mentioned several kinds of "wildman" in the 51st volume of his monumental work *Compedium of Materia Medica*.

Even today, in the area of Fang district, Hubei Province, there are still legends about "maoren" (hairy men) or "wildmen." A local chronicle, about 200 years old, says that the Fang mountains are precipitous and full of holes, in which live many giant maoren. They sometimes come down to steal chickens and are liable to attack men. A lantern on which there is depicted a "maoren" figure was unearthed in this area during an archeological excavation. It has been dated as being 2,000 years old.

There are fairly widespread folk tales about Yeren among the peoples of China. One of the most well-known says that there was a long-haired "wildman" living in the depths of a mountain forest. When it saw people, it would smile, grab their two arms tightly, and then faint with laughter. Once recovered, it would kill and eat them. Thereafter, when people entered the mountains, they took a pair of hollow bamboo poles with them. If by chance they met a "wildman," they would put their arms into the poles, and when the "wildman" fainted with laughter, they would break away from him by slipping the poles off their arms, and run away.[2] It is noteworthy in this connection that the hominoid's ability to smile and laugh has also been reported in different regions outside China. As for Yeren's aggressiveness towards man, it seems that folklore, as usual, resorts to hyperbole, just as it makes the gorilla out to be an aggressive animal, whereas in reality it is a rather gentle creature.

## Eyewitness Accounts

These constitute an important source of information for Chinese investigators. They have collected about 300 sighting reports, covering a period from 1940s to 1980s from witnesses including scientists, officials, army servicemen, foresters, hunters, and members of agricultural communes. Most sightings were solitary, but there are also collective sightings.

The above-mentioned anthropologist Zhou Guoxing writes that among these numerous reports there are two worthy of special note, because the witnesses were scientific workers with a good knowledge of natural science. One is Wang Tselin, a biologist, who reported he saw a Yeren killed by hunters in the area of the Qinling mountains in the southwest of Shanxi Province in 1940. It was a female, with large breasts, whose whole body was covered with thick grayish-brown hair, which was longer and matted on the head. The "wild woman" had deep-set eyes, prominent cheek bones and out-jutting jaws. Except for its hairiness and greater height, the slain Yeren was very similar to the plaster model of a fossil female "Peking Man" exhibited at the Peking Natural History Museum. According to the locals questioned by Wang Tselin, there were two of them, probably one male and the other female. They had been in that area for over a month. They were tall, had great strength, and could move as rapidly uphill as on the flat, but did not have a language, and could only howl.[3]

A drawing from the ancient Chinese encyclopedia showing "wildman" (between the 12th and 3rd Centuries B.C.).

The other account, specially noted by Chinese investigators, belongs to geologist Fan Jingquan, who, with the help of local guides watched two Yeren in the mountain forest near Baoji District, Shanxi Province, in the spring of 1950. Fan Jingquan related that one day at the camp of his geological expedition he talked with two elderly local men who were employed as guides for his team. Sharing their knowledge of local nature the men told Fan, among other things, about their encounters with a "wildman." They said they saw him several times a year, especially during the autumn and winter, in the Wild Chestnut Forest. Usually they said the "wildmen" do not attack people, but nevertheless the guides cautioned the geologists. They warned them that if they encountered "wildmen," they should not look directly into the creatures' eyes, they should observe

An ancient Chinese drawing showing a frolicsome wildwoman.

The creature seen here is referred to as "Feifei." The information provided on this drawing states: "Most often, Feifei is portrayed as a creature covered with short hair, long hair on the head, big mouth with bared teeth (and sometimes with feet turned backward). In 'Er-ya tu' [the Chinese source both for Sinsin and Feifei - DB], Feifei is shown just like Sinsin, but with a sword in hand [a sign of fortitude?].

The creatures seen here are referred to as "Sinsin." The information provided on this drawing states: "Sinsin lives in mountainous ravines, resembles an ape, has human face and limbs, head hair is long, the head and face 'are put straight.' [a hint of upright walking?]. Its voice is like the crying of an infant and the barking of a dog.

their activities only from cover, and should not turn around and run, but move away slowly in the opposite direction.

During a month of work there, none of the geologists happened to see the "wildman," and so before the team prepared to move to another survey site, Fan Jingquan asked one of the guides to lead him to the Chestnut Forest where the creatures had been sighted. The guide complied, and the witness next reports:

> While there was still some light a 'wildwoman' actually appeared with a small child beside her (the height of this child was approximately 1.6 metre). ...The 'wildwoman' seemed alert and fearful of us, maintaining a distance of 200 metres from us. But the 'wildchild,' who seemed unaware 'like a newborn calf who does not fear the tiger,' even dared to approach the guide, attempting to eat the wild chestnuts which the latter had already gathered. The mother growled from time to time, emitting a sound like neighing, as if calling the young one to return to her side. The 'wildwoman' and her child appeared and disappeared behind the bushes. As the sun set the guide began to get apprehensive, so we hastily returned to our camp.

Having returned to the Chestnut Forest the next day, the geologist did not see the Yeren, but on the third day he had a lengthy observation. The "wildwoman" did not seem as fearful as on the first day. The "wildchild" was again the first to get very close to the guide, who had stepped in front of the geologist to give him protection. The "wildwoman" followed somewhat later. Writes Fan Jingquan:

> I was kneeling down, pretending to crack chestnuts and did not dare stand up. Curious and fearful I cast a sidelong glance in order to examine the lower part of the mother. I saw her clearly, even to the extent of observing some traces of blood on the hair on both sides of her thighs... This was a tense moment. Then the 'wildwoman' and child walked slowly away, and when they were about a hundred metres from us, I finally stood up and we returned to camp.
>
> On the way back the guide said, with some pride, that he had been observing the growth of the 'wildchild,' who would be seven years old that particular year. He said that the 'wildmen' lived on the mountain in a cave with a small entrance, which was just big enough to be sealed with a large stone, in order to prevent animal attacks. He also described many moving details of his personal experience of such encounters. I did not have any doubts as to the honesty of the elderly man and felt

Gong Yulan shows scientists a tree against which she saw a humanlike beast scratching its back. Hair from the bark was analyzed and determined to be from an unknown higher primate.

> he would not be exaggerating .... Because of the self-confidence of the elderly man in the forest in leading me to the 'wildman' and the ease of the first encounter, I believe that he essentially understood the habits of the 'wildman' in the area of Qinling.
>
> A year later, on my return to Peking, Soviet scholars learned, through various geological reports, of this occurrence. They interviewed me personally to inquire into the details of this incident.[4]

Two detailed descriptions of a male Yeren come from a production team leader and a forest protection officer of Cuifeng Commune, Zhouzhi District, Shanxi Province. Their sightings occurred in June and July 1977 and refer to a male "wildman" (apparently, the same individual in both cases), over two metres tall and very broad in the shoulders. The upper limbs reached beyond his knees. He had large hands and wide feet and was covered with dark brown hair, longer on the head and spreading down to his shoulders. Except for the nose and ears, the whole face was covered with short hair. His orbital ridge was high, his eyes were sunken, the nostrils directed slightly upwards with the tip of the nose looking like a fleshy lump. The lower jaw jutted out, and his incisors were broad, just like a horse's. The creature even while walking uphill had a 1.5 metre stride. The "wildman" uttered inarticulate sounds and was heard to imitate bird chirping, animal noises, and an infant crying.[5] It should be noted that similar vocal mimicking by hominoids has been repeatedly reported by witnesses in other hominoid habitats.

## Footprints and Other Evidence

One of the most important categories of evidence in favor of the biological reality of "wildmen" comes from their footprints. Chinese investigators are reported to have found over a thousand tracks left by Yeren, and to have made photographs and plaster casts of the best prints. On the whole, a Yeren's footprint is similar to that of a man's bare foot, but at the same time it has certain peculiarities of its own, such as a flattened arch or its total absence, and a considerable lateral mobility of the great toe; moreover very often these tracks are much larger than a man's (there are many footprints measuring 36–38 cm in length, while the largest track found in China reached 48 cm).

The discovery of whole sets of footprints made it possible to judge not only the anatomy of the foot of Yeren, but also the peculiarities of his gait, as well as the size and weight of his body. As is the case in other habitats, the "wildman's" footprints are found arranged in a single line, the toes of one foot pointing to the heel of the other, the length of stride varying from 50 to 150 cm. According to press reports, in June 1980 in the mountains of Jiandao on the border of Sichuan and Hubei provinces a string of huge footprints was found. Each print was 47 cm long, with a distance of 95 cm between them. From the length of the footprints, the distance between them, and the depth of their impressions in the earth, Chinese investigators estimated that the "wildman" was about 2.6 meters tall and weighed about 250 kg. Beside the footprints, they found a "nest" made by intertwining bamboo stems growing in a cluster, a process which demands tremendous physical strength. There was no sign of activity by humans, or any other animals, in that place.[6]

Over 100 hairs have been collected that are said to belong to Yeren. Some of these were collected by eyewitnesses; the rest were collected by researchers in June 1976 from a tree that a "wildman" was believed to have rubbed against. The hair was studied by the Hubei Provincial Medical College and the Institute of Paleoanthropology and Vertebrate Paleontology in Peking. The general consensus is that the hair belongs to a higher primate.[7] However, scientists cannot say definitely that the hair belongs to Yeren because they do not have an approved sample of Yeren hair.

In August 1977, members of a scientific expedition in the mountains of Shennongjia, Hubei Province, following the trail of a Yeren, discovered that the latter had searched out insect larvae from under the bark of trees. Nearby excrement was found (different from that of predatory and hoofed animals), which contained large amounts of cocoon skin. In other cases, excrement, ascribed to Yeren, contained bits of wild fruits, chestnuts, and bamboo shoots.[8]

Previously it was reported that the mummified hands and feet of an unknown primate, killed by peasants in 1957 in Zhejiang Province, had been handed over to scientists by a local teacher of biology. These remains were studied by anthropologist Zhou Guoxing, who writes as follows: "The eyewitnesses thought that they had belonged to a 'wildman,' or a 'strange manlike animal,' but after examining the specimens, I established that they were not the hands and feet of a 'wildman.' They might possibly belong to an enormous monkey .... There is no denying the possibility that they came from an unknown primate in the Jiolong Mountain area."[9]

Thus, judging by publications, as a result of widespread investigations during the second half of the 1970s, Chinese scientists obtained a lot of circumstantial evidence in favor of the existence of Yeren, namely: historical and folklore data, eyewitness reports, footprints, hairs, faeces, and hairs. However, in contradistinction to the successful photographing of Bigfoot-Sasquatch of North America, Chinese researchers have been unable to photograph Yeren or obtain irrefutable physical remains of "wildman." This comes as no surprise to hominologists (students of hominoids) in other countries because science has only recently set about coping with this unusual and difficult task.

And what is the opinion of China's scientific community in regard to this question? Anthropologist Zhou Guoxing in his article in *Cryptozoology* says that there are two completely different views on "wildman" in Chinese scientific circles. The majority of scientists reject the existence of animals of human shape on ecological and zoogeographical grounds. They hold that the legendary "wildman" merely represents some known animal, such as a bear, a monkey, or a gibbon, or that reports of "wildman" might be due to hallucinations, or even deliberate fabrications. A smaller number of scholars are of the opinion that the existence of "wildman" should not be rejected, and that "wildman" might be a living species yet to be identified by science. Even among the latter, there are different views on the classification of the unknown creatures. Some think they are surviving descendants of *Ramapithecus* or *Australopithecus* (especially *A. robustus*), both remote relatives of modern man. Others assume that "wildman" belongs to the ape family, and is possibly a living descendant of *Gigantopithecus* or the orangutan, which thrived in southern China in the Pleistocene period (1,000,000–500,000 years ago).

Zhou Guoxing notes that in most of the areas where legendary "wildman" is reported in modern times, there are still found primeval forests that contain quite a few surviving species of trees of the Tertiary period. *Gigantopithecus* was the dominant member of the *Ailuropoda-Stegodon* fauna thriving on the mainland of China in the middle and later Pleistocene period. Later, most of the members of this fauna disappeared because of geological changes. However, there are still quite a few survivors, including the tapir and orangutan, not to mention the giant panda, which changed its habits and characteristics and remained in the middle and western parts of China. Therefore, it is not impossible that *Gigantopithecus*, as the dominant member of this fauna, could also have changed its original habits and characteristics and survived to the present day. It may have evolved into the large "wildman" now reported in China, or into the "snowman" (Yeti) in areas in the southern part of the Himalayas, and might even have crossed the Isthmus of Bering to become the Sasquatch (Bigfoot) of North America.[10]

It should be noted that some Soviet researchers, in line with the ideas of the late Professor Porshnev, hold a different view on the origin and classification of relict hominoids,[11] but it is not within the scope of this article to compare hypotheses on this question.

## References

[1] Zhou Gouxing, "The Status of Wildman Research in China," *Cryptozoology,* Vol, I, 1982.

[2] *Ibidem.*

[3] Yuan Zhenxin, Huang Wanpo, "A Challenge to Science: the Mystery of the Wildman," *Hua Shi,* No. 19, 1979.

[4] Fan Jingquan, "I Witnessed a 'Wildman' Mother and Child in the Chestnut Forest," *Hua Shi*, No. 23, 1980; *Fortean Times Occasional Paper,* No. 1, 1981.

[5] Yuan Zhenxin, Huang Wanpo, "A Challenge to Science: The Mystery of the Wildman," *Hua Shi*, No. 19, 1979.

[6] Zhang Huimin, "Now Scientists Join Hunt for the Wild Man," *China Daily*, June 30, 1981.

[7] F. Poirier, Hu Hongxing, Chung-Min Chen, 'The Evidence for Wildman in Hubei Province, People's Republic of China," *Cryptozoology,* Vol. 2, 1983.

[8] Yuan Zhenxin and Huang Wanpo, " 'Wildman—Fact or Fiction?," *China Reconstructs*, No. 7, 1979.

[9] Zhou Guoxing, "The Status of Wildman Research in China," *Cryptozoology*, Vol. 1, 1982.

[10] Zhou Guoxing, *Op. cit.*

[11] Dmitri Bayanov and Igor Bourtsev, "On Neanderthal vs. Paranthropus," *Current Anthropology*, Vol. 17, No. 2, June 1976.

(Originally published in *Asia and Africa Today*, No. 2, 1985)

SECTION 4

# Homins Down Under

## The Case for the Australian Hominoids

According to Grover Krantz:

> Some enthusiasts have gathered evidence of hairy bipeds from all continents except Antarctica. Outside of limited areas of North America and Eurasia, this evidence is very scanty, though not necessarily incorrect just for this reason. A single such species with virtually world-wide distribution would be very unlikely. Only man, his domestic animals, and his parasites have such a wide distribution. It is often thought that the more widespread the evidence, the more true it is likely to be. Actually the opposite is the case here and it only increases the likelihood of human mythology as the more obvious explanation. If there are more than one species, then a wide distribution is more acceptable. But if one 'unknown' higher primate is difficult to believe in, three or four would border on the absurd. Those who take this subject seriously would be well advised to postulate no more than is absolutely necessary. To suggest more only makes one sound a bit like a fool. (Krantz 1984, 297–98)

However, the question is: How many hairy bipeds is "no more than is absolutely necessary" for a serious student? I know eyewitnesses in the Caucasus who believe that the "wild man" has survived only in that area and nowhere else. I know a very serious person who claims to have seen a family of the Almas of Mongolia and believes the hairy bipeds inhabit only that country, whereas reports from elsewhere are nothing but folklore. The implication is "my story is true but your story is dubious." In my view, those who take this subject seriously would be well advised to keep clear of such "wild man" patriotism.

All hominologists agree on the hominoid's status in time: "everlasting." What about his status in space? At the risk of sounding "a bit like a fool," I say: ubiquitous.

First, a theoretical consideration. "The capacity for bipedal walking is primarily an adaptation for covering long distances" (Washburn 1960). Being a biped, *Homo sapiens* initially reached distant lands simply by walking. Where lands were separated by too wide a stretch of water, man waited some millennia for the sea level to drop and then marched on to new horizons over a land bridge. But *Homo sapiens* is not the only bipedal primate known to science. So are *Homo neanderthalensis* and *Homo erectus* and even *Australopithecus*. Deduction: Any land colonized by a walking *Homo sapiens* can and, in fact, must have been colonized by an earlier hominid. Hence such a land can, with luck, harbor what we now call relic hominoids.

I must admit there is a lot of hindsight in this theorizing. It is one thing to write about the hominid conquest of the earth as I did (Bayanov 1976, 312) and quite another to add overnight a whole continent to the "empire" of the troglodytes[1]. And what a continent: Australia. So, when the news broke out in 1976, I reacted with wonder and disbelief.

The media advertised the Yowie—the wild man of the land of the kangaroo—as Australia's answer to the Abominable Snowman of the Himalayas and the Bigfoot of America. And the advertisement could be swallowed for what it was when reading about such details of the Australian biped as "the face . . . being like that of an ape or man, minus forehead and chin, with a great trunk all one size from shoulder to hips, and with arms that nearly reached to its ankles" (Joyner 1977). All right, quite a familiar figure for the Yeti or Bigfoot. But then you looked at the year of the report, which was 1912, and you asked yourself in amazement: But aren't the Yeti and Bigfoot an answer to the Yowie?

My present opinion that Australia is as valid a habitat of the hominoid as Eurasia and America is largely due not to the news media but to a little booklet, *The Hairy Man of South Eastern Australia*, Canberra,

---

[1] *Troglodytes recens* is a term proposed by B.F. Porshnev (1974, 449) to indicate living hairy bipeds generally called by us relic hominoids. The word "hominoid" is used here in its etymological sense of "man-like" and indicates any higher bipedal primate which is not *Homo sapiens*.

1977, compiled by policy analyst Graham C. Joyner. Let me simply share with the reader some of the information in the booklet which "did it" in my case. And please note in each case the date of the report.

(6)

"Superstitions of the Australian Aborigines: The Yahoo," *Australian and New Zealand Monthly Magazine*, 1842, pp. 92–96.

> The natives of Australia have, properly speaking, no idea of any supernatural being; at the same time, they believe in the imaginary existence of a class which, in the singular number, they call Yahoo, or, when they wish to be anglified, Devil-Devil... On the other hand, a contested point has long existed among Australian naturalists whether or not such an animal as the Yahoo existed, one party contending that it does, and that from it scarceness, slyness, and solitary habits, man has not succeeded in obtaining a specimen, and that it is most likely one of the monkey tribe.

(7)

Mrs. Charles Meredith, *Notes and Sketches of New South Wales during a Residence in the Colony from 1839 to 1844*, London, 1844, p. 95.

> I never could make out anything of their [the aborigines'] religious ideas, or even if they had a comprehension of a beneficent Supreme Being; but they have an evil spirit, which causes them great terror whom they call 'Yahoo,' or 'Devil-devil': He lives in the tops of the steepest and rockiest mountains, which are totally inaccessible to all human beings. . . The name Devil-devil is of course borrowed from our vocabulary, and the doubling of the phrase denotes how terrible or intense a devil he is; that of Yahoo, being used to express a bad spirit, or 'Bugaboo,' was common also with the aborigines of Van Diemen's Land, and is as likely to be a coincidence with, as a loan from, Dean Swift.

Is it really and merely a coincidence that the hominoids of Australia and those described by Jonathan Swift in *Gulliver's Travels* bear the same name? This fascinating mystery is still to be explored. What we know at present is that the famous book was first published in 1726 while the first British settlement in Australia, according to my information, is dated 1788. To compound the riddle John Green (1973, 11) cites a J.W. Burns, who explored the mystery of Sasquatch in British Columbia in the 1920s, writing in *MacLean's* magazine, April 1, 1929 in an article titled "Introducing B.C.'s Hairy Giants – A collection of strange tales about British Columbia's wild men as told by those who say they have seen them," as follows: "The old man said that she [Sasquatch female] spoke the words 'Yahoo, yahoo' frequently in a loud voice, and always received a similar reply from the mountain."

But to return to Australia and Joyner's booklet:

(9)

*The Goulburn Herald*, May 24, 1881, p. 2.

The *Cooma Express* relates that the Jingera hairy man has again turned up. It was seen on Saturday last by Mr. Peter Thurbon and one or two others. This is the first appearance for some considerable time past. The animal, if such it be, has the appearance of a huge monkey or baboon, and is somewhat larger than a man.

(12)

*The Qucanbeyan Observer*, November 30, 1894, p. 2.

The Braidwood *Dispatch* says that on the 3rd of October last young Johnnie McWilliams was riding from his home at Snowball to the Jinden P.O. When about halfway the boy was startled by the extraordinary sight of a wild man or gorilla. The boy states that a wild man suddenly appeared from behind a tree, about thirty yards from the road, stood looking at him for a few seconds, and then turned and ran for the wooded hills a mile or so from the road ... The boy states that he appeared to be six feet in height and heavily built. He described it 'as a big man covered with long hair.'. . . When running it kept looking back at the boy, till it disappeared. It was three o'clock in the afternoon, and the boy describes everything he saw minutely. The boy is a truthful and manly young fellow, well acquainted with all the known animals in the New South Wales bush, and persists that he could not have been mistaken. For many years there have been tales of trappers coming across enormous tracks of some unknown animals in the mountain wilds around Snowball. Of course, these tales were received with doubt, and put down as clever romancing on the part of the 'possum hunter,' but the story of Johnnie McWilliams is believed by all who know the boy as a true tale. The proof of the existence of such an animal in New South Wales should be of some interest to the naturalist.

(13)

John Gale, *An Alpine Excursion* (Queanbeyan, 1903) pp. 85–89.

Swift, in his introduction to the travels of Gulliver, speaks of a tribe of animals with certain human attributes, passions and vices. The work mentioned is, we know, fiction-satire. But is there in reality (as in fiction) such a creature?... I had hesitated to refer to the story of the yahoo here; but on fuller consideration I arrive at the conclusion, that not only because it is readable matter, but because the allegation I am about to narrate ought to be known in the interests of natural sciences and the zoology of Australia. My informants—the gentlemen who (amongst others whose unsupported statements might be taken *cum grano salis*) aver they have seen this wild, mysterious creature — are not ignorant persons or superstitiously inclined; they are strongminded, experienced, and educated men ... Mr. Cox was relating what had befallen him when camped alone in the ranges of Brindabella about two years ago on a shooting expedition. He was, he said, enjoying his billy of tea in the afternoon, when his attention was drawn to an enraged cry, between a howl and a yell, in the thick shrub of a gully close by. He instantly seized his rifle and looked in the direction whence the sounds proceeded. There he saw a huge animal in an erect posture tearing through the undergrowth, and in a moment it was out of sight before he could bring his rifle to his shoulder. He distinctly heard the crashing of the undergrowth in its flight, and he followed after it. Its speed was greater than that of its pursuer; but as it fled its howling and yelling continued. That it was no creation of an excited imagination—(and from what I know of Mr. Cox, he is not likely subject of wild hallucinations; but, on the contrary, a remarkably cool, intrepid fellow, too well enlightened and educated to magnify a simple fact into a chimera)—is confirmed by this, that in his pursuit he met several wallabies tearing up the gully in such alarm that, though passing close by, they took not the least notice of him. These were followed presently by a herd of cattle similarly scared. Further pursuit was vain, for the thing had now gone beyond sight and sound.

The sources of a second story were Joseph and William Webb.

They were out in the ranges preparing to camp for the night. Down by the side of a range to the eastward, and with only a narrow gully separating them from the object which attracted their attention, they first heard a deep guttural bellowing and then a crashing of the scrub.

Next moment a thing appeared walking erect, though they saw only its head and shoulders. It was hirsute, so much of the creature as was visible, and its head was set so deep between its shoulders that it was scarcely perceptible. It was approaching towards their camp. Now it was in full view, and was of the stature of a man, moving with long strides and a heavy tramp. It was challenged: 'Who are you? speak, or we'll fire.' Not an intelligible word came in response; only the guttural bellowing. Aim was taken; the crack of a rifle rang out along the gully; but the thing, if hit, was not disabled; for at the sound of the shot it turned and fled. The two gentlemen, filled with amazement and curiosity, but not alarm, went to where they had seen and shot at this formidable-looking creature, and sought for its tracks in verification of what had happened. There were its footprints, long, like a man's, but with longer, spreading toes; there were its strides, also much longer than those of a man; and there were the broken twigs and disordered scrub through which it had come and gone. They saw no blood or other evidence of their shot having taken effect.

(15)

A letter published on the subject in the Queanbeyan *Observer*, August 7, 1903, p. 2. (A postscript to the letter mentions an interview with a Harry Williams, who saw the aborigines kill such a creature below the junction of the Yass River with the Murrumbidgee.)

There were a good many blacks at the killing of it, and he saw two blackfellows dragging it down the hill by its legs. It was like a black man, but covered all over with grey hair.

(17)

*The Sydney Morning Herald*, October 23, 1912, p. 15.

...on Sunday (October 12), I heard that George Surnmerell, a neighbour of mine, while riding up the track which forms a short cut from Bombala to Bemboka, had that day, about noon, when approaching a small creek about a mile below 'Packer's Swamp,' ridden close up to a strange animal, which, on all fours, was drinking from the creek. As it was covered with grey hair, the first thought that rose to Surnmerell's mind was: 'What an immense kangaroo.' But, hearing the horse's feet on the track, it rose to its full height, of about 7 ft, and looked quietly at the horseman. Then stooping down again, it finished its drink, and then, picking up a stick that lay by it, walked steadily away up a slope to the right or eastern side of the road, and disappeared among the rocks and timber 150 yards away.

Surnmerell described the face as being like that of an ape or man, minus forehead and chin, with a great trunk all one size from shoulder to hips, and with arms that nearly reached to its ankles.

Hearing this report, I rode up to the scene on Monday morning. On arriving about a score of footprints attested the truth of Surnmerell's account, the handprints where the animal had stooped at the edge of the water being especially plain. These handprints differed from a large human hand chiefly in having the little fingers set much like the thumbs (a formation explaining the 3–1 series of scratches on the white gum tree).

A striking peculiarity was revealed, however, in the footprints; these, resembling an enormously long and ugly human foot in the heel, instep, and ball, had only four toes—long (nearly 5 inches), cylindrical, and showing evidence of extreme flexibility. Even in the prints which had sunk deepest into the mud there was no trace of the 'thumb' of the characteristic ape's 'foot.'

Beside, perhaps a score of new prints, there were old ones discernible, showing that the animal had crossed the creek at least a fortnight previously. After a vexatious delay, I was able, on the Wednesday afternoon, to take three plaster of Paris casts—one of a footprint in very stiff mud, another in very wet mud, and a third of the hand with its palm superimposed on the front part of the corresponding foot. These I have forwarded to Professor David, at the university, where, no doubt, they can be seen by those interested. Anyone acquainted with the nature of mud will not expect to find a cast taken therein three days after imprint as technically perfect as a casting from a regular model, but I believe that any reasonable being will be satisfied by an inspection of these three casts that something quite unknown and unsuspected by science remains to be brought to light.

Since this matter has made such a stir that people in this district have felt that they could attest their experience without further fear of ridicule, an astonishing number of confirmatory cases have come to my knowledge ranging over the country between Cape Howe and Wee Jasper. Such of these accounts as seem of significance I hope to collect.

As Graham Joyner informs us, the author of the above highly commendable report, which shows that Australia was leading, not following, other countries and continents in hominology, is poet Sydney Wheeler

Jephcott (1864–1951), who "spent most of his life in the bush as a cattleman and dairy-farmer .... Neither the casts themselves nor any accounts Jephcott may have collected seem to have survived."

(19)

*The Sun* (Sydney), November 10, 1912, p. 13. The following account was given by a Mr. Charles Harper, described as a licensed surveyor from the Sydney suburb of Leichhardt:

> For many years past vague and mysterious rumours have been current of an Australian gorilla, or 'hairyman,' or some such animal, seen on and in the wild uninhabited mountains and gorges forming the Currickbilly Range, which runs parallel to the south coast of this State, from the head of the Clyde River to North Gippsland ... Scientists assert that this animal, like the bunyip, is a myth, and such animals do not, and never did, exist in this continent, although the old generation of aboriginal natives assert the contrary in both cases.
>
> In various parts of the southern district of this State on the coastal slopes, and at various times, extending over a very long period, I have met men (and reliable men at that) who unhesitatingly assert that they have seen this hairy man-shaped animal at short distances. They were so terrified at the apparition and the hideous noise it made when it saw them that they left their work as timber-getters, and at once cleared out from the locality, leaving their tools and work done behind them. The description of this animal, seen at different times by different people in several localities, but always in the jungle, invariably coincided.
>
> At the risk of being considered by your readers the reincarnation of Ananias or the late Thomas Pepper, I will describe this animal as once seen briefly as possible. I had to proceed some distance into the heart of these jungles for a special purpose, accompanied by two others, and two large kangaroo dogs with a strain of the British bulldog in each.
>
> [Joyner's interjection: On the night of the second day, having just turned in, they heard a most terrifying sound which utterly demoralised the dogs. One of the men was induced to throw dry kindling on the fire, illuminating the scrub for some distance around.]
>
> A huge man-like animal stood erect not twenty yards from the fire, growling, grimacing and thumping his breast with his huge hand-like paws. I looked around and saw one of my companions had fainted. He remained unconscious for some hours. The creature stood in one

position for some time, sufficiently long to enable me to photograph him on my brain.

I should say its height when standing erect would be 5 ft. 8 in. to 5 ft. 10 in. Its body, legs, and arms were covered with long, brownish-red hair, which shook with every quivering movement of its body. The hair on its shoulder and back parts appeared in the subdued light of the fire to be jet black, and long; but what struck me as most extraordinary was the apparently human shape, but still so very different.

I will commence its detailed description with the feet, which only occasionally I could get a glimpse of. I saw that the metatarsal bones were very short, much shorter than in the genus homo, but the phalanges were extremely long, indicating great grasping power by the feet. The fibula bone of the leg was much shorter than in man. The femur bone of the thigh was very long, out of all proportion to the rest of the leg. The body frame was enormous, indicating immense strength, and power of endurance. The arms and forepaws were extremely long and large, and very muscular, being covered with shorter hair. The head and face were very small, but very human. The eyes were large, dark and piercing, deeply set. A most horrible mouth was ornamented with two large and long canine teeth. When the jaws were closed they protruded over the lower lip. The stomach seemed like a sack hanging halfway down the thighs, whether natural or a prolapsus, I could not tell. All this observation occupied a few minutes while the creature stood erect, as if the firelight had paralysed him.

After a few more growls, and thumping his breast, he made off, the first few yards erect, then at a faster gait on all fours through the low scrub. Nothing would induce my companions to continue the trip, at which I was rather pleased than otherwise, and returned quickly as possible out of reach of Australian gorillas, rare as they are.

Who can vouch for the veracity of such a bizarre account? Yet, the creature's foot in this account has surprising correlations with what has been postulated for the foot of Sasquatch from the evidence of footprints. Also certain characteristics and proportions of its body and legs are corroborated by the hominid fossil evidence. The creature's stomach hanging like a sack "halfway down the thighs" is something utterly new and enigmatic to me. Supposing the sighting was real, could it be that the creature was sporting some other animal's skin around its groin?

(21)

*The Sun*, November 24, 1912, p. 13.

Mr. A.B. Walton, of Granville, remembers years ago, hearing the blacks of the Braidwood district speak of 'big feller devil' which they called a 'Yahoo.' They describe it as being taller than a man, and covered with hair. They had seen it climb trees. On one occasion when disturbed by a party of blacks it seized a girl by the throat and strangled her on the spot. The other natives ran away in terror.

According to the aborigines, the 'Yahoo' was only rarely seen, but generations testified to its existence.

'I was only a boy at the time,' says Mr. Walton, 'and never saw the creature myself, but I have no doubt from what the blacks said that it is not a myth.'

(24)

*The Sydney Morning Herald*, June 8, 1935, p. 11.

Though there are few full-blooded aborigines still remaining on the south coast of New South Wales, you can still hear all about the 'Douligah,' or the hairy man, from the few who do remain. Every aboriginal on the south coast firmly believed in the evidence of the douligah, and they dreaded him as children dread the bogeyman. They describe the douligah as a man of powerful build, capable of tearing down small trees and lifting great rocks. He had hair all over his body, and though he remained in the mountains during daylight, he frequently visited the abos' camps at night, and sent them scampering for shelter in caves.

(25)

Roland Robinson, a passage on the doolagarl in "3 Aboriginal Tales," *The Bulletin*, October 13, 1954, p. 27: "A doolagarl is a gorilla-like man. He has long spindly legs. He has big chest, long arms. His forehead goes back from his eyebrows. His head goes into his shoulders, no neck. They live now on Cockwhy and Polawombera Mountains."

(28)

Henry Lawson, "The Hairy Man," in *Triangles of Life and Other Stories,* Melbourne, 1913, pp. 210–23.

But the Hairy Man was permanent, and his country spread from the eastern slopes of the Great Dividing Range right out to the ends of the western spurs. He had been heard of and seen and described so often and by so many reliable liars that most people agreed there must be something. The most popular and enduring theory was that he was a gorilla, or an orang-outang which had escaped from a menagerie long ago. He was also said to be a new kind of kangaroo, or the last of a species of Australian animals which hadn't been discovered yet. Anyway, in some places, he was regarded as a danger to children coming home from school, as were wild bullocks, snakes, and an occasional bushman in the d.t.'s. So now and then, when the yarn had a revival, search parties were organised, and went out with guns to find the Hairy Man, and to settle him and the question one way or another. But they never found him.

So much, in essence, of what positively impressed me in the 1977 issue of Joyner's booklet. In 1980 he added to it some pages of "More Historical Evidence for the Yahoo, Hairy Man, Wild Man or Australian 'Gorilla,'" from which, for lack of space, I'll quote just one account because it gives a new angle or viewpoint on the matter.

William Telfer, *The Early History of the Northern Districts of New South Wales* (c. 1898), University of New England Archives A147/V213, pp. 32–34. Joyner's comment: "Telfer was born at Calala near Tamworth on 26 July 1841. From his own account he seems to have spent his life as a shepherd or a drover. When he died at Gosford on 8 December 1923 he was described as a boundary rider by one of his children."

Here is Telfer's description of an encounter with a hairy man—Joyner retains, and so I after him, the spelling of the document:

> ... i had an Experience myself of this gorilla or hairy man in the year 1883 i was making a short cut across the bush from Keera to Cobedah ... i heard a curious noise coming up the creek oposite the camp over the creek i went to see what it was about one hundred yards away he seemed the same as a man only larger the animal was something like the Gorilla in the Sydney museum of a darkish colour and made a roaring noise going away towards top Bingera ... i was thinking how Easey this animal could Elude persuit travelling by night camping in Rocks or Caves in the daytime.

But the most interesting passage, which introduces a new element into the subject, is this [emphasis is mine—D.B.]:

> ...then they [the aborigines] have a tradition about the yahoo they say he is a hairy man like a monkey *plenty at one time not many now* but

> the best opinion of the kind i heard from old Bungaree a Gunedah aboriginal he said *at one time there were tribes of them and they were the original inhabitants of the Country before the present Race of aboriginals took possession of the Country he said they were the old Race of blacks* ... he said the aboriginals would camp in one place and those people in a place of their own telling about how them and the blacks used to fight and *the blacks always beat them but the yahoo always made away from the blacks being a faster runner* mostly Escaped the blacks were frightened of them a lot if those were together the blacks would not go near them as *the yahoo would make a great noise and frighten them with sticks*. He said *very strong fellow very stupid the blacks were more Cunning* getting behind trees *Spearing any Chance one that Came near them* this was his Story about those people.

And what a magnificent story it is! An aboriginal through a shepherd gives the world in a nutshell an anthropological truth of the first magnitude: How *Homo sapiens* colonized a land already inhabited by an earlier form of hominid. The comparative description of the aboriginals' and the yahoos' physical and mental powers, their arms, and patterns of behavior leaves no doubt on this score in the reader's mind. While science deliberates, with its nose to the ground and its eyes on the fossils, the live tradition of the people puts science wise in this respect, telling what really happened on earth in prehistoric times. And such traditions are to be found all over the world, where *Homo sapiens* once confronted his evolutionary predecessor and, gaining the upper hand, gradually reduced him to the state of a relic. Here is, for example and comparison, how this tradition manifests itself in the Caucasus. The story is taken, as can be expected, not from scientific but artistic prose—an autobiographical book telling of the author's life as a boy in a Georgian village in pre-revolutionary Russia. The village boys, including the author, went on a hike to the mountains and found their way into a secluded cave wherein they discovered the following [emphasis is mine—D.B.]:

> From cracks in the ceiling ghostly vines trailed, and a pale green light filled the room. The floor was covered with bones, bleached white, huge bones, ribs that would have made a cage for a hen and all her chicks, hips wider than an ox yoke, skulls like wine jugs *with eye sockets our fists could go in, bulging foreheads, huge jaws*.
>
> I pointed at the teeth. 'As big as a horse's.'
>
> '*But these aren't horses*,' Teddua said, 'or *cows*.'
>
> 'Not water buffalo and not stag,' Bootla said.

'I know that. Bones from those animals we saw dozen of times. These are men, I think. *Some kind of men.*'

'Men like us,' Bootla said. 'Look. They had long arms with five-fingered hands and feet with toes, not paws.'

'And not hoofs.' Teddua picked up a leg bone. It was half as tall as he. Carefully, he laid it back. 'These men could take apples from the top of the highest tree. Without stretching.'

'And cover six feet in a single step,' Bootla said.

The light from the roof was fading. 'I think we'd better go,' Teddua said. We took a last quick look around. The eyes in the skulls watched us as we went through the passage and swam to the shore.

Walking home, we could talk of nothing else, of course, but the men in the cave. Who were they?

Where did they come from? Why were they buried there? How long ago? We had no answer.

'We will have to ask the Old Men,' Teddua said. 'If anybody knows, surely it must be one of them. . .'

When we came home, my father was there, and we told our story to him and to the parents of Teddua and Bootla. They, as puzzled as we, agreed that we certainly must talk to the Old Men.

So we told our story again while the Old Men listened carefully. When we were through, they asked us questions.

*'Were their swords and daggers beside them?'*

*'Any bowls or tools in the cave?'*

*'Pieces of jewelry or scraps of cloth near the bones?'*

*'No,'* we said.

'And the upper leg bone was how long?'

Teddua marked two feet on the floor.

'And the lower one?'

He added another two.

'I think,' Miriani said, '*you found the Place Where the Giants Came Home to Die.*'

'Were they men?'

'Yes,' Miriani said, '*different from us, but men. Long ago they lived on earth.*'

'Some call them Narts,' Vachtang said. 'Some say they were the first men. We came afterward.'

'What happened to them?' Bootla said.

'Slowly, one by one, they disappeared. Nobody knows where or how, for they left no graves, no bones.'

'But I think you found their resting place,' Miriani said. 'Perhaps *as some animals do, the Giants knew when their time had come, and when they grew old or sick they went to their cave to die.*'

'Did you ever see one of the Giants?' Teddua asked.

'No, but my great-great-grandfather did. At least, he saw a Giant's footprint in the snow up above Mleti.'

'Are all the Giants dead?' Bootla said.

Otar shrugged. 'So they say, but who knows?'

'I have heard *a few, a very few still live at the top of Mount Kazbeck,*' Vachtang said. 'But unfortunately I never had the pleasure to meet one. At least so far.' (Papashvily 1973)

This story of artistic prose should be of great interest not only to the anthropologist and the historian but to the hominologist as well, because it gives him an idea what happens to the skeletons of relic hominoids for which the skeptics are clamoring.

We (Bayanov and Bourtsev 1976, 314–15) said that ancient Greeks and Romans developed their correct ideas of anthropogenesis proceeding from the actual knowledge of contemporary ape-like cave dwellers (troglodytes), i.e., relic hominoids, which were referred to at the time under an assortment of mythological names. As the above examples tend to show our thesis is becoming ever stronger with more evidence coming in.

But back to Australia. An attentive reader may have noticed that all Australian references, cited so far, mention "hairy men," not women or children. For an outside biologist this may be a puzzling obstacle to recognizing the biological reality of the creatures. But not for the hominologist, because this is the right pattern of such sightings all over the world: males are spotted more frequently than females, and females more frequently than little ones. This has nothing to do with their numbers but says a lot for their caution. With other primates, such as gorillas and baboons, it is the male who is the first to confront any danger to his family, and so it seems to be, but to a much greater extent, with relic hominoids.

As will be shown below, the present-day evidence does indicate hominoid females and young in Australia. But it is necessary to stress here that the present-day evidence would never be regarded as seriously and attentively by the hominologist as it is now if not for the historical evidence, such as presented by Graham Joyner's booklet. The author has dug up and given a new lease of life to some of those "photographs on the brain," to those quests for the truth in this problem which had been made by inquisitive and intelligent people in the 19th century and the beginning of the 20th and then quietly buried for decades in the archives and libraries. This also happened in our country and in America. That is why I wrote Joyner: "I find your hominological 'digs' in the archives as valuable to hominology as the Leakey digs to paleonanthropology."

Now let us look, in the light of historical data, at the present-day record of the Australian yahoo, whose name has changed to Yowie. Incidentally, regarding this change Joyner tells me the following:

> The word 'Yowie' has become popular since May 1975 so that it is now, like Yeti or Sasquatch elsewhere, the common word associated with the phenomenon in Australia—but even so it is little known. I have never come across the word at all before 1975 in the area I have studied, or indeed anywhere, but it is possible that it has been used in certain rural areas, perhaps by some people of aboriginal descent, perhaps by some Europeans. It may be the corrupt form of some aboriginal word, or a variant or corruption of Yahoo.

Here is the Yowie's portrait as presented in the media: "It is tall, hairy with head set deep into broad shoulders. It walks on two legs with a loping gait and has a foul smell. Normally it is placid, conversing in grunts, but when provoked or alarmed gives a high-pitched shriek" (Attwood 1979, 9). A Yowie family mentioned:

> Shooters have flushed Brisbane Valley's first and only Yowie family out of their refuge in the ranges around Kilcoy. Sightings of the three strange beasts in recent days suggest the Yowies are moving towards Somerst Dam ... They found three separate tracks—one went along the creek, one kept to higher ground, and the other was in between. All three tracks later converged on one point, but all three tracks were different. 'This led us to believe it was a family group,' Warren said. 'There was a big heavy print, a smaller one and then a smaller one still.'
>
> 'It looked like they were migrating,' he said. 'They appeared to move at night and rest up during the day.' Warren speculated that the yowies may have been chased out of the Stanley Creek area 16 kilometres away by the large number of shooters operating in the area. (Parfitt 1980, 8).

A Yowie female sighted:

> It was April 13, 1976. The bush of the Grose Valley, near Katoomba,

NSW, has been hard for the five bushwalkers to push through, and it was thick, muffling the usual bush noises. But something else was crashing about the scrub, coming towards them. Suddenly it burst into view, and stood, only metres away.

The five got the shock of their lives. In front of them stood a creature they estimated at 1.4 m tall and 1.2 across the shoulders. Pendulous breasts showed through the dark brown hair that hung several centimetres long from most parts of its body. The five startled men ... just stared. The creature did the same, and later they recalled that it stooped a bit, and seemed more inquisitive than dangerous. Its face looked almost human, and it smelled foul. (Boyd 1978, 82).

## The attitude of the aborigines:

The manager of the Woodenbong Aboriginal Reserve, Mr. Terry Roberts, said that as soon as he mentioned the Thel Crewe sighting to his father, the older man knew exactly what it was. He said the creatures are a carefully guarded secret of the tribal elders, and live in the district's mountains. (Ibid., 83).

## Just sightings:

A 52-year-old man told the Lismore Northern Star newspaper he saw a Yowie in 1935 on his grandfather's dairy farm on Three Chain Road, South Lismore. As he stood on the verandah at 9 o'clock on a moonlit winter night, he saw a 'man' walking in from the hills. As the 'person' approached the horses started making a commotion. The 10-year-old witness went inside and told his grandfather who, when he saw what it was, blew out the lamp, grabbed his rifle and with the rest of the family watched from a small kitchen window. The Yowie was eight metres away, clearly visible in the moonlight. He said it was thick-set and hairy, but the head had no neck and was sitting straight on the shoulders. The Yowie seemed to have a hunched back but was standing upright. The Yowie walked past the house, hesitated near the sulky shed, then went on out of sight behind the dairy. As it walked it seemed to be dragging its feet. The witness's grandfather said it was the same creature he had seen a few years earlier when he'd ridden up a gully to pick some guavas. The Yowie came down one side of the gully, crossed a creek, then climbed the other side of the hill, making the horse play up badly (Ibid., 83).

In July 1976 a Mr. Jackson of Sydney hiked out to Ruined Castle. As he came around a bend in the track he saw a tall hair-covered man-like thing moving from nearby scrub onto the track a few metres in front of him. They both stood and stared at each other. Mr. Jackson said, 'The creature, a male, just stood there facing me. I could see his facial features clearly, almost human-like – there was a look of inquisitiveness in his face. He suddenly leaped off the track and disappeared down the embankment and into the trees.' (Ibid., 83)

In 1971, a team of RAAF surveyors used a helicopter to examine the Sentinel, an inaccessible mountain overlooking the rugged terrain near Glen Davis. When they landed their chopper and had a look around on the summit of the Sentinel they were stunned to find huge man-like footprints in the mud (it had rained the previous day) that were far too big for any human being. Later, another military helicopter crew flew over the mountain and at treetop level spotted a man-sized hairy beast moving swiftly through the scrub below. It eluded them among rocks. (Ibid., 83)

The next story, recorded nowadays but referring to the past, also has exact parallels outside Australia and deserves to be quoted at length:

A century-old conspiracy of silence involving the birth of a half human half yowie child on the north coast of NSW has finally been broken. The secret is shared by only a handful of people still living, three of whom agreed to tell their version of the story exclusively to research specialist Liz James.

Two are women, one 91 years old, the other 66. The third is an elderly Aboriginal male, who believes he is about 80 years old. All three claim to have seen the mutant—the offspring of a white woman raped by a yowie.

The yowie girl was allegedly born in 1874 or 1875. News of the abomination spread rapidly through the remote farming community, causing a wave of fear.

The 91-year-old woman learnt of the rape from her mother. 'We were all so fearful it could happen again,' she told Liz James. 'Otherwise, such a thing would never have been mentioned.'

The attack occurred during the husband's absence. Immediately after the creature was born, the mother demanded it be drowned, but she was overruled by her spouse. 'He was a highly religious man and, after searching his conscience, decided it would amount to murder and be contrary to the teachings of his church,' said the informant.

'My mother told me it was like any other baby girl, except for soft, fluffy down all over its body. The infant was reared on goat's milk, and reasonably well cared for, but locked away.'

The yowie girl's mother never became pregnant again because she was convinced she had been contaminated by the Devil.

The 66-year-old woman says the 'hairy girl', as everyone called the creature, had grown into an adult when she first saw her. 'You know what kids are,' she confided to Liz James. 'For a dare, a couple of us hid in a tree overlooking the farmyard and waited for a glimpse of the 'thing.' She looked like a girl alright, only she was covered with this long, thick, orange-coloured hair. A bit like the colour of a Jersey cow. She was more animal than human, though, and ate like a pig, grabbing at her food with her long, hairy hands.'

Liz James says the Aboriginal, whom she identified as Mosek, spoke with respect of the 'Great Hairy Men from another time, beyond the Dream Time.' He, too, remembered the yowie girl well, and reacted angrily when discussing the way she was incarcerated. 'She was treated worse than most animals - boxed up in an outhouse. The only kindness she was shown was the regular supply of food and water she received.'

Mosek also told of a close encounter he had with a yowie out in the open. Man and beast came face to face in a clearing near a lagoon; both, it seems, had gone there to pick berries. Mosek was rooted to the spot, but the 'Great Hairy Man' stretched out a hand, stroked the Aboriginal youth on the side of the head, grunted, then loped off. The 'stroke' was none too gentle, however, and Mosek was obliged to apply a pad of leaves to his ear.

Liz James, who is now working full-time on yowie and related research, says the 'yowie girl' story has scientific significance. 'It means the yowie must be of human descent, close enough in genetic structure to be able to mate with a human and produce offspring.'

What became of the yowie girl of the North Coast? The unfortunate captive vanished into the bush one day, and was never seen again. The man of the house, who survived his wife by several years, apparently set the creature free when he felt his own life was drawing to a close (*People Magazine* 1979, 10).

It is most improbable that the old people, who divulged the "yowie secret" in 1979, had ever read what we wrote in 1976, which has a lot to do with that secret:

We think that much of the mystery and deification or condemnation

> of the creature in historic times is due to the fact that he has been a potential, and sometimes actual, 'diluter' of the human race. Thus, if no permanent natural barriers to hominid interbreeding existed — at least in the late hominids – then there appeared an artificial, social barrier with the advent of *H. sapiens*. Like any social barrier, it was not absolute, and there must have been a certain amount of interbreeding between *H. sapiens* and the pre-*sapiens* creatures (Bayanov and Bourtsev 1976, 313).

More light is shed on the nature of the Yowie and the humans' attitude towards the creature by the following observation of Rex Gilroy, the foremost Yowie investigator:

> There are so many stories of hunters and farmers who have been armed at the time they claim to have seen the Yowie, yet they have never attempted to shoot the beast. Actually, whenever they have been asked why they did not fire on the beast they have always answered that they had been unable to pull the trigger because they felt that they would be committing murder. In other words, to an experienced bushman the Yowie is more man than beast (Gilroy 1976, 24).

"Replicas" of the above are to be found all over the world. As, for example, this one in North America:

> I levelled my rifle. The creature was still walking rapidly away, again turning its head to look in my direction. I lowered the rifle. Although I have called the creature 'it,' I felt now that it was a human being and I knew I would never forgive myself if I killed it (William Roe, quoted in Green 1969, 12).

This explains, on the one hand, why relic hominoids are usually not hunted as common things of wildlife and, on the other, why, if a specimen has been killed after all, the trophy is not readily available to science.

And the final "replica," pertaining to social psychology and showing that the investigator in Australia is in the same plight as his counterparts in other lands: "Rex Gilroy's main grudge against Australian academics is not that they don't believe in Yowies but that they condemn his theories without considering his research or evidence." Says Gilroy:

> There are scientists who have gone out of their way to ridicule me. I had an argument with a Sydney University zoologist and said,

'Look, if I got a Yowie and plonked it on your desk, what would you say then?' He said, 'Well, in the case of the Yowie, of course according to the Wallace Line, no member of the primate family exists below Borneo.' So you're bashing your head against a brick wall, even if you've got the evidence in front of them. If you could get a Yowie to tap-dance with hat and cane in front of one of these people, there is still no guarantee they'd accept it (Abraham 1979, 24).

Yes, the biologist accustomed to dealing with hides and bones usually finds the information presented in this paper, "beside the point" and of no use to his science. This abominable shortsightedness was overcome in hominology by applying the methodology of social sciences in the field of the natural ones, as stated by Boris Porshnev:

> It is precisely the use of nontraditional methods, such as the comparative analysis of mutually independent evidence, that has made it possible to establish the existence of this relic species and to describe its morphology, biogeography, ecology, and behavior. In other words, fact-finding methods have been used in biology that are usually employed by historians, jurists, and sociologists. (Porshnev 1974, 450).

The comparative analysis of mutually independent evidence regarding relic hominoids in Australia not only validates the creatures' existence in that continent but their presence in other areas of the world as well. Because such consistency and coincidence in all aspects of the matter in different lands can only be explained in a sane way by the existence of a real animal. As Ivan Sanderson (1961, 442) put it: "Coincidence is a strange thing, but it eventually runs out, statistically, and simply by the law of averages."

But isn't Australia, zoogeographically speaking, a special case? Yes, it is. This land of marsupials was originally inhabited, as we thought, only by two species of placental mammals (except flying mammals of the order Chiroptera): *Homo sapiens* and his dingo dog. Now we have to add the Yowie. No other land mammals, such as apes, tigers, deer, made it to Australia while man and hominoid did. What does it mean? It means what we already know: that physically and ethologically, and in a way even psychologically, the hominoid stands closer to man than to ape. It means that a treeless land bridge, interlaced with shallow water barriers, that surfaced in the previous epochs to connect Australia with Asia, was not negotiable by any mammals except the higher bipedal primates, i.e., landroving hominids.

So, hairy bipeds seem to be truly everywhere except Antarctica. For our indicator is a walking (and swimming) *Homo sapiens*, and, as far as we know, a walking man has never reached the frozen continent. The nearest he got was the southern end of South America, i.e., Tierra del Fuego. Do we have any evidence of "wild men" in Tierra del Fuego? Yes, we do. But here it is not only the story that matters, but also, if not mainly, the horse from whose mouth we got it. For our source of information this time is none other than Charles Darwin! In Chapter X of *A Naturalist's Voyage Round the World*, he says the following:

> Jemmy believed in dreams, though not, as I have said, in the devil: I do not think that our Fuegians were much more superstitious than some of the sailors; for an old quartermaster firmly believed that the successive heavy gales which we encountered off Cape Horn were caused by our having the Fuegians on board. The nearest approach to a religious feeling which I heard of was shown by York Minster, who, when Mr. Bynoe shot some very young ducklings as specimens, declared in the most solemn manner, 'O Mr. Bynoe, much rain, snow, blow much.' This was evidently a retributive punishment for wasting human food. In a wild and excited manner he also related that his brother, one day whilst returning to pick up some dead birds which he had left on the coast, observed some feathers blown by the wind. His brother said (York imitating his manner), 'What that?' and crawling onwards, he peeped over the cliff, and saw 'wild man' picking his birds; he crawled a little nearer, and then hurled down a great stone and killed him. York declared for a long time afterwards storms raged, and much rain and snow fell. As far as we could make out he seemed to consider the elements themselves as the avenging agents: it is evident in this case, how naturally, in a race a little more advanced in culture, the elements would become personified. What the 'bad wild men' were has always appeared to me most mysterious; from what York said, when he found the place like the form of a hare, where a single man had slept the night before, I should have thought that they were thieves who had been driven from their tribes, but other obscure speeches made me doubt this; I have sometimes imagined that the most probable explanation was that they were insane. (Darwin, 1928, 216)

I am much inclined to think that the creatures described as "wild men" by the savages of Tierra del Fuego were not *Homo sapiens* but *Troglodytes recens ubiquitous*. Realizing that Darwin himself may have been so close to a live object of our long and tortuous research, undertaken in the light of his great and revolutionary theory, I can't help feeling sort of elation mixed with wonder. It is intriguing to conjecture what course anthropology might have taken had Darwin happened to see the "bad wild man" whose sleeping place he was shown.

## References Cited


Abraham, Matt. 1979. Yowie. *The Advertiser,* 23 May, 24. Sydney, Australia.

Attwood, Alan. 1979. The great white Yowie hunter. *The Age*, 17 May, 9.

Bayanov, Dmitri, and Igor Bourtsev. 1973. "Preliminary report on the Patterson film." In *Sasquatch*, by Don Hunter and René Dahinden. Toronto: McClelland and Stewart.

———. 1976. On Neanderthal vs. paranthropus. *Current Anthropology* 17:312–18.

Boyd, Don. 1978. Zowie! Where's the Yowie? *Outdoors*, June, 82.

Darwin, Charles. [1845] 1928. *A naturalist's voyage round the world.* London: John Murray.

Gilroy, Rex. 1976. Apemen in Australia. *Psychic Australian*, August, 24.

Green, John. 1969. *On the track of Sasquatch*. Agassiz, B.C.: Cheam.

———. 1973. *The Sasquatch file*. Agassiz, B.C.: Cheam.

Joyner, Graham. 1977. *The hairy man of south eastern Australia*. Canberra: National Library of Australia.

Krantz, Grover. 1974. Letter to *Current Anthropology* 15:363.

———. 1984. "Unknown hominoid research in North America." In *The Sasquatch and other unknown hominoids*, ed. Markotic & Krantz. Calgary: Western Publishers.

Papashvily, George and Helen. 1973. *Home and home again*. New York: Harper and Row.

Parfitt, Neville. 1980. Yowie shot! Brisbane schoolchildren flush out Yowie family. *People*, 21 February, 8.

*People* Magazine. 1979. Yowie secret. 4 January.

Porshnev, Boris F. 1974. The troglodytidae and the hominidae in the taxonomy and evolution of higher primates. *Current Anthropology* 15:449–50.

Sanderson, Ivan. 1961. *Abominable snowmen: Legend come to life*. Philadelphia: Chilton.

Washburn, Sherwood. 1960. Tools and human evolution. *Scientific American* 203:62–75.


(Originally published in *The Sasquatch and Other Unknown Hominoids*, ed.V. Markotic and G. Krantz, 1984)

SECTION 5

# The Struggle for the 1967 Bigfoot Documentary

## Vancouver Sasquatch Conference

Anthropology of the Unknown: Sasquatch and Similar Phenomena,
a Conference on Humanoid Monsters,
University of British Columbia, May 9-13, 1978

The following two papers by Dmitri Bayanov, Igor Bourtsev, and Rene Dahinden are parts of a single report, entitled "Analysis of the Patterson-Gimlin film and of some Footprints ascribed to the Sasquatch: Why we find them authentic," read in absentia at the 1978 Vancouver Sasquatch Conference. Though praised at the conference, the report was not published by the conference organizers. It was published by Vladimir Markotic and Grover Krantz in 1984, shortened and divided into two papers.

### EYEWITNESS REPORTS AND FOOTPRINTS: AN ANALYSIS OF SASQUATCH DATA

by Dmitri Bayanov Igor Bourtsev and Rene Dahinden

To be scientific, an analysis must be profound, comprehensive and systematic. The specifics of our material suggest that we match it with the evidence of paleoanthropology. Both the sighting reports and the Bigfoot film present a creature which, judging by its appearance, occupies an intermediary position between ape and man, and since science is in possession of fossils of such creatures whose outward features and movements have been surmised or reconstructed, it stands to reason that we compare our evidence with these data. Logic demands that we undertake a comparative analysis of our material to see if it satisfies, both in parts and as a whole, the following three criteria:

1. Distinctiveness
2. Consistency
3. Naturalness

Distinctiveness implies uniqueness, originality, difference from everything else. Consistency means coherence, absence of contradictions. Naturalness stands for things natural as opposed to artificial or man-made.

We agree with those who say that the hominoid problem can finally be solved only through the presentation of physical evidence, be it a live or dead specimen, or its skeletal parts. What needs to be stressed, however, is that science can only solve those problems which it is prepared to solve. In their day, luminaries of anthropology rejected or doubted the real nature of Neanderthal and *Pithècanthropus* fossils because they were not conceptually prepared for those discoveries. The Piltdown case, on the other hand, shows that even "hard evidence," i.e. skeletal material, can be falsified.

Thus, an eventual delivery of traditional zoological material in hominology will resolve the hominoid problem not just by virtue of that fact alone, but also because science will have been prepared to accept the discovery by the entire course of the research, including hopefully, this paper.

## Eyewitness Evidence

The role of the witness is often overlooked because it is taken for granted. To prove that this kind of evidence "works" not only in court, we can refer to such disciplines as history (much written history originates from eyewitness accounts), geography (take such witnesses as Columbus or Cook, for example), astronomy (observations in the past centuries of such phenomena as solar eclipses, comets, supernovae), physics (observations of such a rare natural phenomenon as ball-lightning).

Closer home, in zoology, discoveries of a number of species of animals were preceded by eyewitness reports, as was the case, for example, of the giant lizard of Komodo island, or of the mountain gorilla.

Thus noting the legitimacy of using eyewitness accounts in hominology we point out that the credibility of sighting reports is supported by the following considerations: First, the coincidence of descriptions of the creature in question in different places and at different times in the main traits of their anatomy and behavior, coupled with many differences in details occasioned by the particulars of this or that encounter, the differences of age, sex, or local populations of the hominoid.

Second, the coincidence of the creature's descriptions both by the indigenous peoples of this or that region and by the newcomers or new settlers. In North America it is the coincidence of descriptions by the Indians and by the whites.

Third, the stories of encounters with the hominoid by hunters, fishermen, loggers, prospectors and other outdoorsmen, unacquainted as they are with the niceties of human evolution, contain such details of the creature's anatomy which only make sense to a student of paleoanthropology.

*Portrait of Sasquatch From Eyewitness Evidence* Physical anthropologist, Grover Krantz, draws the following sketch of Sasquatch from a file of sighting reports he has analyzed:

> Presently available descriptions are a mixture of ape and human characteristics, like a gorilla that walks comfortably on two well developed legs. . . The adult animal in these reports stand about 2.5 meters tall and probably weighs from 350 to 400 kilograms. . . The body is covered with hair of 5 to 10 centimeters in length except on the palms of the hands, soles of the feet, and most of the face. Its legs are the same proportion of the stature as in a man, about half of the total, but the arms are often described as being relatively longer than a man's and especially more massive. The shoulders, over a meter wide, are so well muscled to the base of the head that there is no visual constriction at the neck. The face is flat in a human-like manner, but there the resemblance ends. The nose has little projection and the eyes are deep set under prominent brow ridges. The teeth are sometimes seen and noted as being of human appearance. . . The color is generally dark brown or black.
>
> Probably no single observer of one of these animals has testified to all of the above characteristics, but some are included in all descriptions, and most of them have been repeated hundreds of times (Krantz 1983: in this volume).

## Appraisal of Eyewitness Evidence

*Distinctiveness.* Sasquatch, as described by eyewitnesses, is an original figure, not to be confused with any other creature. It is not a human being, that is, at least no modern man, *Homo sapiens,* for it is too ape-like for that. Nor is Sasquatch an ape, for it is too man-like for that. Nor is there need to argue that it is not a bear or any other well known animal.

*Consistency.* Hundreds of reports by observers of North American hominoids, collected over a period of more than a hundred years, "hang together" to form a coherent and impressive picture.

*Naturalness.* We can put forward the following arguments for the naturalness of Sasquatches.

1. There is nothing unnatural about their physique from the naturalist's point of view. If the hominoid, such as the Sasquatch, had a tail, or horns, or hoofs, that would be unnatural.
2. They appear in two sexes and in different sizes, corresponding apparently to different age groups, which is as it should be in a natural population.
3. The reaction of domestic animals, such as dogs and horses, to the Sasquatch is very special and characteristic, the like of which is never displayed in front of humans in disguise.
4. They leave various traces and signs of their presence in nature as one would expect of natural creatures: odors, hairs, leavings, droppings, lairs, and above all tracks which we will discuss later on.
5. The contrast between the "dramatic" appearance of a Sasquatch and his so frequently undramatic or simply modest behavior in the presence of humans betrays a natural being eager to keep out of trouble.
6. There is good reason to believe Sasquatches to be natural beings because the existence of similar "ape-men" is well known to science from the fossil record.

Thus on all counts, the eyewitness evidence has every right to be included in the Sasquatch case.

## The Evidence of the Footprints

In a detective story like this one, witnesses are fine, but footprints are even better. Indeed, it is hard to imagine what would happen to all that fine testimony by the winesses if Sasquatch did not stand on feet big enough to leave giant footprints. The point is simple enough: if the purported witnesses were not day-dreaming or spinning tales, if they really happened to see something peculiar, big and material, standing or walking on the ground, then sometime, somewhere this something is bound to leave tracks for anybody to see and photograph. And this has been going on for a long time in North America. The Sasquatch investigators have assembled an excellent collection of footprint evidence.

## Assessment of the Footprints from the Functional and Adaptive Points of View

*The toes.* These are five in number, from big to small, as in man. But the length of the toes is a matter of controversy. Don Abbott mentions "unduly short toes"; according to Grover Krantz "toes are relatively short", while John Napier says that "the Sasquatch's toes. . .are much longer, more ape-like, than in man" (Napier 1973:121).

In most tracks the toes show a very distinct and characteristic stance: pressed together, their tips pointing downward, they form a perfect arch which leaves in footprints a ridge, sometimes of untouched earth, under all the five toes, of the kind that is never observed in human tracks.

It is this position of the toes that is responsible for the wrong idea that they are short and equal-sized. This is easy to demonstrate with prints made by human hands and fingers.

As for the functional and adaptive explanation of this stance we have to think of the following: the feet must have enough friction on and resistance from the ground to prevent slipping and sliding. The quadrupeds, whose life, by dint of their four legs, is less precarious in this respect, all have either hoofs or claws for the purpose. As for man, making it for the hills he changes his footwear and puts on boots outfitted with spikes. The Sasquatch has only one pair of soles all his life and thus had to adapt his bare feet for the task of successfully negotiating slippery surfaces at different angles which is the everyday terrain of his habitat.

It stands to reason that the most adaptable and most suitable parts of the foot for this purpose are the toes: the Sasquatch walks as if grasping the earth with his feet, and to make the grasp firm and the push-off strong he has to press the toes together, bend them up in an arch, forming a veritable scoop whose forward edge is thrust into the ground. Such adroitness of the toes seems to be as natural and necessary for the creature's survival as the hair on his body.

One logical result of this use of the toes should be the wear of the toe nails and, indeed, we don't detect marks of the nails even in the clearest of Sasquatch footprints showing imprints of the toe tips.

Lastly, we would like to mention that an increased mobility of the toes, especially in the upward direction, is noted in the Kiik-Koba Neanderthal foot in Crimea (Bonch-Osmolovsky 1954:171).

*The metatarsal region.* This part of the Sasquatch footprint is distinguished from that of a human imprint on an unyielding surface by two characteristic features: 1. It is relatively wider than in man and bulges out both on the inner and outer sides of the foot (whereas in man only on the inner side). 2. It shows the so-called double ball (Green 1968:30; 1973:31). The latter feature will be discussed a little later and as for the former, to understand its function we have to look at a set of tracks showing the peculiarity of the creature's striding (Green 1968:1; 1973:50, 52, 53).

A man normally walks with his feet pointing outward, while a Sasquatch walks with his feet pointing straight ahead or even turned in a little. It means the creature's enormous feet find no trouble walking paths which will be to narrow for the feet of his tiny civilized brother.

Placing his feet in a line while striding, the Sasquatch needs some device to prevent swaying sideways. Hence the unusual widening of the feet, especially in the metatarsal region, as an anatomical means of coping with the problem (plus swinging and balancing with his long and heavy arms as a biomechanical auxiliary to the same). Human feet being not as wide as in the Sasquatch, humans cope with the problem by turning their feet out.

The metatarsal area of the Kiik-Koba Neanderthal foot is also much wider than in modern man, with the fifth metatarsal, and especially its proximal end jutting out in quite an inhuman manner. As for the length-to-width ratio of the Kiik-Koba fossil foot, there are closely matching examples among Sasquatch footprints.

*The double-fall feature.* In our opinion, this feature is caused by a flexion furrow on the sole homologous to human *plica transversa flexeria hallucis.* This furrow in the human foot leaves no sign of its presence in the footprints. Why is it so pronounced in the Sasquatch that it leaves a noticeable mark in some tracks? It is because the Sasquatch foot has much more mobility in the toes and much more flexion in the metatarsophalangeal joints than the human foot. Besides the function of preventing sliding and providing a reliable push-off in locomotion, the great mobility of the toes in Sasquatch is caused by the apparent necessity for Sasquatch to raise his heels unusually high when taking long strides on bent knees.

This is clear from a comparison with the dynamics of a human foot in cross-country skiing: the skier can only make long steps on bent knees by raising his heels high and bending his toes up sharply in the metatarsophalangeal joints.

Something similar must happen in Sasquatch locomotion, resulting in greater than human stretch of the tissues of the sole in the area of these joints. When, during the heel strike, the toes assume the "grasping" stance, their tips brought closer to the sole, the tissues contract and form a furrow which leaves a mark in the footprint.

The "dynamics theory" of the double ball is supported by the fact that the said furrow is revealed only in footprints with bent toes, whereas if the "statics theory," suggested by Krantz, were correct the double-ball feature would be present in footprints irrespective of toe position, which is not the case.

Finally, it can be remarked that the double-ball feature seems to be doubly correlated with the surmised locomotor habits of the Sasquatch: first with high "activism" of his toes, and second, with the necessity of highly raising the heels in striding which in its turn correlates with his bent knees and turned-in feet.

*The instep region.* In a normal human foot this is where the arch of the foot is highest, so it is appropriate at this juncture to mention the question of flat feet in Sasquatch.

We want to stress the flat feet in Sasquatch are quite different from flat feet in man. In tracks on an unyielding surface the difference appears in that a human foot with a flattened arch leaves "waistless" footprints, whereas Sasquatch tracks, despite flat feet have a waist, due to the widening of the metatarsal area. But the main difference concerns locomotion.

In man, flat feet are a pathological condition usually resulting in various discomforts and awkward gait. Anatomists point out that a normally arched foot plays the role of a spring, absorbing or softening shocks in locomotion. People with flat feet have difficulty in running and jumping and, what is especially interesting, tend to turn their feet out in walking more than is usual for people with normal feet.

The Sasquatch, as it follows from available evidence, is a superb walker and jumper, and strides with his feet turned in. Hence flat feet are his natural condition and cause him no complaint. How does he manage to keep his brain, and liver, and kidneys from receiving nasty shocks in his bumpy strolls? Very simple: by always bending his knees.

Apes are flat-footed, and it is agreed in anthropology that the normally arched human foot appeared in anthropogenesis as late as *Homo sapiens.* All human infants have flat arches at birth. The Kiik-Koba Neanderthal foot, which we use for comparison, has a flattened arch, which correlates with the specimen's surmised walking on bent knees.

A supposition can be made that the Sasquatch foot is less rigid than the human foot and that a lifted Sasquatch foot may not be as flat as it looks from some tracks impressed with the whole weight of the body.

*The heel.* Krantz (1972), using some theoretical considerations and a cast of a track made by the deformed foot of the Bossburg Sasquatch, has reached the conclusion that the heel bone in Sasquatch must not only be relatively wider but also longer than in man.

Krantz's prediction that the heel is lengthened and the forefoot is shortened, which "means the ankle joint must be set relatively farther forward along the length of the foot" (Krantz 1972:96) seems to us to be theoretically valid and practically confirmed by the Patterson-Gimlin film.

As for comparison with paleoanthropological material, the Kiik-Koba fossil foot graphically illustrates the points made by Krantz, namely, a larger heel bone and the ankle joint set relatively farther forward in comparison with the norm in modern man. The characteristics of this foot are explained as "adaptations caused by the need to support great body weight (Bonch-Osmolovsky 1954:178). The same reason is offered by Krantz to account for the peculiarities of the Sasquatch foot. Since body weight in Sasquatch must be even greater than in fossil Neanderthals, the Sasquatch foot can be supposed to have developed the said characteristics even farther.

## Assessment of the Footprint Evidence

*Distinctiveness.* Even a casual observer of clear Sasquatch tracks *in situ* (and there are plenty of those investigated on the spot by researchers and documented by photographs) cannot confuse them with the tracks of any other animal. Superficially they resemble outsized human footprints, but closer examination and careful analysis show a number of overt and subtle differences from the normal human anatomy and an excellent agreement with certain characteristics of fossil material.

Of special significance is that the footprints not only differ from those of *Homo sapiens,* but indicate a different manner of walking from that of modern man.

*Consistency.* The footprints indicate a foot which is perfectly adequate and consistent from functional and adaptive points of view.

*Naturalness.* Objective observers of Sasquatch footprints, such as Don Abbott, for example, have noted "the apparent naturalness" of the tracks (Hunter and Dahinden 1973:42). Their conclusion is based on such arguments as the depth of impressions, their location and range of distribution in space and time. When we add to this the arguments of anatomical, biomechanical and paleoanthropological nature, cited above, the naturalness of the Sasquatch footprints turns from "apparent" to absolute.

Hence our verdict: the photographs and casts of those prints in the North American collection which reveal a number of typical characteristics, such as the bent-up toes with a ridge of untouched soil under them, or, conversely, fully extended toes; the double ball; the flattened arch coupled with a "waist"-all those represent genuine Sasquatch footprints.

*Conclusion from Eyewitness and Footprint Evidence Combined* We find the eyewitness and footprint evidence to be in complete agreement and testifying jointly to the existence of creatures which are:

1. Habitually bipedal.
2. Man-like, i.e., hominoid in anatomy.
3. Larger than man in size.
4. Heavier than man in weight.
5. Met in the wild and adapted to wilderness conditions, such as the cold (hair on the body) or difficult terrain for barefoot locomotion (special adaptations in the anatomy and mechanics of the foot).
6. Characterized by a different walk from man.
7. Both kinds of evidence point out a creature which has many characteristics of primate forms which, in evolution, preceded modern man, *Homo sapiens.*

Hence our general conclusion: North America or, to be exact, its North-West, is the habitat of a species of higher primates distinct from modern man, *Homo sapiens,* and known by the name of Sasquatch or Bigfoot.

## References Cited


Bonch-Osmolovsky, G.A. 1954. Paleolit Kryma (The Palaeolithic of Crimea). Moscow.

Green, John. 1968. On the Track of the Sasquatch. Agassiz, B.C.: Cheam Publishing.

———. 1973. The Sasquatch File. Agassiz, B.C.: Cheam Publishing. Hunter, Don and Rene Dahinden.

———. 1973. Sasquatch. Toronto: McClelland and Stewart.

Krantz, Grover. 1972. Anatomy of the Sasquatch Foot. Northwest Anthropological Research Notes 6(1):91-104. Reprinted *In* The Scientist Looks at the Sasquatch, Pp. 77-93. Roderick Sprague and Grover S. Krantz, eds. Moscow, Idaho: University Press of Idaho,1977; 2nd ed.,1979.

———. 1983. Research on Unknown Hominoids in North America. In this volume. Napier, John 1973. Bigfoot. The Yeti and Sasquatch in Myth and Reality. London: Jonathan Cape.

# ANALYSIS OF THE PATTERSON–GIMLIN FILM, WHY WE FIND IT AUTHENTIC

by Dmitri Bayanov, Igor Bourtsev and Rene Dahinden

## Introduction

Our knowledge of the present-day existence of relic hominoids is based on three kinds of evidence:

1. Eyewitness reports;
2. Hominoid footprints;
3. The Patterson-Gimlin film.

The three kinds of evidence are considered by us not in isolation but in connection with the historical data on the relic hominoids and are interpreted in the light of the theoretical concepts of Professor B.F. Porshnev (1974:449-456) and Bayanov and Bourtsev (1976:312-318).

Some say the Patterson-Gimlin film is of no significance since we don't physically possess what is portrayed in it. This argument seems invalid to us. Do the close-ups of the surface of Mars or Venus lose value because man has not yet touched samples of rock from these planets? Or, citing a strictly earthly example, does a film showing a suspect on the scene of a crime become less important because the suspect is still at large?

Undoubtedly, the film is of the greatest importance to science, on condition of its authenticity, and that is why we deem it imperative to establish the truth in this matter. (See Plate 13)

Shot in 1967, and thanks to the efforts of Rene Dahinden, the film has been studied and found authentic in the Soviet Union by Bayanov, Bourtsev and Donskoy (Hunter and Dahinden 1973). Dr. Don Grieve has expressed the impression of some scientists who have seen the film:

> My subjective impressions have oscillated between total acceptance of the Sasquatch on the ground that the film would be difficult to fake, to one of irrational rejection based on an emotional response to the possibility that the Sasquatch actually exists. This seems worth stating because others have reacted similarly to the film (Napier 1973:220).

The photographic evidence, as is known, may exist in two forms: in the form of snapshots and in the form of a movie. To take a good picture of Sasquatch you have to be close enough to the creature, lighting conditions should be favorable enough and the hand must be steady enough-all of which have a very low probability of occurrence.

Not so with a movie camera. What is more, a movie image of the

Sasquatch can show how the creature moves about, which random snapshots cannot do.

To test the film we have to examine it in three aspects:

1. The footage itself from a technical point of view.
2. The subject it shows from a morphological point of view.
3. The subject's movements from a biomechanical point of view.

## Technical Characteristics of the Movie Footage

It is a 16 mm color filmstrip, 715 cm long in the part showing the scene in question, which numbers 951 frames, most of which contain the image of the creature under discussion.

No "special effects" have been detected in the film; that is to say it shows what actually happened in front of the camera. The footage was shot without a break from the moment the creature appeared on film and to the end of the roll.

Good frames alternate with not so good ones and in some the view is blurred altogether. According to Patterson, he shot part of the film while pursuing, and, indeed, in part of the movie objects of the scenery, such as trees and logs, appear and disappear in a consecutive order, showing that the film-maker was moving along.

In many frames the image is in good focus. The movie was shot with an ordinary lens and the camera was never near enough to the filmed subject for a close-up. Photographed at a distance of 40 m at the nearest, the creature left a 1.2 image on film which loses sharpness when blown up the size of a screen for the mere reason of enormous enlargement.

Nonetheless, an attentive observer can discern a surprisingly great number of details in the film's subject. Even a more detailed portrait of the creature than can be seen on the screen was obtained by us through printing the frames on photographic paper with different characteristics.

One can be of two minds regarding the implication of the photographic quality of the Patterson film. On the other hand, if the first documentary of a purported Sasquatch were of the quality of a feature film shot in a studio, it would elicit nothing but distrust.

*Framing speed of the movie.* The exact knowledge of this characteristic is necessary for judging the speed of movement of the film's subject. Patterson told John Green that the film must have been shot at 16 frames per second but he was not absolutely sure. To our mind, there is nothing suspicious in that, being an amateur photographer and in the heat of the event, Patterson did not make sure what speed he used in filming, while his avowal of this definitely speaks in his favor.

The speed of the subject's movements is such an essential characteristic so that, regardless of Patterson's word, it demanded an objective evaluation. In solving this problem we were guided by the following idea: when Patterson pursued the creature, his hand-held camera must have made vertical oscillations in rhythm with his steps and, since the camera was shooting all the time, these oscillations must have affected the film. The frequency of these oscillations must correspond to the frequency of Patterson's steps. If we can find this frequency in the film, we will know the frequency of Patterson's steps and, since the possibilities of human performance in this respect are well known to the discipline of biomechanics, Patterson's own speed can be used as a standard to judge the speed of his camera. To find and study the said oscillation in the film each frame was viewed on a screen with a network of coordinate lines on it, allowing the fixing of the position of an object (a twig, for example) in the foreground of the frame.

A detailed study of the frames with this technique made it possible to draw up diagrams of alterations in the position of objects in the frames which revealed vertical oscillations in that part of the film which was shot on the move. The frequency of these oscillations was found to be in the range of one per four frames to one per seven frames. As to horizontal oscillations, these were not found in any regular pattern.

The inference from this is that the more frequent oscillations (one per four frames) correspond to Patterson's running, while the less frequent ones (one per seven frames) were caused by his walking. Finally, an absence of oscillations in a certain part of the footage indicates that it was shot from a standing position. If the movie was shot at 16 frames per second (fps), then in running Patterson made four steps per second (sps) (16:4) which compares favorably with the sprinter's frequency of 4.3 sps. In walking Patterson must have made 2.3 steps per second (16:7) which is a little faster than the norm for walking (2 sps) and below the figure for long distance running (2.8 sps).

If we suppose that the movie was shot at 24 fps, then in running Patterson must have made six steps per second (24:4) which exceeds the sprinter's performance and therefore has to be definitely excluded.

Thus is has been established that the framing speed of the Patterson movie was indeed 16 fps. In solving this important problem Patterson's oft regretted filming on the run turned out to be a blessing in disguise.

## Morphology of the Creature Shown in the Film

*General appearance.* "Roger Patterson's filmstrip shows a hairy man-like creature, walking erect, having well developed breasts and buttocks.

The last three points. . . indicate that it belongs to the Hominid, not the Pongid (apes) line of evolution of higher primates" (Hunter and Dahinden 1973:173). This is also the opinion of Dr. Osman Hill, Director of Yerkes Regional Primate Research Center: "The creature portrayed is a primate and clearly hominid rather than pongid" (Sanderson 1968:29).

*Body size and weight.* To determine the creature's size we used the formula H/L=h/f, where H is the height of the creature, L is the distance between the creature and the camera, h is the height of the creature's image on film in a certain frame, and f is the focal length of the camera. But first the place of the camera on the filming site had to be established for that particular frame, which was done with the help of the same formula and the measurements of distances obtained by Rene Dahinden on the filming site.

According to our calculations, the nearest point of the camera to the creature was at a distance of 40-42 m; the image on film shot from that distance was a height of 1.2 mm; the focal length of the camera lens used by Patterson, as established by Dahinden, was 25 mm. Using these figures in the formula we find the creature's height to be 190 cm. If we take into account the fact that the creature is photographed in a forward-leaning stance with its legs bent, we can suppose it reaches a height of 200 cm when straightened up as much as its anatomy can allow.

A height of 190-200 cm is in agreement with the figures received by other investigators using different methods of computation: 200 cm, (Green 1968:73) or 196 cm. (Grieve in Napier 1973:94) and corresponds to the creature's stride length of 106 cm as measured by Patterson and Gimlin on the sandbar where tracks were found, for in certain frames the stride length appears to be approximately half as long as the creature's height.

All investigators note the unusual breadth of the creature, which we estimate to be 40% greater than the average for a man of that height. Using average data for human beings we calculate the weight of a man 200 cm high and 40% broader than average to be in the vicinity of 220 kg. Since the film creature's shoulders are higher than a man's and it has a barrel-like torso, its weight is expected to be in excess of that figure.

And now to some details of the subject's morphology.

*The face.* "The movie is not sharp enough to show facial detail. . ." wrote John Green (1968:52). This was only true until 1973 when the best frame, number 344, showing the creature's profile was "discovered" in Moscow by printing the stills on photographic paper of suitable characteristics (Bayanov and Bourtsev 1976:40)

Frame 344 gave us a quite detailed view of the face, almost in profile, with its low forehead and protruding brow ridge, a black hole of the right eye socket, a wide nose with a low bridge, the jutting jaws with a thin line of the mouth and a heavy chinless mandible. Also in relief are the right cheekbone (os zygomaticum) and the powerful chewing muscle (musculus masseter). The ear is not clearly in view being covered with hair which forms a bulge at the place where the ear should be. The rest of the face is either hairless or covered with such short hair that it does not conceal the features.

It follows from the above that the face has all the classic features of a *presapiens* hominid. What is more, the profile gives a good idea of the relation in size between the facial part and the brain part of the skull in the sagittal plane, and this proportion is of a definitely non-human i.e., *ipresapiens* character, though not as different from human as in apes.

At the same time the creature's portrait perfectly fits the eyewitness descriptions. On the whole, we learned from frame 344 that the subject's face is more human than simian, whereas prior to the discovery of this frame all the investigators had the impression from the film that the creature's face was more ape-like than human-like, which seemed to contradict the extent of human qualities in the other part of the creature's anatomy.

Finally, a few words are in order concerning the light this episode sheds on Patterson's role in the whole affair. In our analysis we employ an objective approach, as befits science; that is, we examine the objective qualities of the film, regardless of its authorship. Strictly speaking, the latter cannot be ascertained in an objective manner, and we accept Patterson's authorship simply because we have no reason to doubt his and Gimlin's word plus there are no claims from other people in this respect. We have as yet to present other objective arguments for the authenticity of the film, but to counter William Montagna's (1976:7-9) charge that "Patterson and friends perpetrated a hoax," we can offer at this juncture an argument of a psychological nature. Patterson died in 1972, while the best frame from his film, showing the face of the filmed creature in a true light, became known to investigators in 1973. If we are to believe Montagna, that Patterson was able to perpetrate a hoax of that complexity, is it conceivable that he would have failed to his death to present the best frame of his film which could have explained away a contradiction exploited by the movie's critics?

*The legs.* These, apart from their hair covering, appear to be human in shape and proportions though, of course, very big and massive in an absolute sense. What is really different from human is the character of their movements in locomotion which we shall discuss in the next section. Still, there is at least one difference from human anatomy which is of paramount importance. In a certain phase of the stride a bulge is noticeable on the thigh which appears and disappears in a regular fashion and in rhythm with the steps. There can be no doubt the bulge corresponds to the tone of the big thigh muscle—m. rectus femoris—in man, but this muscle in humans never attains such prominence. What is also very significant is that this feature has never been reported by witnesses. And no wonder, for it is noticeable only when viewed at a certain angle, and then it is not such a striking characteristic for a casual observer to behold as, say, an absence of the neck. The significance of this feature will be explained in the next section, when we deal with the specimen's locomotion.

*The feet.* What is telling about the feet is this:

1. The sole is actually seen to be "hourglass" shaped! (see frame 265).
2. There is a hint of an arch in the foot, or at least the sole is not as flat as it appears in footprints (frame 265), corroborating our idea that the Sasquatch foot is less rigid than in man.
3. The heel is actually seen to be sticking out in an inhuman way in some frames, suggesting an unusually large heel bone (calcaneus) as has been predicted by Grover Krantz using theoretical considerations and the evidence of the footprints. That the heel of the filmed subject is really unusual is testified to by the fact that this feature was independently discovered in Moscow and Ottawa. In Moscow it was seen by Bayanov and Bourtsev as "an omen of the creature's reality" (Hunter and Dahinden 1973:178). In Ottawa a scientist of the Canadian National Museum rejected the film because he saw "a human heel under the skin of the creature as it walked. . ." (Ibid.:124). The "skin of the creature" in this case being a euphemism for the notorious "monkey suit," we infer from this statement that for a human heel to stick out from its normal dimension to the extent shown in the film there must be something terribly wrong either with the "monkey suit" or with the heel. Since nobody has yet detected any flaw in the former, the latter seems the only alternative. Thus it is up to the viewer to make up his or her mind whether the protruding heel, as seen in frame 324, is an abnormal human heel, or a normal Sasquatch heel. It is worth pointing out also that this peculiarity has never been reported

by eye-witnesses because it appears only for a fleeting moment when the Achilles tendon is not tight in a certain phase of the stride.

## Some Conclusions

The creature of the Patterson movie by its size and appearance perfectly fits our notion of the Sasquatch. In fact, almost all parts of its body, which are visible in the film, match the data of the eyewitness and footprint evidence. The few features that have not been reported by witnesses are equally significant for at least two reasons: for one, they are consistent with the overall morphology of the creature and the mechanics of its locomotion (the latter will be shown below); for two, they are inconspicuous enough to have escaped the attention of chance observers, but prominent enough to catch the scrutinizing eye of film analysts.

The specimen's morphology is in agreement not only with the witness and footprint evidence, but also with the data of paleoanthropology.

## The Specimen's Movements

The great advantage of a movie is in its ability to "catch" and render movement. The whole "scenario" of the Patterson film consists of one short scene wherein the film star never stays put. This happy circumstance gives us no less information as to the nature of the analyzed creature than its morphology does.

The method which we used to find the framing speed of the movie simultaneously allowed us to fix and analyze every step and movement both of the creature and the filmmaker.

The finding of the filming speed (16 fps) also allowed us to establish the time length of the filmed encounter with the creature: the number of frames in the film (951) divided by the figure of the framing speed (16) equals 60 seconds.

In the course of these momentous 60 seconds the creature is seen to perform a variety of movements, most of which can be embraced by the notion of walking. One way of distinguishing walking from running is: in walking the body keeps in touch with the ground either with one or both feet, whereas in running it is on and off the ground with every stride. The frames showing the feet of the creature testify to its walking, not running. The specimen's walk has been analyzed by two specialists in the discipline of biomechanics, Dr. Don Grieve in London and Dr. Dmitri Donskoy in Moscow, whose findings we use in this analysis. "It seems smooth and resilient like that of a big quadrupedal animal" (Bayanov and Bourtsev in Hunter and Dahinden 1973:174). "The stride is actually much smoother than a normal man's, because the knee is bent as the weight comes on it. A walking man bobs up and down as his body goes over the top of his straightened leg. The Sasquatch in the film moves in a flowing fashion, with her leg being bent at all times" (Green 1968:55).

The cadence of the walk is unhurried: according to calculations, the creature made 1.5 sps (as against Patterson's 2.3 sps) strolling at a human speed of 6.6 km per hour. As for the smoothness and fluidity of movements, Dr. Donskoy provides the following elaboration and adds new observations:

> In the swinging of the leg, considerable flexion is observed in the joints, with different parts of the limb lagging behind each other: the foot's movement is behind the shank' which is behind the hip's. This kind of movement is peculiar to massive limbs with well relaxed muscles. In that case the movements of the limbs look fluid and easy, with no breaks or jerks in the extreme points of each cycle... The gait of the creature is confident, the strides are regular, no signs of loss of balance, of wavering or any redundant movements are visible... The movements are harmonious and repeated uniformly from step to step, which is provided by synergy (combined operation of a whole group of muscles)" (Hunter and Dahinden 1973:190, 191).

Now for some particulars in the movement of the limbs and body.

*Foot angulation.* William Roe said of his Sasquatch that "when it walked it placed the heel of its foot down first. . ." (Green 168:11). The filmed creature does the same. But the striking and unreported feature is not the way the heel is placed on the ground, but the manner in which it is raised off the ground. We wrote of "the apparent necessity for Sasquatch to raise his heels unusually high when taking long strides on bent knees" (Bayanov, Bourtsev and Dahinden 1983: in this volume). The filmed subject demonstrates this point in a marvelously clear-cut fashion (frames 265, 301, 311). The necessity of highly raising the heels in Sasquatch locomotion was one of the causes of the double-ball phenomenon. The evidence of the film gives graphic confirmation to this idea, leaving no doubt in the analyst's mind that the foot has an unusually high degree of flexion in the metatarsophalan-geal joints which causes much stretching in the tissues of the ball (frames 265, 311, 312).

Lastly, a reference to paleoanthropology. "A relatively long trochlea tali in the Kiik-Koba specimen indicates an increased range of forward-backward movements of the foot in the ankle joint in comparison to modern man" (Bonch-Osmolovsky 1954:171). This tallies very well with the angulation of the filmed creature's foot.

*Knee bending.* "After each heel strike the creature's leg bends, taking on the full weight of the body, and smooths over the impact of the step acting as a shock-absorber" (Donskoy as quoted in Hunter and Dahinden 1973:190). As we have pointed out, the shock-absorbing function of knee bending correlates with a flattened arch of the foot. "In normal human walk such considerable knee flexion as exhibited by the film creature is not observed and is practiced only in cross-country skiing. This characteristic makes one think that the creature is very heavy and its toe-off is powerful, which contributes to rapid progression" (Ibid).

Grieve wrote of "the 30° of knee flexion following heel strike" and "The considerable (46°) knee flexion following toe-off" (Napier 1973:218, 219).

Eyewitness evidence: "Its knees were always bent as it walked. . ." (Green 1973:55), sighting of a creature, eight to nine feet tall, by Dick Brown, a high school music teacher near The Dalles, Oregon, 1971.

Evidence of paleoanthropology: "The structure of the long bones of the Neanderthaler's lower extremities shows that the knee joints were not fully straightened" (Nesturkh 1959:249-250). This means, of course, that other bipedal primates, preceding Neanderthals, also had their knee joints not fully straightened.

Thus all our sources of information indicate directly or indirectly a high degree of knee flexion in bipedal primates other than *Homo sapiens.* But what all these sources, *except the film,* are silent about is the implication of this peculiarity for outward morphology of such knee flexing bipeds.

As is widely known, stepping out on bent knees is one of the hardest physical exercises. And this is for *Homo sapiens* with his, on the average, moderate weight in comparison with the average for Sasquatches. If moving about in this manner is not part of a session in callisthenics, but daily routine for a creature as heavy as Sasquatch, the strain on its thigh muscles defies the imagination. Nobody had ever thought of that prior to the Patterson-Gimlin film. But Nature had to take care of that, if her creatures, the way she designed them, were to walk about. Hence the unheard of prominence of the thigh muscle—m. rectus femoris—whose presence in the filmed subject is revealed by the enigmatic bulge on the thigh, appearing and disappearing in rhythm with the steps. Thus, offering the answer ahead of the question, the film made the investigator think of and see the power source of the creature's walking on bent knees.

On the whole, we can't but agree with Dr. Donskoy's pronouncement:

As a result of repeated viewings of the walk of the two-footed creature in the Patterson film and detailed examination of the successive stills from it, one is left with the impression of a fully spontaneous and highly efficient pattern of locomotion shown therein, with all the particular movements combined in an integral whole which presents a smoothly operating and coherent system" (Hunter and Dahinden 1973:189).

Finally, we would like to single out and stress the following points made in this section:

1. Both the integral whole and the particulars of the creature's movements are unmistakably different from those of a human being.
2. They are in accord with the creature's anatomy.
3. They indicate a very massive and heavy body.
4. They agree with the sighting and footprint evidence and the data of paleoanthropology.
5. They reveal facts quite novel to science.

## Assessment of the Film and its Subject

*Distinctiveness:* There are two alternative opinions on the subject of the Patterson-Gimlin film. One is that it is a real female Sasquatch, the other—a man in disguise. Nobody assumes it is a robot, or a bear, or an ape, or a man going about his business in his Sunday best. Thus all agree that within these two possibilities the subject is quite distinctive.

*Consistency:* The consistency of the first version has been amply demonstrated in the course of this analysis. It has been shown that on the whole and in details the filmed specimen fits as well as can reasonably be expected the data of the other kinds of evidence. Besides, it is intrinsically consistent, that is to say all parts of its body are in cohesion with one another, and the movements are in harmony with the physique.

Not so with the second version, whose proponents have never bothered to explain how a man in a monkey-suit could have staged such a convincing performance despite the problems of size, weight, outward appearance and unusual movement.

*Naturalness:* A person, unacquainted with elephants, can have difficulty making head or tail of the creature, because of its trunk, and find the thing alarmingly unnatural. The same goes for a person looking for the first time at. a female Sasquatch with her hair-covered breasts on a muscular body which moves in a manly gait.

As for the advocates of the second version, there is nothing easier for them than to utter "a man in a monkey-suit," but they are sure to get into trouble if they ever try to flesh up this utterance. To do so, it is not enough to understand about man and monkeys, as physical anthropologists and primatologists do. It is also necessary to understand about suits, or clothes in general, on the one hand, and their relationship with a person's body and movements, on the other.

One person who is qualified to judge in this matter is a distinguished Moscow sculptor, Nikita Lavinsky, to whom we applied for advice on this problem. One aspect of Lavinsky's job in the course of a 40-year-old professional career has been modeling posthumous sculptures of outstanding personalities (who died in the revolution or war) using photographs of them for the purpose. To be successful in this endeavor, Lavinsky has to be an expert not only in human anatomy, as all sculptors must be, and the way a person's figure is revealed or concealed by clothes, but he is also a first-class expert in interpreting photographs of dressed-up people in terms of anatomy, angulation and movement.

Having viewed the Patterson-Gimlin film and studied its frames, sculptor Lavinsky stated in no uncertain terms the authenticity of its subject. His argument goes as follows: the creature is different from a man both in its anatomy and movements, yet its anatomy looks true to life and the movements spontaneous. What is more, the movements follow from and are in perfect harmony with the anatomy. These conditions cannot be met simultaneously by a man in any costume. The better a costume from the anatomical point of view, the worse it would be from the viewpoint of biomechanics. A clever costume on a moving hoaxer would *expose, not conceal* a fraud. "All talk of a man in a monkey-suit can only come from laymen who know nothing about the relationship between figure, movement, and costume," said sculptor Lavinsky to us.

Quite independently, a similar point was made by Dr. Don Grieve: "A man could have sufficient height and suitable proportions to mimic the longitudinal dimensions of the Sasquatch. The shoulder breadth however would be difficult to achieve without giving an unnatural appearance to the arm swing and shoulder contours" (Napier 1973:220).

The film was shown at the Moscow Research Institute of Orthopaedics and Artificial Limb Construction who specialists were unanimous in finding the movements of the filmed specimen natural and spontaneous and at the same time not identical with those of modern man.

Dr. Donskoy's conclusion is to the same effect:

> On the whole the most important thing is the consistence of all the above-mentioned characteristics. They not only simply occur, but interact in many ways.

> And all these factors taken together allow us to evaluate the walk of the creature as a natural movement, without any signs of artfulness which would appear in intentional imitations. At the same time, with all the diversity of human gaits, such a walk as demonstrated by the creature in the film is absolutely non-typical of man (Hunter and Dahinden 1973:192).

And again Dr. Grieve: "The possibility of fakery is ruled out if the speed of the film was 16 or 18 fps. In these conditions a normal human being could not duplicate the observed pattern, which would suggest that the Sasquatch must possess a very different locomotor system to that of man" (Napier 1973:220). Since we have established the speed of the film to be 16 fps., Dr. Grieve's conclusion can be stripped of its "if" modality and put to its rightful place in this research.

* * *

We have subjected the film to a systematic and many-sided analysis both in its technical and biological aspects. We have matched the evidence of the film against the other categories of evidence and tested its subject with our criteria of distinctiveness, consistency, and naturalness. The film has passed all our tests and scrutinies. This gives us ground to ask: who other than God or natural selection is sufficiently conversant with anatomy and biomechanics to "design" a body which is so perfectly harmonious in terms of structure and function?

Further research may correct some of our findings, but it seems most improbable that the positive result can be voided. Hence we confidently give this verdict: *The Patterson-Gimlin movie is an authentic documentary of a genuine female hominoid, popularly known as Sasquatch or Bigfoot. filmed in the Bluff Creek area of Northern California not later than October 1967, when it was viewed by Rene' Dahinden and other investigators.*

Until October 1967 we had lots of information on relic hominoids but they remained inaccessible to the investigators' sense of vision. We were dealing then with the underwater part of the "iceberg," as it were. October 1967 was the time when the fog cleared and the tip of the iceberg came into view. True, we still can't touch or smell this "tip," and have to be content with viewing it in the film and photographs obtained from the film. But in this we are not much different from the physician who studies a patient's bones without ever meeting that particular patient—just looking at the x-rays. Or from the geologist, who studies the geology of Mars by looking at the close-ups of its surface.

The difference is, of course, that in the geologist's case seeing is believing and, besides, he has all the might of modern science at his disposal. Those close-ups cost a couple of billion dollars and nobody dares to treat them frivolously. The Sasquatch investigator, on the other hand, offered his photographic evidence to be studied by science for free and the evidence was not taken seriously. According to Dr. Richard W. Thorington, Jr., of the Smithsonian, ". . .one should demand a clear demonstration that there is such a thing as a Bigfoot before spending any time on the subject" (Hunter and Dahinden 1973:124). If by a clear demonstration Dr. Thorington means a live Bigfoot be brought to his office, then it would be more of a sight for the layman than for the discriminating and analytical mind of a scientist.

## Conclusion

The relic hominoid research is of special, potentially unlimited value for science and mankind. Thanks to the progress of this research, we know today that man-like bipedal primates, thought long extinct, are still walking the earth in the second half of the 20th century. We can also see how such a biped looks and how it walks.

We are indebted for this breakthrough to the late Roger Patterson who filmed a relic hominoid in Northern California in 1967, but who, to our sorrow, was not destined to witness the full triumph of his achievement.

The marriage of Russian theory and American practice in hominology proved to be happy and fertile. By joining forces we **have** established not only the authenticity of the film but also that the Sasquatch is part of the natural environment of North America, **end** its most precious part at that. May we offer this conclusion as **our** modest contribution to the cause of growing friendship and cooperation between the peoples of the Soviet Union and North America. The search for humanity's living roots is a cause for all **mankind** and this makes us look forward to new international **efforts** in this intriguing investigation.

## References Cited

Bayanov, Dmitri, and Igor Bourtsev. 1974. Reply to the Troglodytidae and the Hominidae in the Taxonomy and Evolution of Higher Primates, by B.F.Porshnev. Current Anthropology 15 (4):452-456.

———. 1976a. On Neanderthals vs. *Paranthropus.* Current Anthropology 17(2):312-318.

———. 1976b. Tainstvennyi dvunogii. (The Mysterious Biped) Nauka i Religija 6:38-43. Moscow.

Bonch-Osmolovskii, G.A. 1954. Skelet stopy i goleni iskopaemogo cheloveka iz grota Kiik-Koba. (Skeleton of foot and shank of fossil man from the grotto Kiik-Koba). Paleolit Kryma 3. Moscow-Leningrad: ANSSSR.l.
Green, John. 1973. The Sasquatch File. Agassiz, B.C.: Cheam Publishing.
Hunter, Don, and Rene Dahinden. 1973. Sasquatch. Toronto: McClelland and Stewart.
Montagna, William. 1974. From the Director's Desk. Primate News 14(8):7-9 September. Seattle.
Napier, John. 1973. Bigfoot. The Yeti and Sasquatch in Myth and Reality. New York: Dutton.
Nesturkh, Mikhael. 1959. The Origin of Man. Moscow: Progress Publishers.
Porshnev, B.F. 1974. The Troglodytidae and the Hominidae in the Taxonomy and Evolution of Higher Primates. Current Anthropology 15 (4):449-450.
Sanderson, Ivan T. 1968. First Photos of "Bigfoot," California's Legendary "Abominable Snowman." Argosy. 336 (2):23-31, 127-128. February.

# Moscow Conference to mark the 30th anniversary of the Patterson-Gimlin film

Darwin Museum, October 21-22, 1997

## Opening Address

(*Translated from the Russian*)

Dear Friends, Colleagues, and Guests of the Conference,

I greet you in this remarkable place whose connection with our conference is both symbolic and legitimate. The great name of Darwin, who has been called a Newton of biology, is inseparably connected with our research, with the discovery of relict hominoids by modern science. The Darwin Museum is our Alma Mater, the only institution that provided shelter to hominologists when the Academy of Sciences and the Society for the Preservation of Nature showed them the door.

We are assembled today to mark the 30th anniversary of the Patterson-Gimlin film and I want to pronounce the names of our teachers and comrades who have not lived to witness this event and whose memory lives on in our hearts and minds. It is Roger Patterson himself, who shot the film, his teacher in research Ivan Sanderson, fine bigfoot researcher

George Haas, of Oakland, California, anthropologist Carleton Coon, the experienced investigator Robert Titmus. In this country it is Boris Porshnev, Pyotr Smolin, Alexander Mashkovtsev, Vladimir Pushkarev, Maya Bykova, Igor Tatsl. If I failed to mention someone I ask my colleagues to do this when they take the floor.

I see the main purpose of this conference in bringing home to the public what we learned 25 years ago, namely, that the Patterson-Gimlin film is an authentic documentary of a bigfoot or sasquatch of North America.

Here is in brief the film's origin and study. It was taken on October 20, 1967, at Bluff Creek in northern California, by Roger Patterson, in the presence of his assistant Robert Gimlin. Patterson died in 1972 from Hodgkin's disease, Bob Gimlin is alive and fully confirms what we know about the shooting of the film. The movie was seen by scientists at the prestigious Smithsonian Institution in the USA, who concluded the film was a hoax. Doubting the correctness of this conclusion, Canadian investigator René Dahinden brought a copy of the filmstrip to Moscow in 1971, and handed it over for study to members of the Smolin Hominology Seminar.

During his four week stay in Moscow Dahinden showed the film at many scientific institutions and editorial offices of newspapers and magazines. The only institution whose director (Valery Yakimov) forbade the showing was the Institute of Anthropology.

Professional interest was aroused by the film amongst the workers of the Central Scientific Research Institute of Prosthetics and Artificial Limb Construction, whose director, Professor N.I. Kondrashin, stated the following in a letter to the chairman of the USSR Committee on Cinematography: "The Patterson film was viewed and discussed by our Institute's specialists in man's locomotion. The film contains sufficiently clear frames of the walk of a manlike creature, a detailed study of which would undoubtedly be of serious scientific interest."

However, not a single scientific institution did undertake a study of the film nor offer a written conclusion. This task was set themselves by two young members of the Smolin seminar: Igor Bourtsev and Dmitri Bayanov. They were helped considerably by Alexandra Bourtseva. Why was the film studied by young investigators and not veterans? Because 1972 saw the passing of two of our teachers: professors Boris Porshnev and Alexander Mashkovtsev; Pyotr Smolin was of advanced age and in poor health; while Marie-Jeanne Koffmann, engaged in field work, was staying almost year-round in the Caucasus. For these reasons the main work on the study of the film was done by Bayanov and Bourtsev.

For a study in depth it was required to do a lot of preliminary technical work. Patterson, not being a professional photographer, shot the film with a 16 mm cinecamera without a telephoto lens, at a distance of 35-40 m from the subject. As a result the image of the creature on film scarcely reaches 1.5 mm and when it is enlarged to be seen on the screen or even a page it loses distinctness. For this reason it was necessary to enlarge the stills to a certain size and print them on specially chosen photopaper. This work, as well as all mathematical calculations connected with the film (such as the finding of the subject's height and weight, the filming speed, etc.) was performed by Igor Bourtsev. His method of calculating the filming speed (16 f.p.s.) was especially inventive and fruitful because it allowed at the same time the determination of Patterson's every step and movement in the course of shooting. In addition, Igor and Alexandra made sculpture models of Bigfoot based on the film frames. As for my own task, it consisted of comparative analysis, using the data of anthropology, paleoanthropology, and primatology, as well as of committing our study to paper.

The cinecamera caught the Bigfoot's movement, having in this respect a great advantage over the still camera. Movements of man and animals are studied by a special discipline—biomechanics, knowledge of which we were lacking. For this reason Alexandra Bourtseva contacted professor Dmitri Donskoy, the Chair of Biomechanics at the USSR Central Institute of Physical Culture, and invited him to view the film. I am glad to tell you that professor Donskoy is attending our conference and is present in this hall now. He is the only specialist outside our group who offered a written conclusion regarding the Patterson film. He pointed out the naturalness of movements of the film subject and at the same time the difference of its walk from man's walk, which testifies to the film's authenticity.

Having done by that time quite a bit of work on the film, Bayanov and Bourtsev also concluded that the movie was genuine. They shared their views with their American colleagues in 1972; that is, 25 years ago. In 1973, our positive conclusion and Dr. Donskoy's analysis were published in Canada in René Dahinden's book *Sasquatch*, and in 1975 in the U.S., in Peter Byrne's book *The Search for Big Foot*. In Moscow our conclusion was published in 1976, in our article in the magazine *Science and Religion*, No. 6.

In May 1978, the world's first international conference on Sasquatch and similar phenomena was held in Vancouver, to which we had been invited but did not get permission to attend from the Soviet authorities. Still our report on the verification of the Patterson film and Bigfoot tracks, as well as Koffmann's report on the almasty—the Caucasus hominoid—

were read at the conference, having aroused a lively interest among the participants and in the local press. Later on, a short version of our report and Koffmann's report on almasty were published in Canada. But all of that remained largely unknown to the public.

1982 saw the founding of the International Society of Cryptozoology. Its purpose is to discover animals still unrecognized by zoology, such, for example, as the purported denizen of Loch Ness, the mysterious Mokele-Mbembe of the Congo, and, of course, the snowman.

During 10 years in my post of the Society's Board member I called on my American colleagues to undertake a study of the Patterson film, to check our conclusion and present a definitive estimate of that material. With that aim in view, I proposed that a Russian-American commission should be set up. My proposal was received very coldly.

At long last, my call was heeded by one Board member, namely, anthropologist Dr. Grover Krantz, who, I am happy to say, is present here. In his book *Big Footprints* (1992), he presents a detailed analysis of the Patterson-Gimlin film and firmly endorses its authenticity. This happened 20 years after our conclusion and 10 years after the founding of the International Society of Cryptozoology.

Better late than never. The Patterson-Gimlin film, after thorough scientific analysis, has at last been found authentic on both sides of the Atlantic, and the Pacific, if you please. What could be better? The film is genuine, this means Bigfoot is real, and the snowman does exist! Nothing of the sort. As Dr. Krantz wrote to me, "Our scientific community almost unanimously considers the film to be an obvious hoax." Please note it's the opinion of scientists who have never studied the film.

And what's the situation in our country? Two years ago, at the ceremonial opening of this new Museum building, I had a chat on this score with a well-known journalist and writer, Vasily Peskov. I told him that I took to heart his words regarding the role of photography in zoology, published in his newspaper column "A Window on Nature." To quote: "The strength of photographic evidence is enormous. Films and photographs of vultures breaking ostrich eggs with stones made a strong impression on biologists and nature lovers. Scholars studying animal behavior treated the evidence as a sensation and wondered why this avian exploit had not been observed before. But it had been! It was reported by African hunters and by European naturalists ... But anecdotal evidence was treated lightly (Oh, just fables!) and soon forgotten. It took photos and films to realize: what a wonder! So never fail to give credit to the camera." "Doubtlessly, there is still many a mystery in Nature to be seen and confirmed by the camera."

With reference to these words, I mentioned the Patterson film, which opens one's eyes on a great mystery of Nature. Peskov responded that the film is a hoax. "How do you know that," I asked. "Well, it's known that Americans themselves have exposed it", he said and declined any further discussion.

C'est la vie, as the French say. The snowman was banned by Soviet science. In post-Soviet times the truth about him has been engulfed by a sea of mass media rubbish, which has become a market commodity.

Not long ago I was invited to participate in a TV program on the subject of snowmen. My opponent was a scholar who claimed that snowmen are heralds of the extraterrestrials and come here from other dimensions.

I proposed that he expand and deepen his hypothesis by extending it to gorillas, orangutans and chimpanzees. Why disparage them? Why not regard them also as messengers of the extraterrestrials and creatures from other dimensions?

Speaking seriously, we must put an end to this outrage. The truth about our wild cousins must become known to the civilized world, since the uncivilized world does know and has always known this truth. What is needed to break through the curtain of ignorance? First of all, to enlighten the enlighteners, the people of science and education. Let them know that for the hominologist the problem of snowman has long been non-existent. Because for the hominologist the snowman has long ceased being an anomaly, like that of UFOs or poltergeist, for example. The hominologist knows that the relict hominoid is just as natural and necessary a creation of Nature as the gorilla and the chimpanzee. Then why hominology? Why isn't this primate studied by anthropologists, zoologists, primatologists? Because with all its naturalness and necessity, the hominoid is a unique creature, a creature that biology has never dealt with. Anthropology has never dealt with a primate having a human shape but leading a non-human way of life. Zoology has never dealt with a beast having a human shape.

For this reason, information on relict hominoids has mainly been accumulated by demonology and folkloristics, but thereby distorted and mythologized. For the objective study of a subject that is new for the natural sciences, it was necessary to establish a new discipline, a new branch of primatology, which recognizes the reality of higher bipedal primates leading a non-human way of life. What is the meaning and significance of this discipline? The meaning of the ancient dictum "Know thyself" is in the necessity to know human nature. Without a profound knowledge in this respect, is there any point in dreams of a humane society? Man has emerged from the animal kingdom and he has a lot of animal traits. The hominoid is at the top of the animal kingdom, he is a frontier mark and a link between the world of man and that of animals. That is why his study is so important for understanding human nature.

There is no problem of the snowman, but there are problems, and major ones, of the science of him. Hominology is virtually a new-born discipline, and new-born babies' life is never easy. At the initial stage the life of paleoanthropology, the elder sister of hominology, which opened people's eyes on their fossil ancestors, was not easy either. It took science quite some time to recognize the reality of Neanderthal, *Pithecanthropus*, and *Australopithecus*. It's worth reminding our opponents of this. However, the infancy of hominology proved to be even much harder than that of paleoanthropology. Why? Firstly, because bones stay put in the ground and don't run away from the researcher as does the snowman. Pyotr Smolin used a Russian saying on this score: the elbow is close by, but you can't bite it. Secondly, paleoanthropology affects mainly the natural sciences, whereas hominology touches the interests and prerogatives not only of natural sciences but of humanities as well. Being an infant herself, hominology demands revisions and even reforms in some old and venerable disciplines. And revisions and reforms, as we well know, are not easy.

Meeting with my schoolfriends every year, I hear every time the same question: "Have you caught a snowman?" And every time I explain that "to catch" and "to perceive" mean etymologically the same (in the Russian language). To perceive something means to catch it by the intellect. And that's what is most important in science.

The snowman has been caught by the intellect thanks to theoretical and empirical data. The general theoretical foundation is Darwinism, the theory of evolution. Without it all attempts to perceive relict hominoids had failed. The latest and most important of such attempts was the classification of primates offered by Linnaeus in the 18th century. He named the given species *Homo troglodytes* (i.e., cave-man). It was in contrast to troglodytes that our own species was named by him *Homo sapiens*, and not at all because it is really so. Thus, the term *Homo troglodytes* appeared a century before Darwin's theory. Less perceptive or more orthodox naturalists decided that Linnaeus was mistaken and crossed out *Homo troglodytes* from the taxonomy of primates. And the question was closed for two centuries.

The second most important base of hominology is Porshnev's theory, which was conceived in our time on the basis of Darwinism, so it won't be possible to bury it. According to Porshnev, all species of higher primates, preceding *Homo sapiens*, were devoid of language, of human speech, and by this character remained animals. Pushed by the talking primate, *Homo sapiens*, armed with technology and weapons, into high mountains, dense forests, and swamps, the speechless higher primates could no longer embark on a human way of development and have survived to our days as relicts, locally and popularly known as almasty, kaptar, sasquatch, etc.

Their reality at present is testified to by three kinds of data: 1) Eyewitness accounts; 2) Footprints, and sometimes handprints, on the ground; 3) The Patterson-Gimlin film.

Of much importance is the question of methodology. All achievements of hominologists proceed from the methods they use. When science is faced by a subject which defies traditional methods of study, it becomes necessary to devise a new method or borrow one from other disciplines. In fact, scientific disciplines differ not only and not so much by their subjects as by methods of research.

Let me quote Boris Porshnev to support the contention above. In his article, "The Troglodytidae and the Hominidae in the Taxonomy and Evolution of Higher Primates" (*Current Anthropology,* December 1974), he wrote,

> It is precisely the use of nontraditional methods, such as the comparative analysis of mutually independent evidence, that has made it possible to establish the existence of this relic species and to describe its morphology, biogeography, ecology, and behavior. In other words, fact-finding methods have been used in biology that are usually employed by historians, jurists, and sociologists. (p. 450)

In his work "The Struggle for Troglodytes," Porshnev gives a very apt example to explain the difference between the methods used by historians and biologists. There is no need, he says, "to demand that the neck vertebrae of Louis XVI be put on the table to prove that he was guillotined ... This fact is accepted as proven by another, not less scientific method."

And here is what Marie-Jeanne Koffmann said in her "Reply to Professor Avdeyev," published in the magazine *Science and Religion* (1965, no. 4): "I don't believe in the almasty. I possess sufficient data to simply say that he exists!"

Thus, in the first category of evidence the snowman, alias relict hominoid, is caught by the intellect by way of collecting and analyzing sighting reports.

Let us pass on to the second category: the evidence of footprints. Here the best and largest collection of tracks is in the possession of our North American colleagues. And not because Bigfoot leaves more footprints than the hominoids of the Old World, but because the white population of North America have no taboo on information about this creature, and sometimes it reaches the investigator before the tracks are washed away by rain or buried in the snow.

Dr. Krantz has studied Sasquatch footprints and handprints, some even with their dermatoglyphics in view. As it transpires from his article, "Anatomy of the Sasquatch Foot," Krantz also displayed an innovation of method. Using the track of a Sasquatch with a crippled foot, he discovered, in particular, its elongated heel bone. The same feature was discovered by us and confirmed by the frames of the Patterson film; it is also correlated with the fossil material.

His study of footprints enabled Dr. Krantz to write a voluminous monograph, "Big Footprints" (1992), which has no eyewitness accounts. Pointing out the authenticity of the footprints he studied, Krantz winds up his article, "Anatomy of the Sasquatch Foot," with these words: "Even if none of the hundreds of sightings had ever occurred, we would still be forced to conclude that a giant bipedal primate does indeed inhabit the forests of the Pacific Northwest."

So the second category of evidence is also quite reliable in the eyes of the investigator in establishing the existence of relict hominoids.

Let's pass on to the third point, the film whose 30th anniversary we are marking. The advantage of the material of two previous categories is in its abundance, its serial nature, which easily provides for comparative analysis. The Patterson film is unparalleled so far. Cinematographically, we have nothing to compare it with. As of today, there is no other such film in existence. This makes it both immensely valuable and exceptionally difficult to analyze. Eyewitness accounts and footprints are evidence for the specialist, for the hominologist. Thanks to this evidence the hominoid is caught by the researcher's intellect, by his mental vision. A film, on the other hand, enables everybody to perceive its subject and the subject's character by one's proper sense of vision. Seeing is believing. The demonstrability of photography in zoology, as Vasily Peskov justly pointed out, is beyond doubt.

So the crucial question is whether this film is genuine or not. Finding out the truth justified much hard work. Methodological innovations were unavoidable here too, because the work we did 25 years ago was truly unusual. How do you verify the authenticity of a filmed creature if you don't have such a creature outside the film either dead or alive, and if official science deems such a creature non-existent?

First, we compared the film subject with what is known from the sighting reports. Secondly, we made comparisons, taking into account the evidence of footprints. Thirdly, the film subject was checked and compared with the data of paleoanthropology. To reduce our three categories of data—sightings, footprints and the film—to a common denominator, we used three criteria: distinctiveness, consistency, and naturalness. Let me explain what it means. Let's take, for example, such a most distinctive animal as the giraffe. A person who has never seen a giraffe would take an eyewitness account of it to be a joke and its photograph for a hoax. Yet, knowledge of the animal's ecology and way of feeding would convince anyone that there is no contradiction in the giraffe's anatomy, that it is quite natural and consistent with the animal's way of life.

Another example can be provided by a film character—King Kong. The subject is very distinctive and consistent in its anatomy of a giant anthropoid. But it fails our third criterion: naturalness. And this is testified to by the fact that there are no eyewitness accounts of King Kong or his footprints in nature.

A third example is the Piltdown skull. It was presented to the scientific community in England in 1912. It had a human cranium and a simian mandible. Was it distinctive? Very much so. Some scholars were so much impressed that approved of the find. Others pointed out its inconsistency and doubted the skull's authenticity. In 1953 it became evident that the Piltdown skull was a hoax.

So our three criteria, jointly employed, work rather efficiently. Having checked with them every category of our material and compared these categories to one another and to the data of paleoanthropology, we saw that the creature described by eyewitnesses, and the one leaving footprints, and that seen in the Patterson film, was of the same nature and species: distinctive, consistent, and natural. And as for alleged inconsistencies, pointed out by critics who haven't studied the material, as, for example, hair covering, including female breasts, it just confirms the creature's naturalness for those who know its ecology.

In this way we have established the authenticity of the materials provided by our North American colleagues and reported our results at the Vancouver conference, and now in a book to mark the 30th anniversary of the Patterson- Gimlin film.

To attract public opinion to this documentary and use it as a means of enlightenment in hominology, I put forward two proposals:

First, to initiate the inclusion of the Patterson-Gimlin film into the *Guinness Book of World Records*, as the first documentary showing a Bigfoot.

Second, to apply to the International Commission on Zoological Nomenclature with the request to assign the Latin name to the wild bipedal primate filmed by Patterson. The name I suggest is: *Homo troglodytes pattersoni*. Thereby we will give credit to Linnaeus, to Porshnev, and to Patterson.

My third proposal is even more significant. Boris Porshnev divided the science of anthropogenesis into three stages:

The first in time was Darwin's theory of the origin of man from animal ancestors.

The second was confirmation of this theory by the fossil record.

The third is the present and future stage of studying the living relicts of the evolutionary link or links between man and ape.

To make work at this stage and in this direction fully professional and legal, I propose the founding of the Porshnev World Institute of Hominology.

Said Porshnev half-jokingly: "The truth never wins. Its enemies simply die out." Let's do our best to ensure that the snowman shall not die out before the enemies of the truth.

# Bigfoot Adversary Does Verbal Harakiri

As the analysts of the Patterson-Gimlin Bigfoot documentary celebrated its 30th anniversary in Moscow last year (see *The Daily Telegraph*, October 23, 1997), the opposition lost no time in renewed attempts to debunk the famous footage (see *The Sunday Telegraph*, October 19, 1997).

The first attack ended in a spectacular flop, when John Chambers, retired Hollywood makeup artist and the main "witness" for the prosecution, testified to the press that he had never made, nor could he have made, the appropriate Bigfoot suit for Patterson's film.

Now the final debunking is claimed by TV presenter Chris Packham, in the story "BIGFOOT—PROOF or SPOOF?", published by *BBC Wildlife* magazine, September 1998.

What feeds Mr. Packham's desire to strip Roger Patterson and Robert Gimlin of well-deserved fame? The answer is provided by the following statement: "When I first saw the Patterson-Gimlin Bigfoot film years ago, I knew it was a hoax."

To support and sustain this "knowledge," the TV presenter studiously avoided reading the works of the film analysts. Instead, he and colleagues

Chris Packham's pitiful effort to recreate the creature in the Patterson-Gimlin film (left) is compared to an actual film frame.

"talked to special-effects experts, who laughed at the simplicity of the suit. They produced one, and we re-enacted the encounter precisely, to the centimeter. We re-shot it using the same type of camera."

As a result, the Packham film shows with brilliant clarity that, with a monkey suit on, "a man's a man for a'that!"

To complete his mission, the TV presenter used the hospitality extended to him by the late Roger Patterson's wife, Patricia Patterson, for "pawing through the perpetrator's personal effects, his wallet, his phone book, his letters, and photographs."

Packham says that when he found what he sought, "My heart nearly leapt from my chest." But, "In a way, it was a huge anticlimax. I remember walking to the car feeling I had the blood of the Bigfoot legend all over my hands."

What compromising evidence did the TV presenter turned detective find among Patterson's personal effects that killed the Bigfoot legend? He wouldn't say. It's a secret.

This writer and film analyst is sure that a huge anticlimax for Chris Packham is still to come, and that his feeling of triumph in Mrs. Patterson's home was caused by an illusion. A man-made document can never undo a Nature-made fact. As any genuine (not fake) researcher of the Bigfoot film knows, its subject is unassailable and undebunkable, because it is made by Nature, not man. It is as true as the Earth is round.

"A stake protrudes from the bleeding heart of Bigfoot, from cryptozoology itself," boasts Packham. In fact, a stake protrudes from the bleeding reputation of the TV presenter who boldly staged an act of verbal harakiri.

(October 4, 1998)

# Some Initial Thoughts While Looking Over *The Making of Bigfoot*

A copy of Greg Long's *The Making of Bigfoot* was sent by Bobbie Short to Michael Trachtengerts, who gave me the book for a few days. I read the first introductory words: "After nearly forty years of secrecy, the real truth is finally revealed behind the famous Roger Patterson "Bigfoot" film—a hoax that has managed to fool scores of scientists and millions of people around the world."

Stroking the volume's glossy jacket, I thought: What a beautiful tombstone for the future cemetery of Bigfoot debunkers! And what a Titanic of Ignorance, sinking from collision (nay, collusion!) with an iceberg of falsehoods! Here's the intrepid and indefatigable captain Greg Long, and his imposing patrons and well-wishers—Kal Korff, Robert Kiviat, Dr. Kenneth Wylie, and Dr. Dawn Prince-Hughes—the elite passengers of a luxurious liner doomed to perdition. I then read their "Advance Praise for *The Making of Bigfoot*" and thought: Good riddance, gentlemen! You've made your own choice.

I leafed the book and read here and there to see first of all whether it is an unprecedented-in-scale-and-arrogance attack just on the Bigfoot film, or on Bigfoot per se. The latter is the case, I concluded. This follows from the very title of the book and from these words of Introduction (p. 13):

> These giant, hairy monsters grip the imagination of millions of people around the world. But do the monsters exist? A surprisingly large number of people think so. Many of them base their belief on a single piece of evidence: a short, sixty-second strip of sixteen-millimeter color film allegedly taken of a Bigfoot on October 20, 1967, near Bluff Creek in Northern California. (...) No other piece of evidence exists, they contend, which is as compelling, convincing, and indisputable.

This makes it quite clear why the Bigfoot film was chosen as the object of attack on the "monster" itself. So the secret real name of the game must have been, *The Unmaking of Bigfoot*.

I came across and relished these lines (p. 195):

> I wanted Dahinden's final answer. "Do you think you can separate the film from the man, from the photographer?"
>
> "Of course!"
>
> "You think you can separate the two?" I repeated.
>
> "Just examine the f... film!" he yelled. "F... Al DeAtley! F... Roger Patterson! And f... Bob Gimlin! OK? Ignore the human element. LOOK AT THE F... FILM!"

By ignoring the veteran Bigfooter's advice, the author undermined the fate of his book. Actually, he could nicely have connected the film and the man, but only with his merits, not demerits and wrongdoings. The famous film resulted not only from a fluke, but also from Roger Patterson's dedication, courage, resourcefulness, and adroitness. Quite a few other people, with cameras, have chanced to encounter a Bigfoot, but none has managed so far to rival Patterson's achievement. I can imagine any number of people, myself included, in Patterson's shoes at Bluff Creek, coming back empty handed. No doubt, his action was a feat of investigation.

Kal Korff is known to have written (capitalizing his words): "I WANT to see Gimlin PROSECUTED and am WORKING TO ACHIEVE THIS." That is, if Gimlin refused to cooperate with Korff and Long by confessing to his and Patterson's "hoax." In this case, Korff wanted "to give a deposition to the Attorney General's office for CONSUMER FRAUD, specifically Gimlin's comments on [the TV documentary] 'Sasquatch: Legend Meets Science'." So I was interested what Long says about Gimlin in the book. On p. 159 I read the following:

> ... film producer Robert Guenette, who was making a documentary for Schick Sun Classic Pictures, Inc., explored Gimlin's bitterness: "Gimlin's reputation is that of a mild, honest man. I have talked to him several times. He still lives nearby Yakima with his wife, Judy. He has a somewhat embittered attitude about the whole matter; he is angry at the insinuations that he either compromised his honesty to perpetrate a hoax, or indeed was the prime dupe of one. He has repeatedly said that 'there is no question about what was out there ...,' describing the creature and explaining the incident over and over again in detail. In

> all his pronouncements, he has not changed his story. He believes he saw a Bigfoot that October 20th at Bluff Creek. I am only one among many who offered Gimlin large amounts of money to "tell the truth" about what "really" happened that day. His answer to me was, 'I'm already telling the truth'.

Noble Robert Gimlin, I embrace you in my thoughts... How proud I am that you count me among your friends, how lucky I am to have mixed all along with people of your make, not that of Heironimus and his patrons. If only most people stood up for the truth as strongly as you.

I much enjoyed these words of Michaela Kocis on the book's jacket:

> This book is a real EYE OPENER and it is refreshing to see that the lost art of good old investigative journalism is FINALLY BACK. This should set an example for courses on critical thinking and investigative journalism around the world."

No, the book is bound to become not only an all-time example but also an indispensable manual for courses and schools of journalism around the world. Its lesson number one: Don't take your future reader for as big a fool as you are in the matter. Lesson number two: To avoid looking a fool in the matter you start writing about, learn from those who know the subject best.

In investigative journalism of Bigfoot research the top master is John Green. John and I disagree on the nature of Bigfoots and the question of killing, but all insiders do know that his vast investigative work firmly testifies to Bigfoots' reality. Instead of learning from the old hand of Bigfootery, Greg Long questions Green's reputation as "a careful, analytical journalist who checked out all the facts" (p. 180). On p. 377 Long bursts "into hilarious laughter" hearing Dr. Krantz say, "I doubt that any human being could be trained to do that" (walk like the filmed creature). The cock-sure author offers his own conclusion: "It's obvious from my analysis alone that anyone can duplicate the Bigfoot's walk, and with a slightly above-average artistic ability you could build a suit like that" (p. 386). This can be diagnosed as fatal conceit.

The author could also have learned a few things from my *America's Bigfoot: Fact, Not Fiction—U.S. Evidence Verified in Russia*, conspicuously absent in his bibliography. Instead he chose to learn from "the gang who hung out at the Idle Hour tavern" (John Green). Well, the truth was not his purpose.

Long furthermore states, "Perhaps the only 'mystery' left now, is WHY so many millions of people, especially scientists and so-called Bigfoot 'experts' and self-proclaimed 'researchers,' could so easily have been fooled" (p. 10). This is a false question based on a falsehood. Two questions based on the truth are these: Why are the scientific establishment, the government, and the mass media fooling so many millions into thinking that Bigfoot is a myth? And why can they do this so easily?

The first question touches on ideology, business, politics, etc., so I don't know the full answer. The second is less involved, and my tentative answer is this: Bigfoot would long have been recognized real given normal conditions of science in this research. The abnormal conditions are created by those who are affluent, respectable, and "single-visioned" (i.e., one and all close-minded). From academic platforms, through megaphones of mass media, they confidently instruct the public: Bigfoot is a MYTH. Period. The abnormality is helped by Bigfoot researchers who are fundless, maverick, disunited, and quarrelsome. What most of them dare say to the public is: Who knows? Maybe it's not a myth? Question mark. Clearly, supposition has no chance against conviction.

We haven't received yet Chris Murphy's *Meet the Sasquatch,* so I don't know if it is a strong enough grave-digger for *The Making of Bigfoot.* Even if it is, I dream of yet another book on the subject, a marvel of investigative journalism, on a par with the Watergate exposé, with heroes and anti-heroes of the historical Struggle for Bigfoot, presenting positive and negative roles of scientists, presidents, billionaires, institutions, publications, TV shows and documentaries—the whole lot responsible for the present state of the problem. And let the wretched plight of Bigfoot research be not just an eye opener but a MIND OPENER on the integrity (or its lack) in the scientific community, government circles, and the media. I wish David Hancock would find and commission a worthy writer for the task. The title: *Bigfoot: The Biggest Cover-up in Science.*

## Update

Chris Murphy's *Meet the Sasquatch* had to go to press shortly after Long's book was released, so there was no time for an analysis. However, in 2005, Hancock House republished Roger Patterson's book, Do Abominable Snowmen of America Really *Exist?* with a supplement by Murphy on the filming and analysis of Bob Heironimus' claim that he was the "Bigfoot" in the Patterson-Gimlin film. The republished work is titled *The Bigfoot Film Controversy,*

and the contrary arguments Murphy presents were developed by John Green, Rick Noll and others. As a result, we see very clearly that Heironimus does not match the filmed subject's physical stature.

# One More Debunker Exposed

## Comments On David Daegling's Book *Bigfoot Exposed*

Thanks to gifts from Bobbie Short and Roger Knights, the International Center of Hominology now has two copies of *Bigfoot Exposed: An Anthropologist Examines America's Enduring Legend* (AltaMira Press, 2004) by David J. Daegling. As indicated by the subtitle, the purpose of the book is to assure the public that Bigfoot is only a legend. The educational bottom line is this: Bigfoot "is a human invention, and it is reinvented constantly" (p 248). In this respect, it's a repeat of John Napier's message in his *Bigfoot* (1972), and to the question which book is better or worse, I have to repeat Stalin's phrase, "both are worse." Still, Napier's has certain merit over Daegling's, for Napier openly avowed the real reason why he "will happily settle for the myth": otherwise anthropologists "shall have to re-write the story of human evolution" (p. 204).

Daegling is a spokesman of a Knowledge Monopoly, and has written his book accordingly. (See Henry Bauer's article "Science in the 21st Century: Knowledge Monopolies and Research Cartels," in *The Journal of Scientific Exploration*, Vol. 18, # 4). Still, we should be thankful to him and all other monopoly authors, for the more they engage in explaining Bigfoot away, the more they expose their prejudice, bringing closer the day of Bigfoot recognition.

Daegling exposes Bigfoot from the position of associate professor of anthropology at the University of Florida. I am exposing Daegling from the position of science director at the International Center of Hominology in Moscow, Russia. Hominology is a branch of primatology, founded in the middle of the 20th century in science's "no-man's land" between zoology and anthropology. An immediate impulse for its emergence was the Yeti problem, while an underlying historical and scientific reason was the discovery that "wild men" have been known throughout history all over the world. The self-laudatory term *Homo sapiens* was introduced by

Linnaeus in the middle of the 18th century in contrast to what he termed *Homo troglodytes* and *Homo sylvestris*.

Hominology is the science of living non-sapiens hominids (homins, for short), so of necessity it could only come into being after the emergence of the theory of evolution and paleoanthropology. Homins were unknown to modern science because there was no modern natural science to know them. Hominology means a scientific revolution in a number of disciplines, first and foremost in the theory of man's origin (anthropogenesis), as Napier rightly feared. Without considering this crucial factor, it is impossible to understand the attitude of mainstream scientists to the subject of Bigfoot, or any other relict hominids. (Let us note that gorillas, chimpanzees and orangutans are also relict species today). Hominology's evidence comes from natural history, mythology, folklore, ancient and medieval art, eyewitness accounts, footprints, vocalizations, and photography.

Why as yet no living or dead specimen or a part of its body? The shortest answer is that too little time has passed since the birth of hominology. It is a newborn science, devoid of recognition and funding. Even with the greatest funding in the world it takes time to apprehend certain bipeds not wanting to be apprehended, as, for example, Osama bin Laden. There is little doubt that Bigfoot and other homins are not willing to be apprehended and have every capacity to stay at large. The rare cases of their capture are marked in history as special events. On the other hand, the accidental capture of a specimen by apple orchard guards in Russia in 1989 ended in the release of the creature because it threatened to ruin the car in which it was imprisoned. Had the car owner been promised a tiny fraction of the reward for the capture of Osama bin Laden, the situation in hominology would now be different.

In his book *The Locals*, Thom Powell presents the case of a Bigfoot reportedly captured in 1999 in Nevada during a forest fire. The creature was said to have been taken away by the authorities and to have disappeared without a trace. I take the story seriously because of its many realistic details and because we have had similar reports in Russia. Now that the name of the wealthy Hollywood owner of the so-called Iceman has been indicated, I am convinced that Ivan Sanderson and Bernard Heuvelmans were not mistaken when they said that what they saw was an object of biology, not fakery. The corpse was both exhibited and withdrawn for religious reasons.

So a more involved answer to the question why definitive biological evidence in hominology is not available is this. The number of people interested in obtaining such evidence is an infinitesimal fraction of those who are indifferent to the task or are against it for one reason or another. Further, the number of those among the interested who may have a chance to find and recognize such evidence is also an infinitesimal fraction. The negative impact of indifference on one side, and hidden or open hostility on the other, leaves the tiny number of hominologists little chance to quickly obtain traditionally acceptable biological proof. For this reason hominology still finds itself in a cryptozoological phase of development.

And yet of all cryptids in the world Bigfoot is the best documented biologically. We have for it eyewitness accounts, footprints, handprints, a body print, hair samples, scat, recorded vocalizations, and film footage. The progress in obtaining and analyzing so many different kinds of evidence by unfunded volunteers is amazing (see Christopher Murphy, *Meet the Sasquatch,* 2004, Hancock House). For hominologists this is more than enough to take Bigfoot for a reality, but it is not yet enough for mainstream scientists, and this is not only because Bigfoot is not an ordinary primate, but because it is the harbinger of a scientific revolution.

Many hominologists agree with me that it is impermissible for moral reasons to seek the solution of the problem by means of a rifle. We pin hopes on the method proposed by our teacher, the founder of hominology, Boris Porshnev, who wrote:

> If proceeding most cautiously we succeed in conditioning the creature to come and take food in a definite place, that would be a real scientific victory. There is a basis for such prospects, namely, the above-mentioned cases in different geographical areas of local people habituating and even befriending relict hominoids. Scientific work could be launched in such a case even without direct contact of researchers with the specimen, for modern zoology boasts of an excellent means of taking color films with a telephoto lens at a great distance. A relict hominoid would then appear on the screen showing its usual movements and habits against a background of its natural environment. So step by step relict hominoids on earth could find themselves under man's protection and permanent scientific surveillance. At a certain moment it would be possible, of course, to observe the death of this creature. Then the anatomist would get a corpse for autopsy. Thus the perspective of studying *Homo troglodytes* looks as the reverse of zoology's canon: not from dissection to biology but from biology to dissection. (Porshnev 1963, in *Bigfoot: To Kill or To Film? The Problem of Proof*, by Dmitri Bayanov, 2001, pp. 13, 14)

Thanks in part to the Internet, the secrets of habituation are beginning to open up, turning Boris Porshnev's vision into a reality, as indicated by the book *50 Years with Bigfoot: Tennessee Chronicles of Co-Existence*, 2002, by Mary Green and Janice Carter Coy, and by Igor Bourtsev's article "Russian Hominologist in Tennessee" (*Bigfoot Co-op*, December 2004).

Finally, why is hominology scientific rather than pseudoscientific, as alleged by some critics? According to Henry Bauer's *Science or Pseudo-science*, 2001, the main criterion of a scientific pursuit is "connectedness," i.e., "crucial links with the mainstream" (p. 158). "All natural scientists accept and draw on the same laws, facts, and methods" (p. 11). I understand this as follows. The unknown can only be studied and understood by proceeding from and connecting with the known. Magnetism has been known to science since antiquity, while electricity was much of an unknown two centuries ago. Faraday and Maxwell connected electric phenomena with magnetism and thus ushered in the era of electricity. So their work was very scientific.

By this criterion, UFOlogy is not yet a science because so far specific UFO observation reports cannot be connected with or explained by the existing scientific knowledge. Hominology, on the contrary, by the criterion of connectedness seems to be the most scientific of sciences for it provides "crucial links" with and between the theory of evolution, paleoanthropology, mythology, demonology, folkloristics, the history of religion, and the history of art.

In addition, hominology gives a natural answer to the natural question why apes are still with us while brainier apemen or pre-sapiens hominids died out. The answer is they didn't. Their wholesale extinction is the illusion of paleoanthropologists who are as adequate experts on relict hominids as paleontologists were on living coelacanths. Relict hominids are hidden in natural forests and mountains, but above all they are hidden in "the forests of the mind." The task of hominology is to drive them out of those "forests" into the open vistas of science.

Such is the necessary prelude to taking Dr. Daegling on in earnest. Someone declaring nowadays that stones falling from heaven are nothing but a myth would have to refute the science of meteoritics. Similarly, anyone publishing a book declaring that Bigfoot is a myth has to take on the science of hominology in its theoretical, historical, and geographical aspects. As this task proved Herculean for Dr. Daegling, he opted for the simpler job of declaring all the sightings mistaken, all the footprints faked, and the Bigfoot documentary hoaxed. The whole tome of 276 pages consists of nothing but endlessly repeated naysayings. John Green has already challenged Daegling's expertise on Bigfoot tracks:

> People who have never seen any tracks but claim to know more about them than those who did see them are not a rare breed, their number is legion, but for someone to join their ranks waving the flag of "scientific verification" is bald-faced hypocrisy. What the tracks were like may be "anecdotal" to Dr. Daegling, but it is first-hand knowledge to those of us who studied them, photographed them and cast them, and because of our efforts there is plenty of solid evidence available to any scientist who will take the trouble to see if it can be verified or not. Dr. Daegling is not among those who have been prepared to take that trouble. Instead he stayed home and wrote a book (John Green's email "Bigfoot Exposed," Jan.3, 2005.)

As for eyewitness accounts, they, according to Daegling, cannot be trusted, for the following reason:

> Unfortunately, we have been asking the wrong question through the years. What did you see? we ask the eyewitness. If we take the answer at face value, we miss the meaning of the phenomenon. It may be more important to ask the one question the eyewitness may be in no position to answer: Why did you see it? (p. 259)

What a "useful" piece of advice, especially for detectives seeking information from witnesses, or for zoologists interviewing eyewitnesses with the aim of determining habitats of rare animals, or for physicists collecting sightings of ball lightning.

The major part of the author's naysayings is devoted to the Patterson-Gimlin film. This part of the book is of special concern to me and my Russian colleagues because the film was for the first time systematically studied and validated to our own satisfaction in Moscow back in the 1970s. So let us see what the author says about the Russian research and researchers.

It is untrue that "the Moscow Academy of Sciences boasted its own Institute of Hominology" (p. 111). The Institute is even today nothing more than a dream of mine.

It is untrue that Porshnev's first name is Victor (p. 111). It is Boris.

It is untrue that Dmitri Bayanov is schooled in biomechanics (p. 111).

It is untrue that Donskoy's "report ... is thoroughly subjective and devoid of any particulars of argument" (p. 111).

It is untrue that "Up until 1992, (...) there had been no scientific efforts directed at the film that took up the issue from a purely quantitative (and ostensibly objective) standpoint" (p. 119). Daegling's References include our paper, published in 1984, "Analysis of Patterson-Gimlin Film: Why We Find It Authentic." It is based both on quantitative and qualitative analysis and presents quantitative findings.

It is untrue that the film speed "is unknown" (p. 128). Igor Bourtsev determined it in 1973. His method and result stand in black and white in the above-mentioned paper, listed in Daegling's References.

It is untrue that Perez "threw down the gauntlet" (to the mainstream) in the matter of the Bigfoot film (p. 119). This was done by Russian hominologists in their report presented in 1978 at the Vancouver Sasquatch conference.

It is untrue that "The gait of the film subject (...) is easily duplicated by human beings" (p. 147). Mimicked, yes, but not duplicated. Human beings can mimic the walk of different animals, such as bears, camels, elephants, as well as of the film subject. But they cannot imitate it in a natural, uncontrived manner characterizing Bigfoot's gait.

It is untrue that "Skeptical inquiry into the film has made significant strides since 1967" (p. 205). Actually, it hasn't moved an inch. On the contrary, all aspiring debunkers of the film over the past decades have been exposed and defeated, and not a single proof or argument put forward by us for the film's authenticity has been refuted.

Dr. Daegling claims to have found "a glaring anomaly" in the film subject, namely, "the Achilles tendon appears to attach far forward on the heel, where the adaptive advantage of having an elongated heel in the first place is completely lost. (...) A prosthesis explains what is seen in the film; evolution, by contrast, cannot make sense of it" (p. 144). In our paper, published 20 years before Daegling's book and listed in his References, the matter of Bigfoot's elongated heel and Achilles tendon is dealt with as follows:

> The heel is actually seen to be sticking out in an inhuman way in some frames, suggesting an unusually large heel bone (calcaneus) as has been predicted by Grover Krantz using theoretical considerations and the evidence of the footprints. That the heel of the filmed subject is really unusual is testified to by the fact that this feature was independently discovered in Moscow and Ottawa. In Moscow it was seen by Bayanov and Bourtsev as "an omen of the creature's reality." (...) It is worth pointing out also that this peculiarity has never been reported by eyewitnesses because it appears only for a

fleeting moment when the Achilles tendon is not tight in a certain phase of the stride. (In *The Sasquatch and other Unknown Hominoids*, edited by Vladimir Markotic and Grover Krantz, 1984, p. 226)

The film records in some of its frames these fleeting moments. In other words, there is no anomaly with attachment of the Achilles tendon. It is attached in the usual place at the end of the heel, and the impression that it is attached in a wrong place appears only when the tendon is slackened, not tightened. Dr. Daegling hides this fact from the reader by concealing our analysis of the film, described by Dr. Roderick Sprague as "by far the best and most thorough discussion of this classic film" (*Cryptozoology*, Vol. 5 (1986), p. 105).

On p. 211, Daegling quotes Dahinden's phrase "lying by omission." Dr. Daegling's biggest lie by omission is his total silence about my book *America's Bigfoot: Fact, Not Fiction—U.S. Evidence Verified in Russia*, 1997, devoted to our validation of the Patterson-Gimlin film, which is not even listed in his references. A possible reason for the omission is the

Chris Murphy is seen standing in a clearing that was part of the original film site seen in the Patterson-Gimlin film. Dr. Daegling states that the site is "largely overgrown." The truth is that the site is largely missing – about two-thirds of it were wiped out by Bluff Creek, leaving a large gorge. The photo used by Daegling on page 124 of his book appears to be taken from a point further up the creek than directly opposite the film site location. I have also been to the site and can attest to the fact that there is still a clearing.

strength of the case it makes, as indicated by this appraisal by Dr. Henry Bauer, Professor Emeritus of Chemistry & Science Studies:

> Glimpses of the Patterson film in various television shows had left me incredulous that the creature shown in it could be real. This book has made me almost equally incredulous that the film could have been faked, and thus I have become open to the staggering possibility that relict hominids may still be with us in sufficient numbers that we have the chance to learn something about them. I recommend this book heartily as a highly interesting reading adventure. (*Journal of Scientific Exploration*, Vol. 18, Number 3, 2004, p. 533)

On p. 211, we read Dr. Daegling's conclusion that "Poor scholarship is one tell-tale sign of a pseudoscientific approach." This remark applies in full measure to the author. What's more, his book, by its intent and quality, is simply anti-scientific. Its contents do nothing but delude the reader. Fortunately, with the wide means of exposure provided by the Internet, Dr. Daegling's book, unlike that of Dr. Napier, is not destined to delay the search for Bigfoot. The process of undeceiving the public is gathering speed.

# Skeptical Look

## at *The Skeptical Inquirer*

I have never seen *The Skeptical Inquirer* and had taken it for a serious and truthful publication on account of references to it in Henry Bauer's excellent book *Science or Pseudoscience*. So was taken aback by learning that SI published in July 2004 a totally false article concerning the Patterson-Gimlin Bigfoot documentary. Being hominologists and analysts of the film, Igor Bourtsev and I could not stand such a falsehood. To show the editor that we meant business we asked him to publish our rebuttal together with the offer of a reward to anyone who proves the film was faked. We emailed our letter back in July to the SI managing editor Benjamin Radford who has never acknowledged its receipt, in spite of repeated queries. My understanding of his shyness is this: he is afraid both to publish our rebuttal and acknowledge its receipt, not knowing how to explain his refusal to publish. So he opted for silence, hoping in this manner to get away with his misconduct. A vain hope in the age of the Internet. Let everybody know that I find the editor's conduct unseemly because it is dishonest and cowardly. Even doubly cowardly: he fears the truth and fears to be seen fearful. I conclude that the *Skeptical Inquirer*'s skepticism is a one-way street.

It would be comical if it weren't hilarious that while in America Bob Heironimus has triumphantly ousted Ray Wallace from inside the covetable "ape suit," in Britain the late Wallace is still honored as the unrivaled maker of Bigfoot. To wit:

> It's just that, whenever the evidence becomes concrete in one of these more exotic cases, it invariably turns out to be a hoax. Like the legend of the Oregon Bigfoot, a North American yeti which was actually filmed in 1967, widely believed in but which turned out to be the work of Ray L. Wallace, who confessed on his deathbed in 2002, that he was Bigfoot.
>
> It was in 1958, it turned out, that he had first strapped on some wooden feet and made huge foot-prints in the mud near where he worked. It was he who had guided the movie-maker to the best place from which to catch a glimpse of the hairy half-man in the distance. (David Aaronovich, "Big Little Man," *The Observer*, October 31, 2004).

The implication of the Patterson film is a thousand times more important than that of the hominid fossils discovery on an Indonesian island. Thus the *Observer* columnist's entertaining piece is a cocktail of important truths and superimportant falsehoods. How to make the mass media drop falsehoods and stick to facts of hominology is the Big question indeed. Roger Knights came up with the idea of "a commemorative Wall or Hall of Shame" to be constructed in Washington D C "honoring" those skeptics who impeded investigators and refused to credit any evidence. "Pride of place," he wrote, "would go to Dr. Montagna, head of the federal government's primate research lab in Oregon, who said he blushed with shame to think of scientists actually studying the gait of the Patterson Bigfoot creature ..." As a second candidate for the Wall of Shame I proposed the British TV film producer Chris Packham, who falsely claims to have shown that the PGF was hoaxed (see *The BBC Wildlife* magazine, September 1997). Other candidates are, of course, Bob Heironimus, Greg Long, Kal Korff, and now Benjamin Radford.

But that's a promise of punishment in the future, which may not duly impress the culprits. Let's think of an effective deterrent in the present. How about annually awarded titles for the most pernicious Bigfoot debunkers, such as the Liar, or Falsifier, or Impeder of the Year?

Here's our letter to *The Skeptical Inquirer* with attached offer of reward to debunkers which has already been published in FATE and posted on web sites. We dared to make such an announcement not because our Center is rich in funds (the truth is in the opposite) but because through scientific research we do know the film is genuine and therefore irreproducible. Hence our bold truthtelling challenge.

----- Original Message -----
From: Cryptologos
To: letters@csicop.org
Sent: Friday, July 30, 2004 11:17 AM
Subject: Russian response
July 30, 2004

Editor
The Skeptical Inquirer

Dear Sir:

Fully supporting John Green's position and arguments, we are offering some views of our own concerning the latest preposterous attack on the 1967 Bigfoot film and ask you to publish them.

Exposing Greg Long's Attempt To Refute Science With Hearsay

Rebuttal to the article "Exposing Roger Patterson's 1967 Bigfoot Film Hoax" (Skeptical Inquirer, July/August 2004)

The Skeptical Inquirer article by Kal Korff and Michaela Koch fails to mention basic facts regarding the Patterson-Gimlin Bigfoot film, thus presenting the issue in a false light. The film's authenticity has not only been "hotly debated" but firmly established long ago by researchers in Canada, Russia and the U.S. Scientific studies by different specialists of different countries have revealed in the filmed creature a number of non-human characteristics, both in its anatomy and locomotion, thus excluding the possibility of a disguised human being (i.e., Homo sapiens). The analysts of the film had no other motive in investigation than seeking the truth, for which they even risked their reputations.

In contrast, the sole motive of involvement of Bob Heironimus, the source of the alleged exposure, is money. The Korff article says, "it was just a way to make some quick and easy money". Another salient point of the affair is the claim by the debunkers that they know all about the make and origin of the costume allegedly used in hoaxing the film. They claim they know who sold the costume and how it was modified by Patterson. The obvious question is: Why haven't they re-created the costume for all to see in the first place? By failing to re-create the alleged Bigfoot costume and re-enact the film Kal Korff and Greg Long have exposed themselves as sham investigators and presumptuous men.

The truth of the matter is that the truth of Bigfoot, both in the bush and on film, is hidden from the general public by the scientific establishment and the obedient media. We are confident that as soon as the truth prevails, the first Bigfoot documentary is bound to win a grand prize and be placed in the U.S. National Archive. "There are still people on the earth who believe that the planet is flat and not round", says Korff. This rather fits those who still believe that Bigfoot is a myth and its image on film was hoaxed.

We regard Greg Long's book "The Making of Bigfoot" as just another in a series of brazen attempts to refute scientific findings with hearsay. To put an end to such attempts we make the announcement sent in attachment.

Sincerely

| Dmitri BAYANOV | Igor BOURTSEV |
|---|---|
| ICH Science Director | ICH General Director |

ATTACHMENT

Фонд содействия научным исследованиям
и поискам "КРИПТОСФЕРА"

**Международный центр гоминологии**

Россия, 121614, Москва
Осенний бульвар, 12, корп. 3, "Крипто-логос"
E-mail: cryptologos@mtu-net.ru
Тел.:(095) 413-96-05, 413-12-88

CRYPTOSPHERE Fund for Furthering
Scientific Explorations and Searches

**International Center of Hominology**

Crypto-Logos, 12-3, Osenniy blvrd,
Moscow, 121614, Russia
E-mail: cryptologos@mtu-net.ru
Phone: (7-095) 413-96-05, 413-12-88

May, 2004

**$100,000 Offered to Debunkers of Bigfoot Film**

The new discipline of hominology is based on several kinds of data, including the famed Patterson-Gimlin documentary. Taken in 1967 in northern California, it was rejected as a hoax without serious study by the Smithsonian Institution. In 1971 the film was brought to Moscow, where it was analyzed in depth and found authentic by us and Dr. Dmitri Donskoy. Our conclusion, announced at the 1978 academic Sasquatch Conference in Vancouver, reads as follows:

"We have subjected the film to a systematic and multifaceted analysis, both in its technical and biological aspects. We have matched the evidence of the film against the other categories of evidence and have tested its subject with our three criteria of distinctiveness, consistency, and naturalness. The film has passed all our tests and scrutinies. This gives us grounds to ask: Who other than God or natural selection is sufficiently conversant with anatomy and biomechanics to 'design' a body which is so perfectly harmonious in terms of structure and function?

Further research may correct some of our findings, but it seems most improbable that the positive result can be voided. Hence, we confidently give this verdict:

**The Patterson-Gimlin film is an authentic documentary of a genuine female hominoid, popularly known as Sasquatch or Bigfoot, filmed in the Bluff Creek area of northern California not later than October 1967, when it was viewed by René Dahinden and other investigators."**

How we came to this conclusion is described in detail in Dmitri Bayanov's book *America's Bigfoot: Fact, Not Fiction—U.S. Evidence Verified in Russia* (1997). Our verdict has been confirmed, and new arguments for the authenticity of the film have been offered in subsequent studies by our colleagues in North America. Nonetheless, enthusiastic but uninformed debunkers, whom we call "glory vultures," ceaselessly dream of destroying the glory of the first Bigfoot documentary, alleging in the mass media and books that it shows a man in a monkey suit. Fed up with this monkey business, we are now officially offering $100,000 to anyone who can successfully demonstrate to a panel of hominologists and anthropologists that the Patterson-Gimlin film shows a human being in a special suit. We know that people deaf to the voice of reason can hear very well the voice of money. Our offer of one hundred thousand dollars is based on the security of the equipment, vehicles, and intellectual property of our organization.

Dear debunkers, if you mean business and not just a publicity stunt, please send your applications to The International Center of Hominology, Moscow, Russia.

| | |
|---|---|
| Dmitri BAYANOV | Igor BOURTSEV |
| ICH Science Director | ICH General Director |

SECTION 6

# BIGFOOT: To Kill or to Film? The Problem of Proof

## Foreword by Christopher L. Murphy

When I first learned of the contentiousness among Bigfoot researchers on the "kill or film" issue, I gave the matter little or no thought. After all, we have not to this point provided conclusive evidence that the creature even exists. Further, do we really have any control over the issue? Who is to say that the person faced with such a decision would be aware of its ramifications? On the surface, these points provide a good argument that the issue did not merit a great deal of concern.

After reviewing Dmitri Bayanov's manuscript, I arrived at an appreciation of his passionate concern on the issue. In other words, I now see very clearly why he has been insistent to the point of frustration that intentional killing of a Bigfoot must not be condoned under any circumstances except self-defense.

From Dmitri's standpoint, the first question I raised (Bigfoot's existence) is to him a total non-issue. Dmitri is thoroughly convinced that Bigfoot is a natural creature. It is important for the reader to understand this fact right from the outset. The possibility, therefore, of someone being faced with a "kill or don't kill" decision is to Dmitri just a matter of time.

My second point (decision control), is not so easily dismissed. The total number of Bigfoot researchers and enthusiasts hardly compares to the number of hunters and other armed people in Bigfoot's domain. Nevertheless, Dmitri points out that the community of Bigfoot researchers (especially those with a high profile) can influence the decisions of others in all walks of life. Dmitri reasons that a **unified** stand on the issue could definitely "make a difference." The emphasis here on the word **unified** is important. A divided stand (i.e., some for and some against killing) is not much better than no stand at all.

Dmitri, of course, wants an unqualified "no killing" policy to be publicly declared by all high-profile or influential Bigfoot researchers. Certainly, judging by the information on record the major North American Bigfoot researchers are far removed from making such a declaration. However, I have pointed out to Dmitri that some time has passed since he last confronted these researchers with the issue—are we certain they still have the same mind-sets? On this question Dmitri responded that if they have changed their minds, this book will provide them with an opportunity to say so.

I have edited and published this book as a writer and a publisher, not as a Bigfoot researcher. I have not altered the intended message of any information presented, nor have I confirmed or verified the accuracy of any statements, quotations, or facts.

— Christopher L. Murphy

Cover of the original booklet published under this title in 2001. The artwork was performed by Lydia Bourtseva.

## Introduction

This is my third major publication in English dealing with the problems of a new discipline—hominology. The first two, *In The Footsteps of the Russian Snowman* (1996) and *America's BIGFOOT: Fact, Not Fiction* (1997), present and discuss certain evidence for the existence of relict hominoids (actually, hominids) in Eurasia and North America. The books also give the reader an idea why this evidence has been ignored or rejected by the scientific establishment for so long. The science of primatology, as foreseen by Boris Porshnev, is on the verge of a revolutionary change, spurred by the emergence of a new paradigm concerning the origin of modern man—*Homo sapiens*. Within the old paradigm, primatology, anthropology, and paleoanthropology were simply conceptually impotent to deal with the phenomenon of relict hominids. Thus there appeared a division and conflict in science between those who recognized the reality of this phenomenon and those who rejected it.

This work is about a conflict and split among Bigfoot researchers themselves, a source of their weakness for decades. It is a conflict over the eternal problem of ends and means—how to prove the existence of Bigfoot in the face of relentless opposition from the scientific community. Here is the opinion of John Green: "Science will keep its eyes tightly shut until someone produces a body, or part of one, and the more quickly it is done the better. The successful hunter should find it very profitable as well." (*The Sasquatch File*, 1973, p. 71). Grover Krantz's attitude is described in the press as follows: "He is one of the most vocal proponents of killing a Sasquatch in order to prove they exist. 'I want to rub a few faces in the corpse,' he says. His advice is to shoot it, cut off an arm or a leg or anything you can carry and get the hell out." (*Human Behavior*, September 1978, p. 20)

In the mid-1970s, I initiated a "great debate" by correspondence among several hominologists on the "to kill or to film" question—in which Green and I were the main adversaries. The controversy culminated in my rebuttal to Green, "Why It Is Not Right To Kill A Gentle Giant", published in *Pursuit*, Fall 1980. During the 1980s I argued and clashed with Grover Krantz over the issue. He published his pro-kill views and appeals in the article "Research on unknown hominoids in North America" (1984) and in his book *Big Footprints: A Scientific Inquiry into the Reality of Sasquatch* (1992). My counter-arguments were either unpublished or printed in obscure publications, which is reason enough to bring them to light here. The airing is needed not just for argument's sake but because the issue is of great humanitarian significance.

# CHAPTER ONE

## Hunting for Knowledge

The argument on the question of ends and means began not in the 1970s over Bigfoot but back in the 1950s over the Yeti. Loren Coleman, in a book devoted to Tom Slick, a Texas millionaire who mounted two Himalayan expeditions searching for the yeti, has this to say on the issue:

> On March 18 and 19 (1957), the Nepalese government forbade all foreign mountaineers from killing, injuring, or capturing a yeti. ... Tom Slick's expeditions, of course, helped open up the whole debate on the ethics of killing zoological specimens. During the pre-Slick years, the giant panda, for example, was routinely killed and mounted for museums around the world before anyone stopped to consider what effect this action would have on the breeding population. Pandas were killed first to provide evidence that they even existed, then as prized natural history exhibits, and finally captured alive as animals of choice for zoological gardens. Today the giant panda is a symbol of endangered species, and no one would even consider killing one for a museum or any other scientific purpose.
>
> Slick entered the field of cryptozoology when the rules were just beginning to be redefined. His goal, as he saw it, was to collect a specimen, alive or dead, to prove to the scientific world that the yeti existed. By the late 1950s, the era of hunting and collecting was beginning to end. Slick's thoughts on killing began to change, to reflect the times, and to mirror his own notions on the creatures' right to survive. ... Slick received a good deal of bad press about his very "Texan" approach to the yeti hunt, especially from British writers in England and India. (Coleman 1989, 59–60)

If Tom Slick's thoughts on killing began to change back in the 1950s, then the thoughts of John Green and Grover Krantz haven't changed a wee bit over the past decades and present a magnificent relic of the pre-Slick era. Before dealing with this relic in detail, I would like to bring to the reader's attention certain views from the opposite camp. First, the final passage in an article by zoologist Edward W. Cronin, Jr., who returned from a two-year scientific expedition into Nepal convinced that the yeti exists:

Even though I am intrigued with the yeti, both for its scientific importance and for what it says about our own interests and biases, I would be deeply saddened to have it discovered. ... If the yeti is an old form that we have driven into the mountains, now we would be driving it into the zoos. We would gain another possession, another ragged exhibit in the concrete world of the zoological park, another Latin name to enter on our scientific ledgers. But what about the wild creature that now roams free of man in the forests of the Himalayas? Every time man asserts his mastery over nature, he gains something in knowledge, but loses something in spirit. ("The Yeti," *Atlantic*, November 1973, p. 53.)

This passage is remarkable not so much for its message as for its coming from a scientist. If anthropologist Krantz seeks to discover a hominoid by any means, then zoologist Cronin "would be deeply saddened" by the discovery even by the best of means. Cronin is not the only scientist to think so. Apparently, the latter extreme is an inevitable reaction to the former extreme. An ever-growing number of laymen and academics are getting weary of man's mistreatment of animals.

It is fair to say that the inquiry into the snowman problem began by way of a hunt, not science. The coveted objective of numerous Himalayan expeditions by Western explorers and even of the Pamirs expedition by the Soviet Academy of Sciences was to hunt down the enigmatic hairy biped and show the trophy to an astounded world. A hunt mentality is still typical of such North American investigators as John Green and Grover Krantz. A totally different approach, with the mentality and strategy of science (i.e., hunting for knowledge, not trophies), was first introduced into the snowman investigation by Boris Porshnev, so it is worthwhile to look anew at this aspect of his work.

When a translation of Ivan Sanderson's article devoted to the Patterson film was published in a Soviet magazine, Boris Porshnev wrote (in part) the following in his commentary (1968):

The public is much taken by the illusion that the "snowman" problem can only be solved by a sensational breakthrough. A single proof will be obtained and submitted: here you are! No, the process of science is more modest and more majestic. In its course knowledge is accumulated and deepened, new information is added to old information, and its overall reliability increases. A single sensation won't work if only because any sensation can be questioned: photographs and films can be faked, while a live specimen can be declared a rare pathological case, a freak of nature. Science operates, as a rule, not with isolated facts but with series of them. That is why only those investigators who have studied a great number of already known similar facts can judge the validity of the film taken by Patterson and Gimlin. This film can't make a revolution in science. When two, three, ten films are taken, their conclusiveness will increase, if the need for more and more "proofs" does not gradually disappear even for the outsiders. Initially it seemed that some defendants had to submit a "proof" to some judges, whereupon these experts will kindly take into their scholarly hands further progress of the investigation. Now we are clear that it is the defendants who are the only experts and specialists in the matter. Their circle will be joined and enlarged by young biologists who will want to acquire the existing knowledge and take up the torch. And the judges will be dozing away in armchairs in an empty hall. (Porshnev 1968, 52–53)

This commentary shows that back in the 1960s Porshnev well understood what's what and who's who in our research. Free of illusions, he had a clear idea how to push forward. In regard to Patterson's film, Porshnev's words do not mean that he considered the film documentary unimportant or unnecessary. No, he just put that piece of evidence in the context of ongoing research and correctly predicted that it would not resolve the problem—not because it is a film, but because a scientific problem of such magnitude is of necessity resolved in a different way, in a way of accumulating and deepening knowledge. Following Porshnev's train of thought, I wrote in our conference paper on the Patterson film that,

science can solve only those problems which it is ready to solve. Thus, an eventual delivery of traditional zoological material will resolve the hominoid problem not just by virtue of that fact alone, but also because science will have been prepared to accept the discovery by the entire course of the research, including, hopefully, this paper.

In the process of knowledge accumulation, incoming information (including films) must be assessed. Some information will need to be accepted and some rejected. However, who is to do the accepting and rejecting? Who are the experts? Porshnev gives a clear answer, which, regrettably, is still unclear to some people. Of course, hominologists can and must use the expertise of other specialists, but the last word in the appraisal of evidence in hominology must belong to them (hominologists), not to outsiders.

In this connection, it must be said that Porshnev not only proceeded in a scientific manner about the hominoid problem, but happened to found a new science along the way. This fact has to be said loud and clear because most researchers in the field, John Green and Grover Krantz first and foremost, have not yet heard the news. John Napier stated that if Bigfoot is real, then, "as scientists we have a lot to explain. Among other things we shall have to rewrite the story of human evolution" (Napier, *Bigfoot*, 204). Napier shunned the rewriting, Porshnev was inspired by it and, right or wrong, he did it! Thus he was a generation ahead of other evolutionists who haven't even yet realized the necessity of the task.

On page 273 of his suppressed monograph, still existing in 180 rotaprint copies, Porshnev (1963) wrote of "the destinies of the emerging science of relict hominoids." It fell to my lot to name the new branch of knowledge. This happened in 1972, and in 1973 I used the term hominology in my letters. In the second half of that decade the term began to be used in the press. However, was hominology really necessary? Wasn't cryptozoology sufficient to cover all enigmatic animals, including our "charges"? My answer in this regard is as follows. Even with the advent of cryptozoology, hominology is not only necessary but inevitable. The reason is that as soon as an object of cryptozoology (i.e., a cryptid) is recognized by zoology, it stops being a cryptid and comes under the auspices of one or another branch of zoology. Such was the fate of the okapi, the giant squid, the gorilla, and numerous other ex-cryptids. But relict hominoids have no discipline to fall into, except hominology. Zoology, primatology, and anthropology, as we know these disciplines today, have no place for living wild human-like bipeds. The creatures' specifics are not dealt with by any one of the established sciences. Hominology is necessary and inevitable because between zoology and anthropology there is a big gap in knowledge, a no-man's land of science. So the hominologist is a hunter for knowledge in this no-man's land.

Perhaps I should also mention here that in the family of sciences, hominology's elder sister is paleoanthropology. The elder sister deals with fossil ancestors of *Homo sapiens*, while hominology deals with living pre-sapiens bipeds.

As I said in the beginning of this chapter, even the 1958 scientific expedition into the Pamirs was bent on hunting down a snowman, although the expedition members had a very vague idea what kind of animal they were after. Commenting on the results of the expedition in "The Struggle for Troglodytes," Porshnev wrote, "We were obviously not prepared to interrogate nature prior to querying people who have lived for generations in that environment." So five years after the expedition, Porshnev's monograph, *The Present State of the Question of Relict Hominoids*, was the result of querying people, of drawing conclusions from a voluminous file of eyewitness reports gathered by the Snowman Commission of the Academy of Sciences. Chapter 11 of the book, *Preliminary Description of Homo troglodytes*, is devoted to the morphology of the creature from head to foot; Chapter 12 to its biology; and Chapter 15 to the ways and methods of searches. Here are two quotes from the latter chapter:

> If we could find even a single place where tender-hearted inhabitants keep in contact with the animals which interest us, that would give us a key to the practical solution of the whole problem. Why? Because then perhaps it would be possible to repeat and enlarge the practice of giving food to the creatures, to use this method in other places. If we really manage even once to lure and feed a specimen of relict hominoid the whole further perspective of research must be based on repeating and enlarging this practice.
>
> If proceeding most cautiously we succeed in conditioning the creature to come and take food in a definite place, that would be a real scientific victory. There is a basis for such prospects, namely, the above-mentioned cases in different geographical areas of local people habituating and even befriending relict hominoids. Scientific work could be launched in such a case even without direct contact of researchers with the specimen, for modern zoology boasts of an excellent means of taking color films with a telephoto lens at a great distance. A relict hominoid would then appear on the screen showing its usual movements and habits against a background of its natural environment. So step by step relict hominoids on earth could find themselves under man's protection and permanent scientific surveillance. At a certain moment it would be possible, of course, to observe the death of this creature. Then the anatomist would get a corpse for autopsy. Thus the perspective of studying *Homo troglodytes* looks as the reverse of zoology's canon: not from dissection to biology but from biology to dissection. (Porshnev 1963, 390–93)

# CHAPTER TWO

## The Great Debate

Before saying good-bye to René Dahinden in 1972, I gave him a copy of a memorandum with the subject: "Some thoughts on how to get in touch and on best of terms with Mr. Sasquatch," which was an elaboration of the method suggested by Boris Porshnev, as quoted in the previous chapter. "Nobody knows," I wrote, "how the first specimen will be obtained. It may be just a chance event, with no method involved. But I am sure that sooner or later the 'blaring bait' method proposed here (or its modifications) will become standard for the study and preservation of these creatures." I tried to convince René, who craved both money and glory, that by habituating and filming a hominoid, he would be sure to make good money and would avoid all risks and adverse consequences connected with a killing one of the creatures. I was sure then, and I am sure now, that it is the way of filming which may bring one both fame and fortune in hominology. A killing is sure to bring a person *infamy*, and as to a fortune, it is much in question.

Dahinden wrote back that he found the idea interesting and added that he was not sure what he would do (shoot with a cine-camera or with a rifle) if he chanced to meet a Sasquatch, thus implying that it would depend on the latter's behavior. That admission showed that he didn't take my idea seriously enough because the method I suggested excluded the combination of a camera and a rifle.

We continued to touch on the question occasionally, but in a low key, until John Green's publication of *The Sasquatch File* (1973), a paperback which, on the one hand, had some good things to say about Dahinden's visit to Moscow and carried photographs of Porshnev, Koffmann, Bourtsev, and myself. But, on the other hand, he fired a salvo at the opponents of killing. Here are some pertinent quotations from Green's work:

> I have few qualifications as a hunter and no expectation of being the man to bring a Sasquatch in. I do, however, have some very strong words to say on the subject. There seems to be a considerable tendency for people who take an interest in the Sasquatch to weave romantic fantasies of possibilities of communication with them.
>
> There is not one thing in all I have learned of Sasquatch behavior to indicate that their relationship to humans extends beyond their manner of walking. One of the ways that the Sasquatch will be studied is by dissection, and to do this adequately science will require not one cadaver but several, probably dozens at least. All these facts point in only one direction. What is required at this stage is physical evidence. A movie won't help. We already have one. The man with a gun may rightly pause to determine whether he is looking at some idiot masquerading in a fur suit. He may also wisely consider whether the gun he has is adequate to kill a huge animal whose physical capabilities are

unknown. But if he is satisfied on these points, he should not hesitate further. Gun it down, cut off a piece you can carry, and get out of there. We have enough people now who claim to have shot a Sasquatch but can't prove it. There may be hassles to be sorted out afterward, but your first-person story alone is bound to sell for many thousands, and the scientists can collect the second one their way.

> I'm sorry to have to sound so bloodthirsty. It goes against the grain. I don't kill anything much bigger than mosquitoes. But there is too much nonsense being spouted on the other side of this argument. It needs an answer.
>
> The day may eventually come when man decides that it is not his right to do as he chooses with the other inhabitants of the planet, but that is not the situation now. (Green 1973, 70–71)

In fact, the situation was such that among those spouting "too much nonsense" on the other side of this argument, there happened to be Bayanov and colleagues whose photos were published in the book but not their views on the issue. Green's outburst also needed an answer. However, I much appreciated my newly established relationship with Green and did not want to sour it right away by stressing our differences. That is why I wrote him rather evasively stating:

> As for your views in the new book, I find certain weaknesses and contradictions in them, but since you write you are tired of this argument I abstain from offering my comments and counter-arguments for the time being.

Green in reply stated:

> As to killing, it's a problem with people over here, but couldn't you and I agree that the creatures in Russia are likely to be Neanderthal man and shouldn't be shot, but the thing over here is a monstrous ape and a fit specimen for dissection? That's very probably the true situation, and it would save a lot of wasted breath.

To which I responded:

> Unfortunately, I can't bring myself to accept your, "render unto Caesar the things which are Caesar's and unto God the things that are God's," kind of formula in this case. And not because I'm sure that all or any of our homis ("professional" jargon for "hominoid") are

Neanderthal (I am not), but because I'm absolutely positive that your Sasquatch is not an ape. You know that, loosely speaking, that word is even applied to humans, and very painfully at that. Who knows, one day Sasquatches might learn to read and take you to court for calling them that. But seriously, such loose employment of words is no use in our efforts to define the creatures, even preliminarily.

The argument smoldered during 1973 and 1974 and then turned into flame in 1975 when I invited René Dahinden, John Green, George Haas, and Gordon Strasenburgh to participate in a full-fledged debate on the issue. My opponents, Green and Dahinden, are known to the reader through my previous books. George Haas was a staunch ally and deserves a separate chapter. Gordon Strasenburgh was my American colleague who participated in the *Current Anthropology* discussions, insisting that Sasquatch and other hominoids were relicts of *Paranthropus*. The following are some pertinent or more colorful exchanges from the "great debate," along with commentaries. (Dahinden refused to take part in the new round of discussion and the quotes that follow are taken from my earlier exchanges with him.)

**Green (to Bayanov):** In this exchange of opinions we are not trying to change what you do over there, you are trying to change what we do over here.

**Bayanov:** In reply to a grim charge made by John of what amounts to my interference in the internal affairs of another country and mindful that attack is the best kind of defense, I want to say the following. It is not I, but John and René who want to change the situation—the status quo of man's relationship with the homi in every country concerned. Nobody has ever killed a Sasquatch for the sake of science, and if what they advocate should really happen, it would be a precedent—and a precedent is serious business indeed. To cite a dictionary, it's a previous act or action taken as an example or as justification for later acts. Secondly, Sasquatches are not yet (regrettably) Canadian citizens and therefore their defense is still a free-for-all. Thirdly, John mentioned my name and published a photo of me in the same chapter of his book in which he says, "there is too much nonsense being spouted on the other side of this argument," never saying however that I happen to uphold this other side or that I find his side not quite sensible either. Of course, treated as a guest of honor I was not supposed to kick up a row on the pages of John's book, but at least I might have been warned what speeches would be made at the party I was invited to. In view of all the above I plead innocent to John's accusation.

**Dahinden**: I respect living things more than most men, since I see so much of them in the wild. They are busy staying alive and I admire them for it. But there is a place and a time for that. Let me tell you that if we could hear the screams of all the creatures which are killed every minute of the day by other creatures, we would go crazy with the noise. This is nature at work, and it seems to be working well for nature.

**Aside**: Then, referring to my call to learn from nature, René implies there is nothing terribly wrong with the idea of killing a Sasquatch.

**Bayanov:** Yes, René, you're right, animals are busy staying alive, and it is to stay alive that they kill other animals. They never kill for money or glory. To do it nature's way, a Sasquatch killer would have to devour his quarry, which might not be easier than to eat one's hat.

**Dahinden**: The question is not just to shoot or not to shoot a Sasquatch, but goes way deeper than that. The question is, do we have the right to kill any living thing on earth? Anything at all, however small or big it is. Do we have the right to cut down a tree or pick a flower? So, in my view, the Sasquatch is just a small part in the whole question. Let's not single out the Sasquatch for the argument. Since you and others want to single out the Sasquatch for special protection, I don't see why you don't take all the other animals under your wings. It just isn't logical at all, but damn emotional.

**Bayanov**: The reason, René, why I don't take all the animals under my wings is simply because I can't. I wish I had wings as mighty as that. My dream is that mankind will grow itself such wings some day. But I must say that, on the whole, your last argument, René, sounds interesting indeed, and a full answer to it will take a lot of space.

Let me begin by saying that neither I nor any of my allies in the discussion were the first to single out the Sasquatch, or any hominoid for that matter, among the animals. In this case the credit really goes first and foremost to Mother Nature, while humans, including René Dahinden, just followed suit. Is it not Sasquatch, and no other animal, that you have been searching out for twenty years with the stubbornness of a maniac (and I use the word with approbation). Your deeds,

René, speak louder than your words, showing that you do set the creature aside and put special value on it, along with all of us.

Paradoxically, Green also tries to play down the uniqueness of Sasquatches, though he has done a lot to make it known. He believes, "the thing over here is a monstrous ape and a fit specimen for dissection, and when live Sasquatches are caught they will end up in the zoo alongside the gorillas. Or maybe alongside the grizzly bears, since the apes are usually indoors and well heated. Certainly there is no jot of evidence that they are more than animals."

More than what animals? The grizzly bears and the gorillas? There are enormous jots of evidence, John, in your very books to indicate that the Sasquatch are more than **any known animals**. [Examples from Green's books followed.]

**Green:** To take the silly side of the argument (as I see it), if it were decided these things were human they would have to be counted in the census, given the vote, made to send their children to school (and summoned to parent-teacher conferences when the kids couldn't learn their algebra), paid daily allowances, put on welfare, and finally locked up in homes for the mentally retarded. If they were able to understand all that and express an opinion, I expect they would prefer to be apes and take their chances about getting shot.

**Bayanov:** I grant the possibility of Sasquatches deciding to avoid orthodox schooling at the risk of being shot. I certainly would if I had their power of survival in the wild. But the issue is not whether they decide to be shot, but whether we decide to shoot them. There's a hell of a difference between someone deciding to take his life and someone else doing it for him.

Looking at the matter from a different and more optimistic point of view, I can imagine young Sasquatches doing very well in a school specially designed for them. If a chimp brought up and taught by humans can acquire a certain vocabulary, I wonder what heights of scholarship can be attained by an aspiring young Sasquatch under human tutorship. If a human child brought up by animals becomes an animal, I wonder what will become of a homi child brought up by humans.

That's one side of the matter. The other is a reversal of roles, with the Sasquatch becoming a teacher of human boy and girl scouts in the art of survival in the wild. With the present-day "Back to Nature" trend I regard such a possibility as quite feasible. In that case we are bound to have parent-teacher conferences with a somewhat different agenda and composition.

What I have said is meant to show that Sasquatch are not mere animals. Their behavior singles them out among all the known animals, bringing them very close, and let it be stressed we don't know exactly how close, to humans.

Let's take another aspect of Sasquatch uniqueness within the animal world—their physical appearance. Here there are many examples of people taking the hominoid for a human being when it comes to a crucial decision. William Roe's case is typical of a whole series of incidents. Roe, who was placed in a position where he could have shot a Sasquatch, later recalled: "Although I have called the creature 'It,' I felt now that it was a human being and I knew I would never forgive myself if I killed it." (John Green, *On The Track of The Sasquatch*, 1969, p. 12.) Or take the Patterson-Gimlin film, which we now all believe (in fact know) to be genuine. Our opponents among anthropologists and zoologists never say it shows a machine or an ape—they say it's a man in a fur suit. Now is it morally, or let's say emotionally right, to kill a creature which resembles man so much both in behavior and appearance?

Green and Dahinden seem to imply that since the Sasquatch has not been proven to be a human being and is not protected by a corresponding law, he is a permissible target for the gun. But let me stress that the creature has not been proven to be an animal either. It's a hypothesis, which seems plausible, but still a hypothesis—and to pass a death sentence on the basis of a hypothesis is definitely impermissible.

Let me also say that, in my opinion, even if the first killed Sasquatch is proven to be an animal, it would not mean we are free to kill another one. The word "animal" is not a license to kill indiscriminately those whom it signifies. A pet, for example, is an animal, but humans single out pets for special protection. There was a legal case in Moscow in which the public demanded long prison terms for the hooligans who had killed and roasted a swan living in a public pond.

Beside pets, people, as a rule, take a fancy to rare, big and/or intelligent animals rather than to common, small, and unintelligent animals. That applies, René, to your question in a letter to me as to why people are sensitive about the killing of a whale and indifferent to slaughter of herrings. Even your ally in the argument, Green, confirms that this kind of sentiment is rooted in human nature: "I'm sorry to have to sound so bloodthirsty. It goes against the grain. I don't kill anything much bigger than mosquitoes." You say it's illogical and "damn emotional." I agree but insist we should try to have emotions on our side of the problem, not against it. To ignore the reality of emotions in our research is like ignoring the reality of weather on a long and precarious sea voyage. To be on the safe side, we had better use our intellects to understand how emotions work. People walk in bathing suits (and without) on a beach, but try to show up at a banquet in the same apparel and you'll learn what an emotion storm is and that emotions have a "logic" of their own.

The emotional is an integral part of human nature, just as natural and legitimate as the rational. If we had no emotions, we would be computers, not people. A wise policy in any business is not to reject emotions, but to understand their workings so as to strike a balance between the rational and the emotional, and see to it that the emotional help the rational instead of blocking it.

If you still want a reason for man's emotions in our case, I think the explanation is in the fact that man can't be a 'really unbiased' judge in the matters of life and death. Being himself alive and human, he can't help being partial towards anything which is alive and intelligent, or at least looks manlike. Take the abortion problem, for example. It is clear people's sensitivity against abortion at a certain period of pregnancy is caused by the mere fact that at that period the fetus is more alive and looks more like a baby than at an earlier period, and not because it is already a person. And to aggravate our problem, the thing we are dealing with is simultaneously very much alive, very big, intelligent, and manlike.

Finally, to show than the above is not all my invention, let me back up my stand by no less an authority than Konrad Lorenz:

"The scientist who considers himself absolutely 'objective' and believes that he can free himself from the compulsion of the 'merely' subjective should try—only in imagination of course—to kill in succession a lettuce, a fly, a frog, a guinea-pig, a cat, a dog, and finally a chimpanzee. He will then be aware how increasingly difficult murder becomes as the victim's level of organization rises. The degree of inhibition against killing each of these beings is a very precise measure for the considerable different values that we cannot help attributing to lower and higher forms of life. To any man who finds it equally easy to chop up a live dog and a live lettuce I would recommend suicide at his earliest convenience." (*On Aggression*, 1969, Chapter 12, On the Virtue of Scientific Humility, pp. 194–95.)

# CHAPTER THREE
## The Great Debate Continues

**Strasenburgh:** As for the morality of it, to kill an animal (to me, man is an animal) is only to shorten its life.

**Bayanov:** What the hell is this euphemism for? You haven't yet given enough thought to the matter.

**Strasenburgh:** What earthly difference is there between death at the hands of a lucky hunter and at the hands of a (probably intoxicated) logging truck driver? Are accidental deaths unnatural? My preference for a deliberate killing springs from my belief that no living thing can hope for more than a dignified end to his time on this earth. A "chance encounter" with an adequately armed man, a motor vehicle, or old age, for that matter, is neither dignified nor, to use George's word, compassionate. Death at the hands of a hunter who sets out to kill an individual, on the other hand, at least might be compassionate.

**Bayanov:** Unless Gordon (being, I understand, by his own definition, an animal), doesn't mind being killed by someone out of compassion to prevent his possible painful death in an accident or the troubles of old age, or answering the demands of some "ologists," his argument doesn't hold water whatsoever. If he hasn't got such a suicidal bent and wishes to go on living, he is well advised to surmise that other creatures may also want to live. It is to honor this wish and its ramifications for fellow living creatures that morality is about. And one practical asset of morality as against immorality is that it brings about a widely shared sense of security and relative justice in place of insecurity caused by arbitrariness.

Let nobody think that by stressing the moral side of the problem I am laying some special claims in this respect. I am as good a sinner as anybody. But in this particular case, for a change, I find it simply difficult not to be virtuous. I mean normal people are usually pushed to sin not so much by their own will as by the pressure of circumstances. Ours is a rare, perhaps unique, situation wherein we are not really pressed by the circumstances; wherein the circumstances, no matter to what small degree, may depend on us, on our statements, ideas, and actions. Can't we afford to be virtuous if the virtue in question is no more expensive to us than politeness?

**Strasenburgh:** Lorenz's argument about "rising levels of organization" is a rehash of the long-standing European conceit that man is God's glorious capstone to a systematic creation. If he really had something to say, he would not have compared "chopping up a live dog and a live lettuce!" People in the Far East consider dogs very tasty. Monkeys in South America and gibbons in Asia are eaten. And in New Guinea humans are (or until very recently were) eaten.

**Bayanov:** Now about Gordon's jibe at Lorenz. To me the latter's words are not a rehash of anything, but a statement of an empirical fact, confirmed in our case by John Green who says he doesn't kill anything bigger than mosquitoes. What about you, Gordon? Can you really chop up a live cat as easily as a live lettuce? [Aside: I happen to know from his letters that Gordon is a cat person, not a dog person like me.] If yes, I am afraid, you are an exception dealt with by Lorenz in the last sentence of that passage. As for the fact that people in some lands eat anything, including other people, it has nothing to do with Lorenz's statement. You would be right only if you could prove that a cannibal, chopping up a live man for dinner, does it as nonchalantly as when he cuts vegetables.

**Strasenburgh:** If I want a salad I will chop up a lettuce, and if I am bent on answering the demands of the anthropologists, I will kill a homi and present them with a specimen. One action is as purposeful as the other.

**Bayanov:** What's that awful thing about justifying any action if it has a purpose? To take an extreme example, the Nazis' gas chambers were as purposeful as anything, but could there be anything more horrendous?

**Strasenburgh:** Without boring you with an involved explanation of how I arrived at my conclusion, my feeling is that if there is a need for a specimen, then I would prefer to have an individual deliberately shot and killed.

**Bayanov:** The question arises, whose word for the existence of such a bloody need must we heed? The word of God, of John, of René, or of the anthropological community in whose collective skull Gordon invites us to bury our hatchets? So far neither John, nor René, nor Gordon have convinced me of the existence of such a need for human society.

There is also this argument: either we sacrifice one hominoid and save the rest of the species, or we lose the whole damned lot of them. If true, it would be a really serious and valid proposition. If we had to choose between specimen and species there is little doubt what the outcome of the vote would be. But I hope there is no such alternative for the time being. And I'm glad my opponents in the argument over the ocean agree that our "wards" are in no immediate danger of extinction. I do hope we shall make science accept their existence, with methods which **don't go against the grain**, before the creatures reach the point of no return due to their dwindling numbers.

**Strasenburgh:** Of all human tendencies, that of establishing our species as a standard against which all living things are to be judged and valued is the most distasteful to me.

**Bayanov:** I, on the contrary, think that Protagoras of Abdera was quite right when he said, in the 5th Century B.C., that "man is the measure of all things." And not because we are better or higher than anything, but simply because we are men. If we were cats or toothed whales, then man could not be the measure of all things for us.

If John adheres to sapiens chauvinism, then Gordon seems to be heading in the opposite direction beyond all reasonable limits. But extremes converge, and we see that, in principle, Gordon has nothing against a homi killing. If Gordon doesn't like some people's boast that man is the pinnacle of creation, or evolution, then he (Gordon) is free to see man as the nadir of evolution (the notions of "up" and "down" in this connection are largely a matter of taste and convention), but this shouldn't make him overlook the real qualitative difference between representatives of different stages of evolution. I hope I'll commit no indiscretion by quoting from Gordon's letter to me of May 14, 1974:

"We are, in my opinion, on the verge of extinction. If our species is to save itself and our world, some rather drastic changes in our attitudes have to occur. I would hope that the shock and the wonder of discovery of the homi could be turned toward changing certain attitudes. (Now perhaps you will begin to see why I object even to the suggestion that anything 'places man on top' of anything.) No one can doubt that we are a very powerful species. But can we control the power we wield? If we cannot, then our "power" is a negative force, not a positive one!"

Can we control the power we wield? Which at an individual's level may be the power of a hunting rifle. Don't you see, Gordon, that your question is of direct relevance to the problem? And that your other pronouncements contradict what you so well stated last year? Also that your position as stated in the above passage is very close to, if not identical with mine?

**Dahinden:** Scientists say they don't want any footprints, photographs, or movies but the creature itself, living or dead.

**Bayanov:** Strangely enough, this is interpreted by René as a license from science to shoot a Sasquatch and, judging by his oft-repeated mention of it, is his strongest argument. In my opinion, René, it's one of your weakest arguments.

First of all, your relations with the scientific establishment have been so far of the kind which doesn't make your leaning on them, when it suits you, very convincing. One can't denounce someone's attitude in one sentence and use it as a positive argument in another.

Secondly, you like to check people's credentials in our business by asking what such and such have done for our cause. So let me ask you what the scientists you refer to have done in our research and what they know about the matter. Have they studied all the existing evidence and found it insufficient for a definite conclusion? Is that the reason they want more evidence and of a different kind? Otherwise, bearing in mind the prevailing attitude of the establishment which doesn't even recognize the existence of the problem, those scientists' advice to you may have been just an attempt to put you off. The czar in a Russian folk-tale, when he wished to get rid of a good fellow, would order him: "Go there I know not where and bring me back I know not what!" Porshnev, in this connection, used to cite certain detractors of the research into the problem of space travel who said to the pioneers of that research: "You first go to the moon and then we shall discuss your problem."

If the anonymous scientists René refers to did not display the same attitude and were really eager to help us with their considered advice, then let them speak up and give their reasons, let them state in public their approval of homi killing. Neither of us has any illusions about the profession. Scientists are only human, and their opinions can only be of more value than that of the man in the street when it is supported by a detailed scientific analysis.

René tells me that whether he shoots a Sasquatch or not will depend on whether the latter behaves or not when they happen to meet. Do you expect, René, the giant to pose obligingly for a snapshot or let you smilingly handcuff him to earn his life? The trouble is, René, that even if you become the victim of an unprovoked attack by the creatures and kill one of them while defending your own life, there's little chance the public will believe your story after what you've said so many times.

We all expect a lot of good to ensue from our research, while a wantonly killed homi would be nothing but a terrible nasty fly in the ointment, or, as the Russians say, a spoonful of tar in a barrelful of honey.

If my arguments are found to be well reasoned out and valid, I expect them to be accepted by all the parties concerned. The only force to be resorted to, said an eminent and "heretical" Russian biologist, Alexander Lubischev, is the force of reason, and the force of reason is no violence. And if these arguments are accepted, I expect all of us hominologists to agree on the "rules of the game" and make the agreement known to all concerned. I mean recognition of the "most favored creature" status of the relict hominoids or a self-imposed ban on homi killing as the main clause of the agreement, and a call to all concerned not to resort to shooting.

I realize how difficult such a decision would be for Dahinden and Green. But still to change one's mind must be easier than change both mind and action, and thank goodness none of the hominologists has fired a single shot yet at the Sasquatch, only a volley of words at each other. It is human not only to get and keep opinions but also to change them. We all at one time thought there could be no such thing as a relict hominoid—now we know better, which shows there's nothing wrong in changing one's mind, and the process can go on as long as one lives. The only mistake so far has been that we all made up our minds individually regarding a problem which concerns us all collectively and therefore must be solved collectively.

On the other hand, if Dahinden and Green or anybody else have valid counter-arguments to my points, then discussion must go on, and should I come to see their reasons are stronger than mine, I shall change my mind and accept their views and method. If we want the whole world to see reason we ought ourselves to set an example.

The worst thing would be, of course, if we have neither agreement, nor valid counter-arguments to my points.

**Green (extracts):** I don't have the slightest inclination to seek a collective solution. I agree entirely with René that the life of a Sasquatch and the life of any other animal weigh the same on the scale. I'll suggest also that in many cases, both in Asia and in North America, the public declaration against killing is merely a means of avoiding difficulties, and that a specimen would be collected without hesitation if the opportunity arose, with excuses being provided afterwards.

I quite agree that the larger and more complex the organism, the more trauma is involved in killing it. But in a world where humans are killed by the millions for the most inconsequential reasons, such as getting from one place to another at a higher rate of speed, or to satisfy various cravings or beliefs of other humans, I am simply not about to get worked up about a few Sasquatches. I wouldn't expect a Sasquatch to excel a chimpanzee in learning sign language. And the ability to learn sign language does not prevent chimps from being shot for various reasons. Moreover I have seen them myself in cages and also stacked in a deep freeze awaiting dissection, the latter a much more disturbing sight than the former, for some reason. As to the Sasquatch becoming a teacher of human boy and girl scouts we are farther apart in opinion than we are geographically. In my opinion the statement would be of equal validity if made concerning a bear.

I agree that the fact that this is an animal does not provide any license to kill it. The decision, however, will be made by humans, and in North America the attitude towards science is such that there will be no hesitation in granting various institutions a license to collect specimens both for live study and for dissection.

I have no illusions about our ability to influence mankind's attitude towards killing either animals or other men. Truly great men have been trying to achieve improvements along those lines for thousands of years, with only temporary successes. Whether or not someone kills a Sasquatch will rest as a grain of dust on the scales.

The odds are overwhelmingly that the actual conclusion will be achieved by a lucky guy with a gun who has never previously contributed anything. My only wish is that he hurry up about it, and I sincerely wish Patterson and Gimlin had done so.

My sole aim, therefore, is to get the message to the multitudes that if they get a chance to shoot one of these things it is important and worthwhile to do so, and in case they should succeed in doing so it is vital to come away with physical evidence of their success. From my point of view, therefore, the person who wants to hunt with a camera is no problem, but the person who publicizes his opinion that it would be wrong to shoot these things definitely is a problem—particularly since his arguments are emotionally attractive to the uninformed. In view of this handicap, I have no hesitation about appealing to those emotions that are favorable to the result I wish to see achieved, even though those are not commendable emotions.

I am not trying to improve human character—at least not in this instance—I am trying to prove the existence of the Sasquatch.

And both of us [Green and Dahinden—D.B.] really are very tired of this particular debate, not because of you [Bayanov] but because we have to live with it over here. And don't try to tell me all we have to do is change sides, we would then just have to argue with the people who now agree with us (which would be even tougher because they happen to be right).

**Bayanov (after discussing some of the above points):** Those are things which demanded immediate attention and I postponed discussion of other points in John's letter because I don't want to impose too much on his time, bearing in mind his words that he and René, "really are very tired of this particular debate." I am pleased to note though that they are tired not because of me, "but because we have to live with it over here," which doesn't tally somehow with John's words, "I don't consider the price (of a killing) at all high." Well, let's give our hawks a good rest so that they don't lose their integrity fighting doves on two fronts, and if in the meantime they feel like changing sides they will be most welcome to do so without fear of attack from their present fellow hawks. I promise to take those on and show that, contrary to what is said in John's letter, they don't happen to be right.

## CHAPTER FOUR

# The Message of George Haas

The fifth participant in the "great debate" was George F. Haas (1906–1978), of Oakland, California. I present his contribution to the discussion in a separate chapter because his words and the memory of him are very dear to me. We first heard of George Haas and his *Bigfoot Bulletin* from René Dahinden, and in March 1972 I began to correspond with Haas. It was from him that we first learned of Roger Patterson's passing in 1972. The *Bigfoot Bulletin* was a new kind of venture in North American hominology, much approved at the time both by Dahinden and Green. One chapter of Green's *Year of the Sasquatch* (1970) is devoted to that newsletter. Green wrote:

One of the most significant developments of the year 1969, at least from a Sasquatch hunter's point of view, was the birth of an unassum-

Trio of Bigfooters (from left to right): George Haas, René Dahinden and Archie Buckley, January 1975.

Jane Goodall, angel savior of nonhuman primates, greeting one of them.

ing little publication called the *Bigfoot Bulletin.* For years people who were spending their spare time and money running down Sasquatch reports have talked about the need for a means of communication, but none ever did anything about it. Some kept in fairly regular contact by letter, but others did little or no writing, only dropping in during their travels or telephoning at intervals that might be years apart. Quite a few didn't know that anyone else was active in the field at all.

The man who changed all that is George Haas, of Oakland, California. George has worked at a variety of outdoor jobs and has been a keen woodsman for many years, but he got into the Sasquatch business via the bookshelf. From correspondence with George I know that his experience included a period as Ranger in Charge of the Calaveras Big Trees State Park in California and six years in Yellowstone National Park where he designed, built, and operated an 18-acre reforestation nursery.

The first issue of the *Bigfoot Bulletin* came out on January 2, 1969. It was just two mimeographed pages. The first item reported the finding of 16-inch tracks in the snow on the Bluff Creek Road, December 2, 1968. Most of the rest of the first page listed published articles on the subject in current papers and magazines, but on page 2 was an article by Jim McClarin, who has continued to be the Bulletin's most prolific contributor. It was the first of a series of old-time stories to reach a modern audience in the Bulletin.

Despite its growth in size and circulation, George Haas has continued to distribute it (the Bulletin) free of charge, as well as handling an ever-growing volume of correspondence resulting from it.

A number of Sasquatch hunters are basically more inclined to compete than co-operate—because they each want to be the first to bring one in. From this point of view some already object to the Bulletin as making too much hard-won information available at no cost and no effort to anyone who comes along. But for those whose main interest is to see the facts brought home to a doubting world, the Bigfoot Bulletin is an undiluted blessing. No one can buy the Bulletin. It is sent only to those who contribute information.

George Haas is also known as the organizer, archivist, and spokesman of the Bay Area Group of Bigfoot investigators. Their aims, methods and activities are explained in the following texts. First comes George's letter to me of April 16, 1975 (somewhat shortened here):

Dear Dmitri:

Thank you for including me in the discussion of whether or not to kill a homi. You have covered the ground so well in your "Discussion" paper that I don't think there is much I could or need add. I fully agree with all the points you have made on why we shouldn't kill one. I will try not to waste time and space in repetition. Instead, I will make a few additional remarks that have a bearing on the subject. Also, I want to take this opportunity to try to clear up a few inaccurate impressions regarding me and other members of our Bay Area Group that seem to be held by some members of the opposition.

Fortunately, there seems to be a growing trend in the West to recognize that animals and other forms of life have rights of their own; that they were not just "put here for the benefit of man" but are fellow passengers on the spaceship Earth through time and space and thus entitled to the respect and consideration due to any fellow traveler. Most of this trend seems to be due to the growing influence of eastern philosophies on our western culture and I hope this trend continues.

As a species we have appointed ourselves trustees of the earth and of everything on it but actually we do not "own" anything. As individuals, as groups, as societies, we, in effect, hold all things in trust for future generations, not only of men but of all other species as well. How we manage this self-appointed trust is the measure of our integrity. If we log off all the redwood groves for the sake of a few jobs, if we exterminate all the coyotes to save a few ranchers' sheep, if we kill off all the eagles for a few souvenir feathers, then our sense of values is warped and distorted and we have failed to live up to our trust. The redwoods, the coyotes and the eagles have rights of their own and unless we can see that, we are in a bad way.

What we must not forget or overlook is that in Bigfoot (and in other forms of relict hominoids) we now have a totally unique opportunity to do something worth while before it is too late: to demonstrate our integrity and to save and protect all the individuals of what we all agree is undoubtedly a rare and unique form of life. We should all work together to accord to **this** species at least the inherent rights it has, the respect it deserves, so that future generations will not look back at us and say "they muffed it." We have the opportunity now to avoid the killing of even one individual for the questionable reasons of expediency, fame, financial returns or supposed medical benefits. Surely we can rise above such fleeting aspirations and do right by **one** species. We may never get the chance again.

It may be, as I have stated before, that the Bigfeet, for example, will eventually be classified as "animals" but they may be something more. I call for **caution** in advocating a killing. It seems to me to be a little reckless to advocate and encourage others to shoot something before we really know what it is. In this connection, let me quote the little Himalayan folk tale from Odette Tchernine's book, *The Snowman and Company,* page 158: "One day as I was walking on the mountainside, I saw at a distance what I thought to be a beast. As I came closer, I saw it was a man. As I came closer still I found it was my brother."

Shooting at the creatures in a given area would certainly destroy all chances of making any kind of friendly contact there in the future. Furthermore, if one were hit and wounded that particular creature might become exceedingly dangerous, not only to the one who shot him, but to other people living in the area and to campers and other forest visitors. This possibility certainly should be taken into account by those who advocate a killing. From time to time we of our group take or send young people, sometimes teenagers, into our particular research areas and we certainly don't want them to be subjected to an angry, wounded Bigfoot. This is another reason why we have always been very reluctant to report any of our experiences and chance having outsiders coming in to shoot up the place. This should be obvious to those who cannot understand why we refuse to share some information.

I think all of us who are doing research in this fascinating subject should be aware of and take into account public opinion. We should understand that all over the world there is a growing awareness of the importance of preserving our environment and all of its various elements. This is not something "emotional"; it is something essential if we are to understand pollution and all the other environmental problems and learn how to deal with them. If we don't learn, we are on the way out. It cannot be denied that there is this growing awareness, especially among youth, everywhere. Almost without exception, I find these people deeply concerned about the Bigfoot mystery and what we, as investigators, are doing about it. They take an exceedingly dim view of the idea of killing or caging one. Our public image of Bigfoot investigators has greatly suffered in the past and the advocating of killing a Bigfoot certainly has not helped. After my "The Present State of Bigfoot" was printed in the Willow Creek newspaper, I received a letter from a woman of that area thanking me for expressing such opinions and saying that she and all her family and friends have always contended that if they ever did see a Bigfoot, they would never report it for fear the "Bigfoot hunters" would be out to kill it. I have received other such letters and I think this attitude is typical of most of the people who live in Bigfoot areas. The same sentiments are held by many Indians also. The real reason why Indians won't talk, why they are reluctant to give information to investigators, is not because they are afraid of being laughed at, as some anthropologists contend, but because they have a long-standing protective attitude toward Bigfoot. In the experience of our Bay Area Group, we find the residents of Bigfoot areas, as well as Forest Service employees, more willing to talk and give information when they find that we go into the woods unarmed and when we explain our protective aims and policies.

In the September 1974 issue of *The Sierra Club Bulletin* is an article titled, "The Rings of Life," by Galen Rowell, a writer and mountaineer. In this article he describes the death of what was undoubtedly the oldest tree in the world, a 4,844-year-old Bristle-cone pine in the White Mountains of California. With the permission of the Forest Service, it was cut down with a chain saw by a "scientist," a dendrochronologist and geographer, from the University of North Carolina, *in order to see how old it was!* What adds to the horror of this atrocity is that this act was completely senseless and unnecessary since the same information could have been obtained by using a borer, taking out a pencil-sized core of the annual rings without injuring the tree. Here was a living tree, probably 1,000 years old when King Tut ruled Egypt, and it might have lived another thousand years. Surely there is something wrong with our sense of values when such atrocities are perpetrated in the name of science. The correlation between this act and the killing of a Bigfoot to prove it exists is obvious.

A Bigfoot left alive in its native habitat certainly should be of more value to science in the long run than a dead one. A corpse, of course, is of value for dissection and for determining its identification, but that's the end of that particular Bigfoot. A live one, on the other hand, is still available for study, for possible contact and, above all, for perpetuating the species.

I think I should dwell a little upon our Bay Area Group and try to dispel the impression held by some that we are an "emotional" group who only want to see a Bigfoot and who spend our time camping in the woods and reading fantasy fiction. My own qualifications, for what they are worth, are fairly well known in Bigfoot investigation circles. I don't hold a degree in any branch of science but I have had a great deal of practical experience in forestry, botany, mountaineering, conservation, exploration, and all-round woodsmanship. During the past 20 years I have been collecting all the information I have been able to find on Bigfoot and related creatures so that now I have what is generally conceded to be one of the largest collections of its kind in the world. I might be called the archivist and spokesman of our group.

One of our leaders is Archie Buckley whose knowledge of ambulation, accumulated through many years of research and practical application, is profound. In him we have an unexcelled expert on footprints and their evaluation. His knowledge of wildlife and his practical woodsmanship are second to none. His success in drawing a Bigfoot into his camp as described elsewhere, using the friendly contact technique we advocate, was, and still is, unique.

We hope:

1. To collect all available bibliographical information.
2. To collect all the information we can regarding the creatures by actual field studies without molestation.
3. To collect all possible information—short of killing a specimen—leading to their identification and acceptance by scientists.
4. To work for legislation for their protection.
5. To work toward getting areas set aside for their refuge and preservation. Perhaps all the National Forests could be declared Bigfoot preserves or reservations. It may come as a surprise to some who read this to hear that some members of the U.S. Forest Service are already interested in this phase of the problem.
6. And, yes, we do want to see one ourselves.

We have publicly stated our aims and policies, outlined our methods of approach and encouraged others to do likewise. We have proposed an alternative course of action. Someday, somebody, using our non-violent methods of making contact, should be able to bring back convincing scientific proof of which there could be no question. For example, anyone—be he scientist or otherwise—viewing Bigfoot films comparable to those made of the gorillas by Adrien Deschryver of eastern Zaire and denying their authenticity, would be a die-hard skeptic indeed.

I also want to correct the false impression evidently held by some that I feel that the spread of information about Bigfoot is harmful to their cause and that we should "leave them alone." There is no doubt the creatures would have been better off if nothing had ever been written about them and if all reports had never been made but it has long been too late for that. Indeed, I have contributed no small share in spreading information about them in the publishing of the *Bigfoot Bulletin* together with two published articles and a vast correspondence. On the contrary, I feel that greater public knowledge about them may lead to a greater understanding of the problem and that in turn may lead to their protection and preservation. A mark of a simian is curiosity and we are simian folk too. Undoubtedly, we have inherited and developed that old simian thirst, curiosity, to a far greater degree than any other primate and I see nothing wrong in it. Curiosity leads to information and information may lead to understanding.

I think all of us in this investigation of relict hominoids are in it because we are a little more open-minded than the average man in the street. We have demonstrated that we have the vision necessary to accept that we are dealing with an unknown as yet unaccepted by the scientific community and we have demonstrated a willingness to investigate. But let us not fall into the trap, peculiar to many scientific disciplines, of constructing a body of concepts, of building up a set of "facts" and of thinking that's all there is, that we have all the answers. Therefore, let us carefully weigh and consider every proposal for the solving of this mystery, no matter how much it differs from our own current beliefs, and not dismiss it without due consideration. This is what "investigation" is. How else are we going to find out anything new? Let us not become "set" in our ways of thinking.

Let me take this opportunity on behalf of all of us in the Bay Area Group to send our greetings and best wishes to our esteemed friends and colleagues in both Canada and the U.S.S.R. and to all those elsewhere who are doing research in this fascinating field. We hope that in spite of differences of opinion and aims, there will develop between us new understanding, mutual tolerance and respect of our different viewpoints.

George F. Haas
April 16, 1975
cc: René Dahinden, John Green, Gordon Strasenburgh

Some more information about the Bay Area Group sent over by George at the time follows:

We have no grants, no funds, nor do we seek any. Members pay their own expenses. We have no profit motive in this Bigfoot research and investigation; we are out to obtain knowledge and information only.

> Most scientists say they need a body but we do not feel that the killing of a specimen for that purpose can be justified in any way whatever. ... Sooner or later the body of one that had died naturally or been killed by accident will be found.
>
> As a group we spend upwards of two or three months in the mountains of northern California each summer on field research and we occasionally make weekend or longer trips into our search areas during the fall, winter and spring. We investigate sightings and footprint finds reported to us but more than that, we spend time in the woods looking for evidence ourselves. We study food plants and their distribution. Since we have evidence that Bigfoot is a deer hunter, we study the deer and their distribution and migration habits. One of our most important finds was a butchered fawn. We feel certain this was a Bigfoot kill from the evidence we found near it.

Some extracts follow from *The Present State of Bigfoot* by George F. Haas:

> Out of all the hundreds of reports of sightings made during the past hundred years and more, probably not more than about 10% are truthful or valid accounts. As in any other field of investigation, especially of Fortean phenomena, Bigfoot research has been pestered with hoaxes, frauds, practical jokers, and mistaken identifications. Inevitably, the field has attracted the publicity hounds, the bounty hunters, the greedy, and the confidence men. The investigator has been seriously handicapped by wasting much time in evaluating or running down such stories, but all should be investigated. However, there remain about 10% of the reports that merit serious consideration. It would only require one valid report to prove the case.

The "hawks" have been inclined to dismiss George Haas and his group as naive and gullible romantics. The above passage gives the lie to the assertion.

> Most of us in our Bay Area Group feel that we are dealing with a creature that is more than a "mere animal." We feel that the Bigfoot can and should be studied in their natural habitat somewhat like George B. Schaller studied the gorillas in Africa, without molestation and certainly without killing or capture.
>
> We could go along—but with strong reservations—with current

plans for a capture providing humane methods were used and providing the creature were turned loose immediately after the examination. However, such a capture might result in imparting to the creature involved various human diseases to which it might not be naturally immune. Also, it would be inevitable that tremendous pressure would be brought to bear upon the capturers against a release.

In proportion to his population, Bigfoot has probably been sighted as often as the mountain lion. He knows all about people, cars, tents, logging equipment and campgrounds because he sees them all the time; he sees campers, hunters, fishermen, and hikers by the hundreds of thousands in the woods every season. While he generally avoids them, he is not particularly afraid of people but he is understandably *cautious*. He is never seen unless he wants to be seen or doesn't care. He is, as Archie Buckley points out, a "master of concealment." He can approach a camp, even in daylight, and never be detected. He uses all available cover and even at night will approach a camp or other human habitation by coming in from tree to tree, using them as cover just as the Indians used to do. One Bigfoot that approached to within 46 inches of the car in which Archie Buckley was sitting did just that; he used four different trees as cover in coming in. Tracks in the mulch told the whole story the next morning.

The time has come to use new methods if we are ever going to solve the Bigfoot mystery, see one for ourselves, get close-up photographs or, more important, make friendly contact. We are never going to learn anything of any zoological or anthropological value about the creatures using present methods. We learned little of value from the big game hunters about the gorillas; that had to wait until a man like Schaller went in to live with them.

We go into the woods *completely unarmed* and in small groups of not more than two or three in any one place. We set up unobtrusive camps, try to blend into the landscape and set out baits and lures in an attempt to entice the creatures in to look for us rather than the other way around. Most important is our attitude. This is a relaxed, friendly attitude, purged of the hate-fear that provokes hostility and aggression in return. Anybody who knows anything about dogs will know what I mean. I am also speaking here from the experience of having spent a whole day with a wild coyote on a mountainside in Wyoming and of days on end spent with a herd of Rocky Mountain Bighorn Sheep taking pictures at a distance of fifty feet.

If we appeal to Bigfoot's evident curiosity or hunger then sooner or later friendly contact should be assured. Indeed, that this system works is attested by the fact that Archie Buckley, with infinite patience and an intimate knowledge of wild creatures, did succeed in drawing a Bigfoot into his camp.

With the increasing awareness and acceptance that Bigfoot may indeed exist and inhabit our western forests, there is, in our experience, a mounting public attitude of: "leave Bigfoot alone." This is encouraging since this attitude will no doubt grow into strong public pressure against any exploitation for private gain should one of the creatures be captured. Also, with increased and more enlightened attitudes toward rare and endangered species in national government circles, it is certain that private exploitation of Bigfoot will not be allowed.

We hope that all Bigfoot investigation and research will take a more human and civilized approach to the solving of the mystery in the years to come. *Patience* should be the watchword for there is no *hurry*. The mystery is not going to be solved next weekend or in the immediate future—the search may go on for years.

January 1973

# CHAPTER FIVE
# Incomparable Sociability

George Haas never read Boris Porshnev, yet he and I spoke with one voice about the way of solving the hominoid problem. If only that voice could have been heard in Green's books, if only George could have been on the ISC (International Society of Cryptozoology, founded in 1982) to help me counter Grover Krantz on the issue.... But fate willed otherwise.

George's contribution to the great debate was so much to my liking that I even entertained for some time pipe dreams of George launching a new publication, called *Current Hominology,* a man and nature magazine, devoted to ecological, ethical, and philosophical problems, with man represented by the hominologist and nature by the homi. Alas, George was not in the prime of life; illness intervened and in October 1977, I received word from Warren Thompson of the Bay Area Group who said, "George Haas has asked me to write to you on his behalf. He has been in the hospital for the past five weeks because he has cancer. It was just disclosed this past weekend that his condition is now considered terminal and that he has only three to four days left to live."

Happily, the next message was from George himself, and said:

> Just a few lines to let you know that my condition which we wrote to you about a few weeks ago has taken a turn for the better. I was released from the hospital a week ago after being there for two months, I am now back home in my apartment. I am slowly recovering, getting stronger and feeling better every day. Well, I just wanted to let you know that I am still here and looking forward to hearing from you again in the future.

Overjoyed, I replied with this message:

> For Bigfoot to survive,
> George Haas must be alive.
> Back home George is!
> We feel gorgeous.
> Long live George the Wise!
> Leave us not for paradise.

He left us for good on February 16, 1978. "With his passing," wrote Warren Thompson to me, "we have all lost a wonderful friend. The field of Bigfoot research has lost one of its major assets. As was his wish, we plan to continue on with the work he initiated."

I continued to communicate with Warren Thompson, who published and distributed the *Bigfoot Bibliography,* initiated by George Haas, but contact grew thinner and some years later ceased altogether. Thus, on the question of ends and means, I was left to face Dahinden, Green, and Krantz without an ally in North America, at least an ally known to me and of a stature and abilities of George Haas. Dahinden was not much of a problem because, unlike Green and Krantz, he didn't widely advocate a "kill Bigfoot" policy, and in the early 1980s he broke relations with us anyway. But Green and Krantz, the outstanding hominologists of North America, continued to remain my "professional" colleagues as regards the reality of relict hominoids while we were destined to be bitter opponents on the "kill or film" issue, and that was somewhat tragic.

I greatly valued Green's literary work, because his three paperbacks, *On The Track of the Sasquatch* (1969), *Year of the Sasquatch* (1970), and *The Sasquatch File* (1973), enabled the researcher to compare North American and Eurasian data on relict hominoids. Even more fruitful in this respect was Green's fourth book, a hardcover volume of 492 pages published in 1978 and titled *Sasquatch: The Apes Among Us*. It was most useful not only because it supplemented information previously published by the author but also because, unlike the previous books, this one had a Sasquatch Index, with such subdivisions as Behavior, Description, and Indicated Diet.

In the opinion of Boris Porshnev, George Haas, myself, and many others, the key to the solution of our problem is friendly contact with the hominoid. Haas writes that Bigfoot "is never seen unless he wants to be seen or doesn't care." Obviously, it is essential for the researcher seeking friendly contact to have an idea about situations in which Bigfoot wanted to be seen or didn't care. And here Green's volume, with its Index, comes in handy. The very first lines of Chapter 1 read as follows:

> You are sitting alone in the house at night when you hear a slight noise outside, and turning your head you are confronted by a bestial black face, more like a gorilla's than anything else, staring in the window at you. An animal looked in out of the darkness to see what was inside your lighted room, but it wasn't trying to get in to eat you, and it felt no hostility towards you. It was just curious and for some reason abandoned its usual caution. In reality you were granted a rare privilege, a chance for a close look at one of the most interesting creatures on earth and one that very few people ever see. It's too bad you weren't in shape to appreciate it. (p. 13)

To appreciate what stands behind those lines, let me quote from Green's book (with the help of its Index) a couple of cases of Sasquatches approaching buildings.

> Another report recently received from New Jersey, although it happened some time ago, takes us back to the Pine Barrens, to the town of Lower Bank. A couple who lived there in the fall of 1966 claimed to have found 17-inch, five-toed tracks outside their house after they had seen a face looking in the window more than seven feet off the ground. Instead of panicking, they began leaving table scraps outside, mostly vegetables, which something ate. The only thing rejected was a peanut butter and jelly sandwich. Then came a night when they forgot to put out any scraps, and a loud banging was heard outside the house. The husband went out to find out what was going on, and saw a 'Bigfoot-like' creature covered with gray hair throwing a garbage can against the side of the building. He fired a

shot in the air, without scaring it off, then fired at it, whereupon it fled and did not return. (p. 269)

There is reason to think the creature was grey with old age and that is why it began to rely for sustenance on garbage cans. Among items indicated in the Diet Index, Green mentions "garbage." Incidentally, the famous Bossburg cripple-foot tracks were found in a place where a Sasquatch was coming to feed out of roadside garbage cans. He must have been elderly too. Such evidence is known in Eurasian data as well.

Here's another case dealing with the evidence of the night watchman at a small sawmill in the forest twenty miles north of Orofino, Idaho:

> The animal was described as being about six feet tall, completely covered with shiny dark hair except for its face, hands, and the nipple areas of its very large breasts. All visible skin was pink. He thinks it may have been nursing a baby.... He and several other employees had seen many tracks in and around O Mill all during the summer of 1969. There were at least three different sets of tracks, one very large set, one human-size, and one child-size. The tracks indicated that they had explored in and around the buildings. They played in the sawdust pile and ate sandwiches he had put near the carriage. He could hear them jabbering among themselves and throwing boards which got in their way. They were in and around the mill most of the summer. (p. 289)

Hominologists generally believe that female Bigfoot (and other hominoids) raise their very young offspring in secure natural sanctuaries, well away from civilization, which is probably true. Yet the female with a presumed baby mentioned above found a secure refuge at a peaceful sawmill in the forest whose watchman "threatened" her with nothing more than sandwiches. Similar cases are found in Eurasian data too.

In her paper sent to the Vancouver conference, M.J. Koffmann said that the almasti of the Caucasus used to be offered food and even clothes by humans. Special sympathy was offered to their females with babies. The creatures take food from man—dairy products, meat, honey, porridge, all sorts of fruits and vegetables. John Green's Sasquatch Diet Index, which includes "stolen items and handouts," lists, among others, the following items of interest to Bigfoot: apples, cattle, chickens, cooked food, corn, fish, flour, garbage, macaroni, oranges, peaches, prunes, rabbits, sandwiches, sheep, table scraps, turnips, and vegetables.

I wrote in *Current Anthropology* in 1976: "Judging by the available data, the American hominoids look more 'archaic' than their European counterparts." The Green 1978 volume has almost erased the difference in my mind. Perhaps Bigfoot are more "pure-blooded" apemen than some Eurasian, especially Caucasian, populations but basically, I think, homins[1] on both sides of the Pacific are of the same Pithecanthropus (i.e., apeman stock). In ecology and ethology (i.e., behavior) North American hominoids (or, in keeping with present-day classification, relict hominids) are no more different from their Eurasian counterparts than the American brown bear is different from brown bears elsewhere. That is clear to anyone who has studied relevant data gathered by Russian and North American investigators, John Green in particular.

As for Neanderthal, in the opinion of Grover Krantz, this form should be put in *Homo erectus (Pithecanthropus)* and the whole group should be treated "as something distinctly less than human" (personal communication). I tend to agree with this view, adding the important reservation, however, that "less than human" is not necessarily tantamount to "purely animal." I think that the "superanimal" qualification would be more to the point. "*Homo erectus* existed for over a million years with relatively little change—a kind of evolutionary plateau—and then was transformed rather quickly into *Homo sapiens*" (Grover Krantz, 1980, *Sapienization and Speech*). So it is logical to surmise that today's wild bipedal primates in Eurasia, Australia, and the Americas are relics of that evolutionary "standstill," which lasted long enough for homins to penetrate and settle whole continents before the advent of *Homo sapiens.* Adapting to local environments they must have more or less departed in their physique from the fossil *Homo erectus* forms presently known to science.

In *A Hominologist's View From Moscow, USSR*, I called upon colleagues to follow the example of Jane Goodall who, having made friends with free-ranging chimpanzees, was able to thoroughly study their way of life in the wild. I mentioned three factors that facilitated Goodall's research: First: the chimpanzees are diurnal animals; second: they lead a community life; third: Goodall's project had financial support. I then referred to the handicaps of our task, having failed to mention one very important point which must facilitate the solution of our problem—the incomparable sociability of the apeman!

Nonsense? Paradox, yes, but not nonsense. No species of the great apes in the wild is known to make such regular and profound contacts

[1] I now use the term "homin" instead of "homi" and as a substitute of the old term "relict hominoid." Hominology is the science of homins.

with humans as our "wards" do. By "profound" I mean that they take not only food from humans but sometimes take the latter along as well, usually causing them in so doing no physical harm. Such cases are on record both in Eurasia and North America. One entry in Green's Sasquatch Index reads, "Carry things: animals, persons."

Why then is it paradoxical to speak of Sasquatch sociability? Because in relation to man it is, as a general rule, one-sided. In those rare-but-repeated cases when a homin is ready for friendly contact, humans are usually not. In my book, *In the Footsteps of the Russian Snowman*, a case is listed involving Anatoly Pechersky, a teacher, who was approached and followed in the mountains of the Kirghiz Range by an old and hungry hominoid. Contact was interrupted when a fearful Pechersky produced a hunting gun. In North America, as follows from Green's writing, similar contact was broken when man not only produced a gun but fired at the creature. The idea of friendly contact with wild hairy bipeds is as strange a fantasy to civilized bipeds of the twentieth century as the idea of friendship with gorillas was to civilized people of the nineteenth century. For them a norm, not a fantasy, was the slaughter of these relict animals by trophy hunters.

The above points are well illustrated by the late anthropologist Carleton Coon in his report at the Vancouver conference entitled, *Why There Has to Be a Sasquatch*. He relates a case in New Hampshire which he investigated personally and reported as follows:

> A man who lived just below the border in Massachusetts had driven his pickup truck, which he had converted into a camper, to a wooded glade along a highway. He stopped there and went to sleep at the wheel, his two young sons likewise snoozing on bunks by a window. At 11:00 p.m., the man awoke. Something was rocking his vehicle from side to side. An earthquake? He stepped out and was immediately grasped on the left shoulder by a seven-foot creature covered with light brown or yellowish hair. Its right hand pushed the camper off his running board onto the ground. It looked down on him, and stuck out its tongue. The man jumped free, the creature stepped back. The man drove as fast as he could up the highway, the creature followed him. Later on, one of the sighter's sons saw the creature's face peering into his bedroom below.

Coon writes elsewhere (in Vladimir Markotic, ed., *The Sasquatch and Other Unknown Hominoids* (1984), pp. 46–47):

> Several elements in this narrative had been recorded elsewhere, in other encounters, by persons who never heard of our New Hampshire actors and vice-versa. Both these encounters took place at night. The animals rocked the vehicles. It touched its human occupant. Its touch was not aggressive, but apparently a clumsy attempt at interaction, what might have been called a pat of affection, or way of saying: "I'm hungry!" or "I need a drink of water." The face-in-the-window syndrome is also on record elsewhere.

On record is also a knock-on-the-door-and/or-window syndrome, as illustrated by the case of Mecheny in Siberia, described in my book. All of that, and many other peculiarities of Sasquatches and their relatives in Eurasia, bespeak their "more than animal," (i.e., "superanimal") qualities.

Knowing this, how should we proceed about "discovering" these creatures? Searching for them in forests and mountains has long been compared to trying to find a needle in a haystack. Grover Krantz has improved upon the simile: "the proverbial needle is furtively moving about inside that haystack and leaving little indication of its previous locations." So what's to be done?

Some clever housewives, instead of groping for a dropped needle on the floor, easily find it with the help of a magnet. A powerful magnet, I suppose, could extract a needle even from a haystack. So if we want to discover a hominoid by way of friendly contact we should use all means that attract him and put aside all things that repel or alert him. That is why George Haas says of his group that they set out baits and lures in an attempt to entice the creatures in to look for us rather than the other way around; that they go into the woods completely unarmed; that most important is their attitude, which is friendly and relaxed, free of hate and fear. Similar advice was offered by me to René Dahinden.

To entice a hominoid, to induce him to respond accordingly to a friendly gesture and offer on our part is one reasonable line of action. It demands, besides clever planning and preparation, immense patience and persistence. Some initial results on this way have already been marked by the Bay Area Group in California and M.J. Koffmann's group in the Caucasus. Our beacon in this approach should be the pioneering work in befriending chimpanzees and gorillas by such primatologists as Jane Goodall, George Schaller, and Dian Fossey.

Another no less obvious and reasonable line is, as Boris Porshnev states, to use the help of those "tender-hearted people" who are already in friendly contact with the creature, having positively responded to his sociable advances. Porshnev pinned special hopes on this tactic, calling it "a key to the practical solution of the whole problem." When he wrote that, we only had information about such cases in Eurasia, mostly the Caucasus, with the problem that local inhabitants there tend to obey taboos which forbid them to "betray" wildmen. Today we know, from the Green volume in particular, that friendly contacts of this kind are also on record in North America, where the population (except the Indians) knows no taboos in regard to Bigfoot and Sasquatch.

Our task then is to:

Inform potential contactees of our interest;
Find actual contactees;
Interest actual contactees in cooperation with us.

Obviously, there is the question of money for us, just as there is for the opposition. Says Grover Krantz on this point: "With enough money, say about half a million dollars, a Sasquatch could almost certainly be obtained by expert hunters."

I realize that permanent friendly contact with a Bigfoot may cause some inconvenience to a household or its neighborhood. But would not one take inconvenience in one's stride for half a million dollars? Would not the man who fired at the creature demanding its daily portion of table scraps have behaved differently had he known that half a million bucks awaited him for continued supply of table scraps to his hairy intruder?

Where to get the prize money? Of course, tender-hearted private funders would be most welcome and may step forward, considering the fact that the dividends would start coming as soon as photographic evidence was obtained. But perhaps the necessary funds could be raised through contributions by people of ordinary incomes around the world if we explain to them the aim and significance of the operation. The next step, then, is to find enterprising and trustworthy people or an organization that would be willing to undertake such a venture.

# Memories of the 1970s

Author afield in Kabarda, North Caucasus, back in the 1970s, the time of the "Great Debate."

Russian–Mongol summit: specialists on *almasty* meeting the expert on Mongol *almas*. Standing: Igor Bourtesev and Alexandra Bourtseva. Seated: Dmitri Bayanov, M.J. Koffmann, and Professor Rinchen of Mongolia. Moscow, May 1976.

# 1997 Moscow Conference to mark the 30th anniversary of the Patterson film

The Darwin Museum new building, where the conference was held, was opened in 1995. It is the headquarters for the Smolin Hominology Seminar which is sponsored by the museum, as well as at present by The International Center of Hominology.

Igor Bourtsev at the podium, while I show photographic evidence to lovely foreign journalists, French on the left and British in the center.

Professor Grover Krantz playing the part of Bigfoot — demonstrating the walk of the Bigfoot seen in the Patterson–Gimlin film. Bayanov interpreting for the audience.

Three Bigfoot veterans with the "Bigfoot Blues." (L to R) Dmitri Bayanov, John Green, Grover Krantz.

M.J. Koffmann querying Grover Krantz regarding some fine points of his talk. Igor Bourtsev, behind his Bigfoot model, is all attention.

Three speakers enlightening an audience of over 100 conference attendees (L) Valentin Sapunov, (C) Vadim Makarov, (R) Leonid Yershov.

A gathering at Alexandra's apartment for a little R & R. From left to right: Vadim Makarov, Dmitri Bayanov, Igor Bourtsev, Marie-Jeanne Koffmann, Grover Krantz, Dmitri Donskoy, Michail Trachtengerts and in the foreground, Alexandra Bourtseva. I am standing beside the painting used on the front cover of my book, *In the Footsteps of the Russian Snowman.* The ancient painting (1545) depicts hominids which we now call snowmen. John Green took the photograph.

(L to R) Igor Bourtsev, Grover Krantz and Bayanov. The cast Igor is holding is from a print found in Pamir-Alai, Tajikistan, in 1979. In discussing the cast, Grover said that he detected certain features in it that indicated its authenticity. Grover kept those indicators secret.

Bigfooter and the Big Gun. Grover Krantz and Alexey Shishkin, a member of our hominology seminar who guided Grover and John Green around Moscow, are seen here beside the huge cannon in the Kremlin that is known as the Tsar-Cannon. It was cast in 1586 by the Russian master A. Checkov. It weighs 40 tons; its length is 5.34 m and its calibre is 890 mm. It was made for the defense of the Kremlin but was never fired. It is now a monument to the art of casting in the 16th century.

Three of the world's most eminent hominologists, (L) Dr. Grover Krantz, (C) Igor Bourtsev, (R) John Green, are seen together here. Igor is holding the sculpture he created after the hominid seen in the Patterson–Gimlin film.

# V.I.P. in Moscow

In close touch with Jane Goodall, world-famous primatologist, UN "Messenger for Peace," accorded warmest welcome in Moscow, June 1999.

My journey to Willow Creek, Northern California, was an adventure unto itself. After arriving in Seattle, Washington, John Green, Chris Murphy and Tom Steenburg met me, and we then picked up Bob Gimlin in Yakima. We had two vehicles, and shared company during the long drive. Seen here left to right, Bob Gimlin, John Green, Chris Murphy and myself. Tom Steenburg took the photo.

One of several stops along the way to enjoy the beautiful Oregon scenery. Sasquatches on the Pacific Coast inhabit some of the most picturesque country in the world.

Bob Gimlin (L) and I inspect Murphy's film site model. Murphy was pleased that the model got the "blessings" of the man who was there. Gimlin pointed to one of the tiny stumps and said, "Hey, that's the stump I jumped from to see how far my boot would sink."

Conference speakers' press interview before the opening session.

At the life-size wooden statue of Bigfoot, carved by Jim McClarin in the 1970s as a gift for the people of Willow Creek.

Greetings from Russia, esteemed Patty Bigfoot!
(Frame 350 of the documentary, enlarged by Scott McClean)

# In Bigfoot Country

At the Bluff Creek clearing famed since October 20, 1967, when it was walked by the Bigfoot caught on film by Roger Patterson. As a souvenir I took home a pebble from the creek.

American technology to back up hominology. Doug Hajicek is at the wheel.

A typical view of forested mountains in Northern California where searching for Bigfoot is like trying to find the proverbial needle which "is furtively moving about inside that haystack and leaving little indication of its previous locations" (Grover Krantz). As the saying goes, you don't find Bigfoot, Bigfoot finds you. The only exception that proves the rule is the case of Patterson and Gimlin.

Witnessing "Bigtrees" in Bigfoot country.

# Our Man in Tennessee

In 2002 we learned with incredulity of the existence of a farm in Tennessee where a Bigfoot family had quietly been hosted and "feasted" for half a century. That was the now famous Carter Farm. After I failed to inspire our overseas colleagues to launch a full-scale investigation and study at that site, Igor Bourtsev flew from Moscow to Tennessee in September 2004 and spent five weeks as a guest of Janice Carter, who broke her farm's long-held secret. Here he is seen with the authors of the mind-boggling and eye-opening *50 Years with Bigfoot*, 2002, Janice Carter (L) and Mary Green (R).

Janice leaving at appropriate height a "wee bit" of food for her furry friends. The gifts would be taken in the dead of night.

Igor showing a tree limb bent down and twisted by Bigfoot in the territory of the Carter Farm. A clear demonstration of the hominid's Herculean strength.

A wooden construction believed to be Bigfoot's "territorial marker." Such constructions are also found by hominologists in Russia and Australia.

Here, for comparison, is a similar "marker" in a Russian forest of the Vyatka area, some 700 km northeast of Moscow.

Here is a Bigfoot bed inspected by Igor on Janice's property.

Igor playing the part of the Bigfoot, called Fox, as the latter emerged from the basement of the house during Janice's abortive attempt to photograph him.

**Encouraged by the Carter Farm case, we are now in touch with several other Bigfoot befrienders and contactees in the U.S. who give us the best hope for a humane and near solution of the problem. We trust people trusted by Bigfoot.**

# CHAPTER SIX

# On the Cryptograpevine: Nessie Meets Bigfoot

"Then it's agreed... I'll believe in you if you believe in me..."

The Nessie-Bigfoot dialogue which follows was prompted by the above cartoon, originally published in Dan Perez's **BigfooTimes**, August 31, 1981. The piece that follows was first published in **BigfooTimes** on January 7, 1985. The message is as applicable today as it was then.

**NESSIE:** Then it's agreed... I'll believe in you if you believe in me.

**BIGFOOT:** Our mutual recognition is no problem, Nessie.

**NESSIE:** And what is the problem, Biggie?

**BIGFOOT:** The problem is our recognition by humans. A 1978 Gallup Poll found that only 13% of the American public believes in you and me, far less than the belief in Extra Sensory Perception (51%), precognition (37%), or astrology (29%).

**NESSIE:** And why should we care about their beliefs, Biggie?

**BIGFOOT:** Because our very lives depend on that, Nessie. While you've been cruising the murky waters of the Loch, a momentous thing has happened on the ground. Some of our most ardent fans have set up the **International Society of Cryptozoology** to prove that you and I are real.

**NESSIE:** And how can anyone doubt that? Thousands of eyewitnesses would testify to our existence.

**BIGFOOT:** Ah, Nessie, you don't understand the ways of civilization. Recognition by a crowd of fishermen doesn't make you a bit real. To be real for civilization you have to be recognized by science, that is to say by the majority of scientists around the world.

**NESSIE:** And what is our standing with scientists, Biggie?

**BIGFOOT:** That's a good question, Nessie. Even though we are on a par as two superstars of cryptozoology, the attitudes of scientists towards you and towards me are quite different. A recent poll of 300 scientists showed that acceptance of Bigfoot is far lower than the acceptance of Nessie, 10.6% and 31% respectively.
**NESSIE:** Oh Biggie, I'm thrice more believable than you are! That's great. How do you account for that?

**BIGFOOT:** One reason may be big differences showed up in the scientists' ideas about the impact the discovery of you and me would have on science. Over 50 percent of the anthropologists thought discovery of me would have a severe impact. But only 3 percent of the anthropologists felt that discovery of you would have a severe impact.

**NESSIE:** Are you saying you're more important to science? How can you prove that?

**BIGFOOT:** According to one enlightened opinion, being a primate and a hominoid, possibly even a hominid, I would be a close genetic relative of man, perhaps too close for comfort, and the legal and moral implications involved could be substantial. You, on the other hand, do not threaten man's elevated status in the animal kingdom.

**NESSIE:** So on some points I am ahead, on others you are. No reason for us to quarrel.

**BIGFOOT:** Not really, Nessie. But still you are more secure or, let's say, less vulnerable than I am. Recently the Vermont (USA) House of Representatives passed protective legislation, saying your kind "should be protected from any willful act resulting in death, injury or harassment." Congratulations, Nessie! I wish I could live to see my folks enjoy such legislation.

**NESSIE:** You sound so pessimistic Biggie. Do humans really hate you so much?

**BIGFOOT:** Well, it's not a matter of hatred, Nessie. I'm afraid it's something worse. There are, you see, three kinds of bipeds on earth: people, androids, and hominoids. People are really not too bad, but androids... well, they are the death of hominoids. Just listen to what one of them has in store for my species.

> My aim is knowledge,
> to know a thing I must probe it.
>
> First I will capture it
> with nets traps helicopters dogs pieces
> of string, hole dug in the ground, doped food
> tranquilizers guns, buckshot, thrown stones
> bow and arrows
>
> Then I will name the species
> after myself
>
> Then I will examine it
> with pins tweezers flashlights microscopes telescopes
> envelopes statistics elastics
> scalpels scissors razors lasers cleavers axes
> rotary saws incisors osterizers pulverizers, and fertilizers.
>
> I will publish the results
> in learned journals.

Then I will place a specimen
in each of the principal zoos
and a stuffed skin
in each of the principal museums
of the western world.

When the breed nears extinction due to
hunters trappers loggers miners farmers
directors collectors inspectors
I will set aside a preserve consisting of:
1 mountain
1 lake
1 river
1 tree
1 flower
1 rock
and 1 tall electric fence.*

**NESSIE:** And why do creatures of civilization turn so bestial at the mere thought of you, Biggie?

**BIGFOOT:** How do I know, Nessie? Maybe because "homo homini lupus est," which translates throughout history as "man is a wolf to man."

**NESSIE:** I see, Biggie, your problem is they treat you like a human being. If you were not hominoid but, say, fishoid, wormoid, or even poisonous snakeoid, then humans, I'm sure, would be as nice to you as they are to me.

Postscript: Recently your author visited the same secret meeting place of Nessie and Biggie and overheard them talk as follows:

**NESSIE:** Biggie, do you know the good news? The Commissioners of Skamania County, in Washington State, say the willful slaying of a Sasquatch would be punishable by up to one year in jail, a $5,000.00 fine or both. Besides, an official of the Federal Office of Endangered Species (OES) said that he hoped that nobody would ever actually shoot a Sasquatch to prove its existence.

---

*Atwood, Margaret (1970), see Bibliography.

**BIGFOOT:** How did you learn that, Nessie?

**NESSIE:** I heard a crytozoologist read that to his mate. It was in the ISC Newsletter, Volume 3, Number 2.

**BIGFOOT:** Is that all you heard?

**NESSIE:** Yes, Biggie.

**BIGFOOT:** Well, Nessie, I overheard my cryptozoologist read something else in that Newsletter: "A Vietnam combat veteran announced plans to hunt down the supposed Sasquatch of the Pacific Northwest and to shoot it. Grover S. Krantz, a physical anthropologist at Washington State University (Pullman, WA), supported the venture and offered the team his advice." Did you ever hear a scientist advocate the killing of a Nessie or a Mokele-Mbembe to prove their existence?

**NESSIE:** No, I didn't.

**BIGFOOT:** Neither did I. So you see, Nessie, as long as "homo homini lupus est," I am in mortal danger.

## Bibliography

**Atwood, Margaret.** 1970. *Oratorio for Sasquatch, man, and two androids*, in *Poems for Voices*. Toronto: Canadian Broadcasting Corporation. (Also in *Manlike monsters on trial: Early records and modern evidence*. Vancouver, British Columbia: University of British Columbia Press, 1980, 306–15.)

**Bayanov, Dmitri.** 1980. Why it is not right to kill a gentle giant. *Pursuit*, Fall, 140–41.

———. 1996. *In the footsteps of the Russian snowman*. Moscow: Crypto-Logos.

———. 1997. *America's Bigfoot: Fact, not fiction*. Moscow: Crypto-Logos.

**Coleman, Loren.** 1989. *Tom Slick and the search for the Yeti*. Boston: Faber and Faber.

**Coon, Carleton S.** 1984. Why there has to be a Sasquatch, in *The Sasquatch and other unknown hominoids*, ed. Vladimir Markotic and Grover S. Krantz. Calgary: Western Publishers.

**Green, John.** 1968. *On the track of the Sasquatch*; Agassiz: Cheam Publishing.

———. 1970. *The year of the Sasquatch.* Agassiz: Cheam Publishing.
———. 1973. *The Sasquatch File.* Agassiz: Cheam Publishing.
———. 1978. *Sasquatch: The apes among us*. Saanichton: Cheam Publishing.
**Greenwell, John.** 1982. Human nature: the dinosaur vote. *Science Digest*, April.
**Greenwell, John Richard and James E. King.** 1980. Scientists and anomalous phenomena: Preliminary results of a survey. *Zetetic Scholar*, no. 6
**Haas, George.** 1969. *Bigfoot Bulletin*, January 2.
**Krantz, Grover S.** (1992) *Big Footprints: A Scientific Inquiry Into The Reality of Sasquatch;* Boulder: Johnson Books.
———. 1980. Sapienization and speech. *Current Anthropology* 21(6).
Lake Champlain monster draws worldwide attention. 1982. *The ISC Newsletter,* Summer, 4.
**Lorenz,** Konrad 1969. *On aggression.* London: Methuen & Co.
Markotic, Vladimir and Grover S. Krantz, eds. 1984. *The Sasquatch and other unknown hominoids*; Calgary: Western Publishers.
**Porshnev, Boris.** 1963. *The present state of the question of relict hominoids* (in Russian). Moscow: VINITI.
———. 1968. The struggle for troglodytes (in Russian). *Prostor* (Alma Ata), no. 4-7.
———. 1968. Comment on the Patterson-Gimlin film (in Russian); *Znanie-Sila* (Moscow), no. 9: 52–53.
Proposed Sasquatch hunt stirs new controversies. 1984. *The ISC Newsletter,* Summer, 1–3.
**Strasenburgh, Gordon R.** 1975. On paranthropus and relic hominoids. *Current Anthropology*, 16(3): 486–87
**Tchernine, Odette.** 1961. *The snowman and company*, London: Robert Hale.

# Why It Is Not Right to Kill a Gentle Giant

## A Rebuttal to John Green's Conclusion in *Sasquatch: The Apes Among Us*

HAVING READ half of Green's 1978 volume plus its last chapter, I sent him the following message of acknowledgement: "It seems to be an admirably good book with a notoriously bad ending." My opinion did not change after I read the whole thing.

Why the book is good is clear to any reader interested in the subject: It is the most complete and detailed presentation of Sasquatch sighting-reports in North America to date. It is as if the author had opened his files for everybody to see and learn. Many thanks to him for that from Sasquatch-lovers all over the world!

Not so, alas, with the concluding chapter of the book. It is bad, not only because of its bad treatment of Sasquatches but also because it is evasive and unfair in the treatment of truth. For this very reason the quality of the last chapter is not apparent to the reader; hence my protest.

On page 150 Green says that "Dmitri Bayanov and I have engaged in a protracted debate by mail in which he contends that no 'Homi' as he calls them should ever be killed." Yet in the space of 492 pages of his book and in its last chapter specially devoted to the question, Green never cites or answers a single argument of mine from that "protracted debate" of ours. Love's labor lost.

Instead of citing my arguments and dealing with them, John Green cites his pessimism regarding the future of mankind, as if saying: Personally I have no grudge against Sasquatch, but why bother about good treatment of animals if mankind itself is going down the drain? To wit:

> In my opinion our species is a blight that the world would have been better off without. Had we realized a century or two ago that we were becoming too numerous, and had we been able to halt human multiplication, everywhere, man might have been the crowning glory of the planet. With his present numbers and continued growth, man can only be considered as a cancer that is destroying the planet. (p. 461)

My answer to that is that mankind's plight has something to do with its ill-treatment of living creatures. To my mind the quality of life depends first and foremost on the quality of man's social behavior, and only in second place on the quantity of population. I'd rather live in a country densely populated with decent people than sparsely with thugs.

The main and immediate problem for mankind is not its numbers but the lack of an ethical balance between the rational and the emotional, between the brain and the heart in man's behavior and institutions.

The trouble with Hitler was not that his policy increased the world's population. It did the opposite. The trouble with Hitler was that he stopped at nothing to achieve his aims. And as far as science is concerned, the trouble is not that there are too many scientists in the world, but that more than half of them are engaged in the arms business, i.e., devote their time and talent, at the taxpayer's expense, to devising ways of destroying life instead of improving it.

But why such generalizations? Why make a fuss over the fate of Sasquatches? Is our research really of so much consequence? Fortunately, in this respect Green makes no mistake: "This is not a game or a fantasy, it is a question of serious scientific research of tremendous importance. It may not have the glamour of moon shots, but in what it can teach us about our origins and our physical potential it may be even more important." (*The Sasquatch File*, p. 70)

On that we agree, so let us return to what is at issue here. It was with great relief and approbation that I learned from Green that "public objection to the killing of any of the great apes for research purposes has risen to the point where projects involving it are not approved, even though there are sufficient animals available." (p. 463) Yet when it comes to Sasquatches, the author is adamant: "Should they be hunted for scientific purposes? Definitely yes ...

They will have to be studied, and study will have to include dissection. ...Sasquatches are not available for study without killing them." (pp. 462–63)

If there is a public objection to the killing of the great apes why wouldn't, or shouldn't, there be to the killing of Sasquatches? Green gives no answer, or, at least, the answer is not to be easily grasped. What he clearly says is that "To give special treatment to one type of animal because it reminds us of ourselves obviously reflects concern for ourselves, not for animals." (p. 462) Does it follow from this that by killing a man-like creature we display our unselfishness? It is obvious that special treatment of man-like animals reflects concern both for ourselves and for that type of animal.

Why is public opinion in favor of special treatment in such cases? Green seeks rational reasons and, to his great satisfaction, finds none. And no wonder, because the reasons are emotional. But does that invalidate them? Much of my protracted argument with Green centered just on this question. Back in 1975 I wrote him that "the emotional is an integral part of human nature, just as natural and legitimate as the rational. If we had no emotions we'd be computers, not people. A wise policy in any endeavor is not to reject emotion but to understand their workings so as to strike a balance between the rational and the emotional. ...Man can't be a 'really unbiased' judge in the matters of life and death. Being himself alive and human he can't help being impartial towards anything which is alive and intelligent, or at least looks "man-like"...To show that the above is not all my invention let me back up my stand by citing no less an authority than Konrad Lorenz who, in his book, *On Aggression*, in the chapter "On the Virtue of Scientific Humility," says the following:

> The scientist who considers himself absolutely "objective" and believes that he can free himself from the compulsion of the "merely" subjective should try—only in imagination, of course—to kill in succession a lettuce, a fly, a frog, a guinea-pig, a cat, a dog, and finally a chimpanzee. He will then be aware how increasingly difficult murder becomes as the victim's level of organization rises. The degree of inhibition against killing each of these beings is a very precise measure for the considerably different values that we cannot help attributing to lower and higher forms of life. To any man who finds it equally easy to chop up a live dog and a live lettuce I would recommend suicide at his earliest convenience. (pp. 194–5)

After reading all that in 1975, Green still continued asking in 1978: "People kill other animals, so what are the grounds for treating this one differently?" (The Vancouver *Sun*, issue of May 15, 1978) Still I am grateful to him for citing in *Sasquatch* such examples of eyewitness reports of the animal, which he proposes to treat no better than frogs, as this one:

> It was half-ape and half-man. I've been reading up on the abominable snowman since then and from articles you get the idea that these things are more like gorillas. This thing was not like that at all. It had hair all over the body as if it was an ape. Yet the face was definitely human. It was more like a hairy human. (p. 194)

In an appeal to the would-be participants in the Sasquatch conference in Vancouver I wrote in December 1977: "Modern *Homo sapiens* represents and symbolizes the achievements and failures of civilization. Modern *Homo troglodytes* represents and symbolizes the top achievement and subsequent retreat of nature. Now for the first time in history the two species are going to meet in the limelight of science. Let us make this meeting a happy one for both. Let it be on the credit side of civilization."

Green, however, sees no alternative to a *bloody* meeting. "To begin with, one [Sasquatch—D.B.] must be presented to the scientists in the flesh in order to establish that such a creature exists at all." (p. 462) Then, "sasquatches are not available for study without killing them." (p. 463)

Let me remind Green that nowadays it is perfectly feasible to establish the existence of life forms without presenting them in the flesh to the scientists. Millions of dollars have been spent on the search for the minutest signs of life on Mars by such a proxy method. Surely a tiny fraction of that sum would suffice to establish beyond any doubt the existence of a creature the size of a Bigfoot here on earth. And why must anybody bring a Sasquatch in the flesh to a scientist in an armchair? It is the scientist who must be brought in the flesh to the mountains to meet and study Sasquatches in their habitat without so much as disturbing them.

Jane Goodall and George Schaller have set an example of this kind in the treatment of higher primates in the wild, and it is the business of those who know what's what in our research to persuade the scientific community to start moving in that direction. Of course, it is easier to bestir the "trigger-happy types" than the scientists, but the easy way is not necessarily the right way.

Since I relied on the reputation of Jane Goodall to argue my point (see *The Scientist Looks at the Sasquatch*, p. 152), Green decided to deprive me of that advantage by citing Goodall to support his own case (*Sasquatch*, p. 466). I corresponded with Jane Goodall and asked her what she thought of Green's stratagem and our controversy. She replied from Dar-es-Salaam, Tanzania, on August 22, 1978:

> Dear Dmitri Bayanov,
>
> Very many thanks for your letter of 11 July. I am answering it more quickly than usual because of the questions in it. Most particularly the one about the killing of a Sasquatch—or any other form of ape-like or human-like creature. I deeply deplore the killing of animals for

> museums. I loathe to see a stuffed chimp or gorilla—or monkey—or lion and so on. A photograph is just as good. A film is better. ...Nor do I look back kindly on what I wrote, which John Green has been able to use to his own ends. I no longer agree with what I said then—I don't think it is at all justifiable to use chimpanzees to find out about a disease which people would not get if they did not eat one another! When I included this (as the book was going to press) I was trying to find out something really useful which scientists had learned from chimps. Something which really would alleviate human suffering. The kuru was a bad example, and got into the final book before I had thought about it properly. Anyway—to shoot a creature just to see what it is—well, that is even worse. Most undesirable ethically—and the mark of a poor researcher.

I wish all of us could be as magnanimous and uninhibited as Jane Goodall in admitting a mistake.

Now let's look at the matter from a different plane. Let us imagine that during one of his lectures on the "rightness" of killing a Sasquatch John Green could tune in on an exchange of opinions between two UFOnauts invisibly hovering in the auditorium.

"You know, mate," says one to the other, "back home on our planet they'd never believe the existence of such bloodthirsty creatures in the universe. Let's vibrocute this one and dispatch his body or part of it to our learned skeptics."

"But that would be murder!" gasped the other.

"I don't think so," continued the first. "Take a good look at him and be rational. Is he really like us? His body is not transparent; he cannot levitate; he doesn't speak the Milky Way lingo; his brain doesn't work on neutrino energy ..."

I wonder what message Green's brain would send back to his would-be executioner. Perhaps this: "Hey there, just let me see a killed Sasquatch and then do with me whatever you like!"

## References Cited

Austin, Edie and Tim Padmore. 1978. Sasquatch: The skeptics want a corpse. The Vancouver *Sun*, May 15, p. A6.

Bayanov, Dmitri. 1977. In *The scientist looks at the Sasquatch*, p. 152. Moscow, ID: The University Press of Idaho.

Goodall, Jane. Personal letter, quoted by permission.

Green, John. 1973. *The Sasquatch file,* p. 70. Agassiz: Cheam Publishing Ltd.
———. 1978. *Sasquatch: The apes among us.* Seattle: Hancock House.
Lorenz, Konrad. 1969. *On aggression*, pp. 194–5. London: Methuen & Co. Ltd.
*Pursuit*, Vol. 13, No. 4, Fall 1980.

(Originally published in *Pursuit*, Vol.13, No.4, Fall 1980)

# SECTION 7
# Hominology and Cryptozoology

## A Hominologist's View from Moscow, USSR

[Editorial lead-in by *NARN*:] The following comments, kindly offered for publication, were received in letter form, July 1976, by one of the editors (RS) in response to copies of *NARN* reprints sent to Dmitri Bayanov and Igor Bourtsev.

Thank you very much for the *NARN* materials you sent to us. Some of the articles—those by Grover Krantz and Wayne Suttles—we received previously from other colleagues and discussed at our seminar, while those by Bruce Rigsby and Gordon Strasenburgh are new to us and are scheduled for discussion.

It is also the first time that I have seen and can appreciate your Editorial, which started it all in *NARN*. I must say it is written in the best and noblest tradition of scientific inquiry. The fruits of science can be used to harm man, but the spirit of science, which is revealed in your editorial, can only be of the greatest service to mankind.

It is also gratifying to learn that you were influenced by Green's reference to the editors of *Soviet Ethnography*, who found it possible to publish an article by Boris Porshnev (1969) on the problem of relict hominoids. As a Russian poet put it: "We aren't given to divine how our word will echo." The echo in this case turned out to be fine and is still reverberating. In fact, it was the only echo of the article that I know of, since those whom Porshnev addressed here with his publications on the subject invariably greeted them with utter silence.

No matter, we have survived, along with the hominoids, and are gaining ground. In this connection, I would like to elaborate on or even correct one statement by Strasenburgh (1975, 282), namely "the surviving Hominoid Problem Seminar, composed of interested laymen who meet at the Darwin Museum in Moscow." His saying that our seminar is composed of laymen implies contrasting it with a group of professionals. I wonder what professionals Strasenburgh has in mind in this case.

We have many members with higher education (our average attendance is 25 and we meet monthly, except summer vacation time), several are biologists, with zoologists among them. The late Pyotr Smolin (who died last September), the founder of the seminar, was one of the most versatile biologists and eminent zoologists of this country. It is true that physical anthropologists are conspicuous by their absence in our ranks, through no fault of ours. That is why we welcome so much and appreciate Krantz's work on the subject. But does it give grounds to say that our seminar is composed of laymen? As Strasenburgh (1975, 281) himself has aptly observed, "the majority of scientists who have been quoted in the media regarding the subject simply do not know what they are talking about." Since they do not, and we do, who are the laymen in the field then?

It is true none of us can work full time on the subject for the simple reason that all our research is done free, and man, alas, cannot live by thought alone. But still there are more appropriate names than "laymen" for such enthusiasts.

My name is "hominologists." Though we often use it in jest, I would like to stress some of its serious implications. In a letter to John Napier, I wrote (in 1973): "The living missing link is 'unknown' to science because there is no science to know it." Indeed, between zoology and anthropology there is a big gap in knowledge, a no-man's land of science, in which the hominoid is securely hidden from the eyes of orthodox scientists. This situation is caused both by the history of science (the young age of the current evolutionary theory, which alone could have shed true light on the nature of hominoids) and by the characteristics of the hominoids themselves (their resemblance to man in appearance and to animals in behavior, their love-hate relationship with man, etc.). The study of relict hominoids reveals not only their reality, but a new kind of reality in general. Since hominoids are different both from the apes and *Homo sapiens*, their study constitutes a legitimate branch of knowledge that deserves a name of its own. Hence the birth of *hominology*, as a branch of primatology, called upon to bridge the gap between zoology and anthropology, which task is in full accord with the work and aspirations of Charles Darwin.

I cannot but agree with Strasenburgh that our studies must go on no matter whether a specimen is brought in or not. In fact, the inquiry into the theoretical and historical aspects of the problem is already so advanced that even if relict hominoids were no longer in existence today, there would be still enough sense and material for hominology to exist as an historical discipline. But they do exist, and this makes me view the situation in a somewhat different light than Strasenburgh does.

I think the hominologist is duty-bound to have the hominoid recognized by science at large as soon as possible. The reason for my impatience is as follows.

There was a time when apes and monkeys were nothing but objects of amusement for man. Today we know better, and the non-human primates have become the subject of the most serious and valuable research. There can be little doubt that the study *in vivo* of non-human bipedal primates from the genetic, psychological, and medical points of view will be even more relevant to man's ever-present task to "Know thyself" than is the study of apes and monkeys. The hominoids may turn out to be as necessary for the health, and eventually the very existence, of mankind as wild plants are necessary for the continued existence of agriculture.

They are potentially the most precious element of the environment, and it is by human-induced changes in the environment that their existence is put in jeopardy. To prove the point it is enough to recall the fate of hominoids in Western Europe, where they disappeared, along with other species of wildlife, in historically recent times. A similar process, but at a much faster pace, is underway in other parts of the globe. Even if this process is stopped and reversed in the future, as has been the case with some threatened species, the disappearance in the meantime from the face of earth of one or another local variety, not to mention species, of the hominoid will be a terrible loss in the inventory of life forms on the spaceship Earth. Hence we must act right now to prevent this from happening.

But having realized this we seem to land in the long familiar vicious circle: for science to protect the hominoid it must first recognize the creature's existence, and the creature's existence is thought to be unproveable without the funds resulting from such recognition. This is how Strasenburgh (1975, 289) sees the dilemma:

I think several million dollars spent toward obtaining a specimen is justified on the basis of the evidence we presently have. Anything less is likely to be a severe handicap. I am, however, aware that such a sum of money is beyond the means of those who are presently interested in the question, as well as those who might be potentially interested in it. So let us turn to the ethnographers and ethnologists and their lack of interest in the wild man myth.

But perhaps the dilemma is more apparent than real. Do we really need millions of dollars to lift the leaden eyelids of skeptics? It all depends on how we propose to go about it.

As my American colleagues well know, I am in principle against not only a killing, but any kind of violence against the hominoid in the name and for the "good" of science. The idea of destroying or crippling anything to prove its existence, moreover such a formidable and awe-inspiring figure as a hominoid, seems unsavory to a great many people; this writer included. Cannot *Homo sapiens* afford to be magnanimous enough to grant Most Favored Creature Status at least to his nearest kin in the animal kingdom? To quote George F. Haas (personal communication, 16 April 1975):

Fortunately, there seems to be a growing trend in the West to recognize that animals and other forms of life have rights of their own; that they were not just "put here for the benefit of man" but are fellow passengers on the spaceship Earth through time and space and thus entitled to the respect and consideration due to any fellow traveler. Most of this trend seems to be due to the growing influence of eastern philosophies on our western cultures and I hope this trend continues.

Given this, the only alternative left to the hominologist, dreaming of meeting the hominoid in person, is making friendly contact with the creature. The feasibility of this hope is based on the following considerations. Both anthropologists and zoologists have rich experience in making friendly contact with objects of their study, from humans still living in the Stone Age to chimpanzees to wolves to crocodiles. No matter how unique the hominoids are, they cannot be so different from all other creatures as to be immune to man's friendly advances.

Moreover, we have indications that the ancients knew ways of making friends with hominoids, and there is reason to believe this "know-how" still lingers on and is used clandestinely by certain indigenous *cognoscenti* here and there in *homi* country (homi is our jargon for hominoid). In the early days of our research, namely in September 1959, a local old man in the Caucasus (in Daghestan) agreed to show a *kaptar* (popular name for the hominoid in Daghestan) to a group of hominologists that included Jeanne Koffmann of our group in Moscow and Yuri Merezhinsky of the Kiev University (Department of Ethnography and Anthropology), by taking them to a spot on a brook where kaptar used to bathe. The offer was made on the condition that no harm would be done to the creature.

The creature appeared as promised, but Merezhinsky did not keep his word and fired a pistol shot at it, which killed nothing but a priceless chance to attempt friendly contact with a hominoid (see Boris Porshnev 1968, 112).

A unique and nearly successful attempt to make friends with a Bigfoot was made in June 1970 by Arthur Buckley of the Bay Area group in California headed by George Haas (personal communication, 16 April 1975). Says Haas of Buckley's attempt: "Nobody else has ever done it. His techniques should serve as models for all future attempts at making contact."

Before trying to answer the question why Buckley was not fully successful and why his, and similar, techniques are slow in catching on, let us put and answer the question: Will contact prove the existence of hominoids? My answer, just as Haas's, is: Yes, undoubtedly so. In his words, "Anyone ... viewing bigfoot films comparable to those made of the gorillas by Adrien Deschryver of eastern Zaire and denying their authenticity, would be a die-hard skeptic indeed." As for physical evidence, such as bones or a carcass, given contact, we are as sure to obtain it eventually as both humans and homis are mortals.

Now, why this delay in putting into practice what Haas and I have been advocating? To better understand the answer first read what David A. Hamburg of Stanford University has to say in the foreword to Jane Goodall's book *In the Shadow of Man*:

> The difficulties of solving the mystery were formidable—many experts thought insurmountable. ...The chimps were not cooperative. They stayed away from her, kept her at a great distance. ...Four years passed before truly abundant observations were possible (Goodall 1971, 13–14).

Goodall's accomplishments are above all praise and are a great inspiration to us. Yet, compared to our problem her task was facilitated at least by three important factors: 1) chimpanzees are diurnal animals; 2) they lead a communal life; and 3) Goodall's project had financial support during those crucial four years.

We can do nothing at present about the hominoid's contrary first two characteristics, namely, his solitary nocturnal life, except intensify the expression of our desire to make friends with him. Let us recall that in Goodall's case the ice was really broken when the chimps accepted offerings of bananas from her. Buckley used fish as the bait, making bigfeet aware of his presence and good intentions with calls "in a friendly, encouraging voice."

On our side, I have worked out a technique, dubbed the "braying bait method," which combines food baits, certain natural (animal) sounds believed to beguile the hominoid, and a live herbivor known to interest and attract the creature. The method has never been tried out, mainly because of the absence of the third factor above: lack of funds.

Considering the handicaps of our problem, it may take more than four years of intensive attempts before contact is made. To increase the probability of being in the right place at the right time we had better have not one but several or even many simultaneous attempts being made by separate investigators in separate places.

And all of them have to be backed financially, if we do not want the baiters to eat their baits.

Thus, in the end, as Strasenburgh foresaw, it all boils down to the question of money. However, in my estimate the program, as I have described it, would cost hundreds of thousands of dollars, not several million. The dividends would start coming as soon as photographic evidence would be obtainable. Yet, how to obtain funds to get the program off the ground?

One answer seems to be through international cooperation. What one country is shy to do on its own, may be bravely done by a community of countries or on a bilateral basis. The United States and the Soviet Union are actively engaged in environmental cooperation, having outlined "a program of research in the preserves which will be set up in deserts, mountains, and forests both in the Soviet Union and the United States. They will be used for the observation of air masses, soil, flora and fauna and the use of water resources." The quote is from a TASS report on a Soviet-American biospheric preserves symposium that was held in Moscow in May of 1976. The inclusion of a staff of hominologists into this program would seem to come in handy and at a negligible increase in its cost, considering the total sums involved. What is needed now is an understanding of the hominoid problem on the part of those concerned and the will to prod things in the right direction*.

In a *Moscow News* article, entitled "Need for international cooperation in search for relic hominoids," René Dahinden said in 1972:

> If by way of international cooperation we manage to determine the quality of the material we possess as well as to evaluate the possibility and probability of the existence of the above-mentioned creatures in certain parts of the world, then we'll make a big step forward in solving the problem, which I consider one of the greatest scientific problems of all time.

I am all for Dahinden's idea of international cooperation in this research and hope that we can not only make a big step forward by joining forces but finally solve the problem.

In conclusion, to reassure those who are tired of waiting for a quick solution, as well as those who insist that all we learn from history is that we do not learn from history, I shall quote the *Encyclopaedia Britannica* on the history of meteoritics, which bears a striking similarity to our case:

> History of Meteoritics. - Since very ancient times men have known about meteorites falling; however, the scientific study of meteorites is hardly older than 150 years. ... It was just this miraculous character of meteorite falls and the favour they found with the churches which, during the period of enlightenment, made scientists suspicious of their reality. In the 18th century members of the French Academy, then the highest authority in all scientific matters, were convinced that such an irregular phenomenon as the fall of a stone from heaven was impossible, and preferred to doubt all the reports of witnesses and to change their statements to conform with acknowledged scientific theories. Following their lead, keepers in many museums of Europe discarded genuine meteorites as shameful relics of a superstitious past. It is an interesting fact that the preservation of the precious iron meteorite of Hraschina, which fell in 1751, is due to the protection given to it by the bishop of Zagreb's consistory; they collected the sworn statements of eyewitnesses and sent the document, together with the iron, to the Austrian emperor. Both came into the possession of the Vienna museum. The influence of this clerical report on the Hraschina fall went even further; it became one of the strongest weapons in the hand of the

* The 1978 Soviet invasion of Afghanistan led to a sharp deterioration of the Soviet-American relations and put an end to environmental cooperation between the two countries.

> German physicist E.F.F. Chladni (1756–1827) when he began his fight against the scientific authorities who ridiculed belief in meteorites. His paper of 1794, in which he defended the trustworthiness of this and of a few similar accounts and explained meteorites as pieces of cosmical matter that enter our atmosphere, marks the beginning of the science of meteoritics. Most of his colleagues remained skeptical, but a shower of stones that fell in 1803 at L'Aigle, not far from Paris, finally convinced the Paris academy and the rest of the scientific world of the reality of the fall of meteorites. From that time astronomers, physicists, chemists and mineralogists of many countries have contributed to knowledge of them (*Encyclopaedia Britannica* 1965).

Since hominoids can not be expected to fall from the blue on their own, as is the case with meteorites, I wish some prankish UFOnaut would dump a load of bigfeet on the heads of skeptics among modern academics.

## References Cited


Bayanov, Dmitri, and Igor Bourtsev. 1976. On Neanderthal vs. *Paranthropus*. *Current Anthropology* 17(2): 312–18.

Dahinden, René. 1972. Need for international cooperation in search for relic hominoids. *Moscow News*, 5–12 February, 5(1100):6.

*Encyclopædia Britannica*. 1965. Meteorites. In *Encyclopædia Britannica*, 15:272–77. Chicago: Encyclopædia Britannica Inc.

Goodall, Jane. 1971. *In the Shadow of Man*. Boston: Houghton-Mifflin.

Porshnev, Boris F. 1974. The struggle for troglodytes. *Prostor,* No. 7. Reprinted in *L'homme de Néanderthal est toujours vivant* by B. Heuvelmans and B.F. Porshnev (29–205). Paris: Plon.

———. 1969. The problem of relic paleoanthropus (in Russian). Soviet Ethnography 2:115–30.

Strasenburgh , Gordon R., Jr. 1975. Perceptions and images of the wild man. *Northwest Anthropological Research Notes* 9(2): 281–98.


(Originally published in *Northwest Anthropological Research Notes (NARN),* vol. 11, no. 1, 1977.)

# Why Cryptozoology?


## Abstract

One of cryptozoology's main concerns is the possible survival of animals thought to be long extinct. Because the geological record is incomplete, the possibility of fossil animals surviving to the present cannot be ruled out. Comparison is made between the objects of study of zoology and the non-animal subjects of other disciplines, including botany, and the conclusion is drawn that the elusiveness of some animals is the main reason for the existence of, and need for, cryptozoology. Given the nature of cryptic animals, cryptozoology has to apply its own methods of research.


## Introduction

Since the article "What Is Cryptozoology?" by Bernard Heuvelmans (1982), published in the first issue of this journal, has not made critics any less vociferous (May 1984; Simpson 1984; Diamond 1985), it seems worthwhile to devote more attention to the *raison d'être* of the discipline. One of cryptozoology's main concerns is the possible unrecognized "persistence of presumed extinct animals to the present time or to the recent past" (*The ISC Newsletter* 1982). Consequently, we should examine the problem of extinction and survival in relation to cryptozoology.

## The Extinction-Survival Dichotomy

Recent attention to the notion of species extinction has somewhat obscured the phenomenon of species survival. Both extinction and survival are *conditio sine qua non* of biological evolution. But, man's intervention apart, dying and living in nature are so propitiously balanced that the variety of life and the number of species on earth have generally been on the increase for millions of years, this being due not only to the ongoing process of speciation, but also to the persistence of old forms.

This being so, how do we know precisely what is extinct and what is extant in biology? According to Charles Darwin ([1859] 1929, 280 & 299), "No fixed law seems to determine the length of time during which any single species or any single genus endures," and "the utter extinction of a whole group of species has sometimes been a slow process, from the survival of a few descendants, lingering in protected and isolated situations." A noted Soviet paleontologist, L.S. Davitashvili (1948, 486), has this to say on the subject:

> It is always necessary to remember the incompleteness of the geological record. The first appearance of a given species in the geological record and its disappearance from the latter can in no way be taken for the dates of its origin and final extinction. The real life span of a species (or a group of species) is usually much longer than the period determined from the geological record. Consequently, the dating of the extinction of a form or a group is not as simple a matter as may appear from the frequent citing in the paleontological literature of extinction dates for various organisms. [Author's translation.]

Put another way, it is a matter of observation, of empirical investigation, whether an organism is extinct or extant. And here there is an interesting difference between specimens and species. If a skeleton is found, it means a specimen is dead for good; and as long as a species (or other taxon) is known only from the fossil record, it is presumed to be extinct. But nothing prevents its "rising from the dead" the moment its living representatives are discovered. This having happened a number of times, there is nothing highly unusual or unscientific in expecting or predicting further occurrences of this kind.

What irritates critics is that some cryptozoologists are willing to acknowledge the possible survival today of dinosaurs and pre-sapiens hominids. Let us note, in this connection, that the wholesale dying out of the dinosaurs had intrigued paleobiologists long before the advent of modern cryptozoology, and that, despite many attempts, no fully satisfactory explanation of the phenomenon has yet been offered. Indeed, who can convincingly explain why such reptile forms as crocodiles, lizards, snakes, and turtles survive today, but dinosaurs—which, incidentally, were of different sizes and had the widest range of distribution—do not. Or, indeed, why gibbons, orangutans, chimpanzees, and gorillas are still living, but not *Gigantopithecus, Australopithecus, Homo erectus, or Homo neanderthalensis.* Survival of any of these forms thought to be extinct would actually solve, not pose, a problem. For we do know that they existed, and we can only speculate why they ceased to exist.

Thus, cryptozoology not only follows established scientific principles, but attempts to answer some topical questions confronting zoology today.

## Hidden Or Hiding?

If the subjects of cryptozoology are "hidden" animals, then many sciences have their "hidden" objects: astronomers attempt to detect unknown celestial bodies "hidden" in the vastness of space (some of which cannot even be observed optically); geologists search for minerals "hidden" in the earth; paleontologists search for "hidden" fossils, etc. Why, then, don't we have cryptoastronomy, cryptogeology, etc.?

Up to a point, the history of exploration in botany is remarkably similar to that in zoology. The age of discovery in both is far from over. In the eighteenth century, Linnaeus described some 7,000 species of higher plants. Today, botanists know over 350,000 species of higher plants, and, according to specialists at Moscow University, not less than 2,000 new species of lower and higher plants are described every year from all over the world. The world flora, just like the world fauna, is characterized by the coexistence of evolutionary old and new forms. "Living fossils" are known among plants as well, as, for example, *Metasequoia glyptostroboides*, which was first described by paleobotanists and later found alive. And botanists also often depend on the testimony of native peoples in botanical exploration and the discovery of new species. Yet, there is no cryptobotany, nor is there going to be. Why not?

The answer is too obvious to be always noticed. There can be spectacular plants "hidden" in the jungle (e.g., *Victoria regia*), but there are no plants *hiding* anywhere. The difference between zoology and all other disciplines engaged in the study of inanimate or non-animal objects is that the subjects of zoology have the *ability* to hide, and thus remain unknown or only vaguely known to science.

Two diverging behavioral strategies seem to have evolved to enhance the survival of species. Many species are highly open and visible—some even moving in herds and flocks—and usually have naturally high population densities; these species are generally well known to man. Other species, however, are less visible, and their population densities may be low; they survive because their secretive ways are adaptive.

For example, crows, pigeons, and sparrows are commonly observed by all, but it is unusual to observe an owl in the wild. Even more interesting are examples of these differing behavioral adaptations among related species: crow/jay; giraffe/okapi; baboon/orangutan.

Of course, we know little or nothing about the population densities or reproductive rates of so-called "cryptids" (i.e., the subjects of cryptozoology), but there seems to be little doubt that the larger ones must be among the rarest of living species. It is reasonable to suppose that giant octopuses are not as numerous as those of moderate size, that sauropods, if extant, are not as numerous as turtles, and that Sasquatches (whatever form of pre-sapiens hominid or hominoid the species may represent) are not as numerous as bears in any given area.

It is difficult to find a rare mineral, it is time-consuming to locate a rare plant, *but it is many times more difficult and time-consuming to discover a cryptid, not only because it is rare, but also because, in contrast with minerals and plants, it attempts to avoid detection.* To borrow Darwin's phrase, cryptids are "lingering in protected and isolated situations," such as deep waters, thick vegetation, or mountainous terrain. Being active and mobile, cryptids inevitably leave, from time to time, the depths of their sanctuaries and come into view of human beings. But they invariably retreat to their natural strongholds as soon as humans come in force to "discover" them. Hence, a cryptozoological situation: an animal is occasionally sighted by a local people or by visiting outsiders, but is absent from zoology's inventories of the Animal Kingdom.

That is the main answer to the question "why cryptozoology?" But there are also auxiliary ones, such as the difficulty of photographing elusive animals in the wild, the difficulty of retrieving specimens from remote areas, and the propensity of native peoples to attach certain beliefs to rare animals.

## Method In Madness

*Webster's Collegiate Dictionary* gives the following definition of science: "Science is systematized knowledge considered in reference to the discovery or understanding of truth."

First, it should be noted that the systematizing of knowledge and the discovery of truth are supposed to take time. Secondly, each discipline uses its own methods of discovering truth, methods that are best suited for its own particular purposes.

Let us spell out some common terms in cryptozoological research: eye-witness; testimony; identification; tracks; footprints; surveillance; reference files; photographs; laboratory analysis. Has all this been consciously borrowed from a criminal investigations department? Not at all. The remarkable coincidence of terms came about quite naturally and spontaneously because criminologists and cryptozoologists utilize similar methods.

The first order of business in cryptozoology is to collect and analyze the testimony of witnesses. There is no end to mirth, scoffing, and solemn warnings from critics in this connection. When preparing for an expedition in search of the Yeti, Charles Stonor (1955, 7) was given this advice by a zoologist: "Do remember that a Native [sic] will always tell you what he thinks will please you and what you want him to say." And here is a warning from none other than George Gaylord Simpson (1984, 12): "Humans are the most inventive, deceptive, and gullible of all animals."

It is noteworthy that such an attitude is generally evinced by individuals whose professions are not concerned with collecting and analyzing human testimony. Of course, all testimony is subject to doubt, but it is one thing to state that men are fallible—or that some people are fond of spinning a tale—and quite another to assert that no man is trustworthy, and no witness can tell the truth.

The role of the witness is often overlooked because it is taken for granted. To show that this kind of evidence is valid not only in court, I shall refer to the physicist's knowledge of a phenomenon called ball lightning. Since there is no generally accepted physical theory for this phenomenon, the reality of ball lightning is doubted by some physicists, who suggest that it is an illusion caused by the effect of a flash of ordinary lightning on the retina of the eye. To determine the truth, a team of Soviet physicists, headed by I. Imyanitov, has collected about 1,500 eyewitness reports of ball lightning worldwide. These reports, both published and solicited, were subjected to computer analysis. The Soviet government newspaper *Izvestia* reported, on September 30, 1976, what Imyanitov stated about their findings: "Collective portraits of the phenomenon made by eyewitnesses of different nations and at different times have coincided, and this in itself is telling proof of the reality of ball lightning." Similarly, in cryptozoology the process of "de-mythification" of evidence starts with the analysis of eyewitness accounts.

The next most important—and for terrestrial cryptids, inevitable—kind of evidence is tracks. This would seem to be perfect zoological material, but many zoologists simply do not accept tracks of unknown animals. Fortunately, on this subject cryptozoology finds support in paleontology, which is concerned, among other things, with the footprints of unknown, albeit extinct, animals. For example, Mossman and Sarjeant (1983, 75) state the following: "The skeletons of extinct land animals in museums suggest that the main fossil evidence for such animals is bones. This is not the case. The bones of the animals are very much rarer than their tracks. Indeed, many extinct land animals are known only from their tracks; their bodily remains appear never to have been preserved." The authors even find certain advantages of this kind of evidence compared with bones: "The animal's tracks and trails . . . are a much more extended and dynamic testament. Studying these clues to an animal's behavior, the paleontologist is able, as it were, to see the animal in action."

In cryptozoology, footprint evidence plays a decisive role in probing the reality and identity of the so-called relict hominoids (Yeti, Sasquatch, etc.). The best studied is the collection of footprints of the North American Sasquatch (Green 1978; Hunter with Dahinden 1973), which occasioned

Krantz (1972, 103) to state: "No matter how incredible it may seem that the sasquatch [sic] exists and has remained uncaught, it is even more incredible to believe in all the attributes of the hypothetical human track-maker .... Even if none of the hundreds of sightings had ever occurred, we would still be forced to conclude that a giant bipedal primate does indeed inhabit the forests of the Pacific Northwest." Demythification of the Sasquatch footprint evidence has been further advanced in subsequent works (Krantz 1983; Bayanov, Bourtsev, and Dahinden 1984a; Buckley 1984).

There is no denying that such footprints can be, and are known to have been, faked. However, the crucial question is not whether some tracks have been faked, but whether some are authentic. The latter question is answered by the investigators involved with a most emphatic "yes." And, to cite Napier (1973, 203), "if one track and one report is true-bill, then myth must be chucked out of the window and reality admitted through the front door."

A major "tool" of cryptozoology is photography, both still and motion picture. Again, the critics maintain that photographs and films are easily faked, and therefore cannot be accepted as evidence. And, again, these opinions come from individuals who do not vitally depend on photography in their professions.

A lot of work has been done on, and much information obtained from, the Patterson-Gimlin Sasquatch film, the study of which I was fortunate to participate in. We have subjected the film to a systematic and many-sided analysis. We have matched the evidence of the film against the other categories of evidence, and tested its subject with our criteria of distinctiveness, consistency, and naturalness. Our conclusion: "The Patterson-Gimlin movie is an authentic documentary of a genuine female hominoid, popularly known as Sasquatch or Bigfoot, filmed in the Bluff Creek area of northern California not later than October 1967" (Bayanov, Bourtsev, and Dahinden 1984b, 232).

Let me stress that, given the nature of cryptids, and barring their chance discovery, their *planned* discovery, which is the hoped-for intent of cryptozoology, is almost certain to be preceded by eyewitness, footprint, and photographic evidence. Thus, the existence of such evidence, and the study and evaluation of it, is a clear indication that cryptozoology is proceeding appropriately as a particular line of scientific inquiry.

## References Cited

Bayanov, Dmitri, Igor Bourtsev, and René Dahinden. 1984a. Eyewitness reports and footprints: an analysis of Sasquatch data. In *The Sasquatch and other unknown hominoids*, ed. Vladimir Markotic and Grover S. Krantz. Calgary: Western Publishers.

———. 1984b. Analysis of the Patterson-Gimlin film: Why we find it authentic. In *The Sasquatch and other unknown hominoids*, ed. Vladimir Markotic and Grover S. Krantz. Calgary: Western Publishers.

Buckley, Archie. 1984. Report on Sasquatch field findings. In *The Sasquatch and other unknown hominoids*, ed. Vladimir Markotic and Grover S. Krantz. Calgary: Western Publishers.

Darwin, Charles. [1859] 1929. *The origin of species by means of natural selection.* London: Watts.

Davitashvili, L.S. 1948. *Istoria Evolutsionnoi Paleontologii ot Darwina do Nashikh Dnei* [History of evolutionary paleontology from Darwin to our days]. Moscow-Leningrad: U.S.S.R. Academy of Sciences Publishing House.

Diamond, Jared. .1985. In quest of the wild and weird. *Discover*, March.

Green, John. 1978. *Sasquatch: The apes among us*. Seattle: Hancock House.

Heuvelmans, Bernard. 1982. What is cryptozoology? *Cryptozoology* 1: 1–12.

Hunter, Don, with René Dahinden. 1973. *Sasquatch*. Toronto: McClelland and Stewart.

*ISC Newsletter, The*. 1982. Formation of the society. 1(1): 2.

Krantz, Grover S. 1972. Anatomy of the Sasquatch foot. *Northwest Anthropological Research Notes* 6(1): 91–104.

———. 1983. Anatomy and dermatoglyphics of three Sasquatch footprints. *Cryptozoology* 2:53–81.

May, Robert M. 1984. Cryptozoology (Science Journals). *Nature* 307:687.

Mossman, David J., and William A. S. Sarjeant. 1983. The footprints of extinct animals. *Scientific American*, January.

Napier, John. 1973. *Bigfoot: The Yeti and Sasquatch in myth and reality*. New York: E. P. Dutton.

Simpson, George Gaylord. 1984. Mammals and cryptozoology. *Proceedings of the American Philosophical Society* 128(1): 1–19.

Stonor, Charles. 1955. *The Sherpa and the snowman.* London: Hollis & Carter.

# Hominology as Part of Cryptozoology, Its Theoretical Basis and Sources of Information*


## *Abstract*

*The paper begins with the origin of the term "hominology." "Relict hominoid" in this research actually implies "relict hominid." The author feels the need for a one-word term to denote the object of hominology, as simple and convenient as "cryptid" is in cryptozoology.*

*Relict hominids being cryptids, this makes hominology a part of cryptozoology. But it's a matter of fact that a scientific committee to study "humanlike hairy bipeds" was set up in Rome by Dr. Corrado Gini in 1962, 20 years before the International Society of Cryptozoology. The paper quotes Dr. Gini's words regarding the importance of the Committee's work.*

*Briefly discussed is Boris Porshnev's theory of the nature of hominids, as being the theoretical basis of hominology. The treatment of this theory in Russia and in the West is mentioned.*

*The unique specifics of relict hominids have caused the emergence and existence of hominology, even with the advent of cryptozoology. Hominology is the most "explosive" part of cryptozoology. It can also be described as a branch and sub-discipline of primatology, currently finding itself in a cryptozoological phase of development.*

***Hominology's Data Base***

### *Historic Evidence*

***Natural History.*** *Briefly presented and discussed is information about "hairy bipeds" in the works of Lucretius, Pausanias, Plutarch, Nizami al-Arudi, Ali Masudi, Johann Schiltberger, Linnaeus, and Bernheimer.*


---

* Presented at the Cryptozoological Conference "On the Track of Mysterious Animals" in Rome, Italy, March 27-28, 1999.

***Mythology and Folklore.*** *The evidence of wildmen is all around us in our culture. The role of "hairy bipeds" in mythology and religion is briefly discussed. Presented are views of theologian Luigi Maria Sinistrari (1632-1701) regarding succubi and incubi, called folletto in Italy, follet and lutin in France, and duende in Spain. According to legend, King Numa Pompilius used wine and honey to bait and capture such creatures, named Picus and Faunus, on Aventine Hill.*

***Ancient and Medieval Art.*** *Corrado Gini was a pioneer in collecting such evidence. Today's collection includes images of "hairy bipeds" across Eurasia, from Western Europe to China. Demonstration of slides showing such images.*

### *Modern Evidence*

***Sighting Reports:*** *This is the first and main source of information on the present-day existence of relict hominids. The paper mentions the works of such analysts of these data as Porshnev, Koffmann, Green, Bindernagel, and Magraner.*

***Footprint Evidence:*** *This allows the researcher not only to classify the track-maker as a hominid, but also to estimate the biped's size, and sometimes its weight. Mentioned are analysts of footprint evidence in Russia, China, and North America. Demonstration of slides showing Sasquatch tracks, and certain footprints depicted in petroglyphs in Russia.*

***Photography:*** *The Patterson-Gimlin film, showing a female Sasquatch in 1967 in northern California, is for the time being the most informative piece of photographic evidence in the whole field of cryptozoology. The author briefly relates the history of its study and shows slides with stills from the film.*

## Introduction

It all began with the Himalayan Yeti expeditions in the middle of the 20th century. They stimulated similar activity in Russia and North America, which is still in progress.

Having been nicknamed "Abominable Snowman," the Yeti gave us the popular name "snowman" in many languages: *snezhny chelovek* in Russian, *uomo delle nevi* in Italian, *homme des neiges* in French, *Schneemensch* in German, etc. But when it came to applying a scientific name, the problem proved to be formidable, because the nature of the snowman was not clear at all.

Scholars in the West began to use the term "bipedal anthropoid."

Anthropoid usually means anthropoid apes. Apes are quadrupedal. "Bipedal anthropoid" seems to be a contradiction in terms. So Russian researchers came up with the term "hominoid."

What they actually meant was "hominid," but at the time nobody dared to say out loud that pre-human hominids could be surviving. The word "hominoid" was used with the adjective "relict," meaning "remnant," or "surviving." Boris Porshnev's monograph of 415 pages, printed in 1963, was entitled *The Present State of the Question of Relict Hominoids*. He referred in it to "the emerging science of relict hominoids" (p.273).

Following Porshnev, I began to call this science "hominology," and defined "relict hominoid" as a higher bipedal primate different from *Homo sapiens*. I was and am still dissatisfied with the term "relict hominoid." I believe we need a one-word term, as simple and convenient as "cryptid" in cryptozoology. In my view, "homin" could fulfill this function. Thus, *hominology* is the science of *homins*, that is living hominids different from *Homo sapiens*. We need a special term for the objects of our research in order to distinguish them from both fossil hominids and modern man.

Since the International Society of Cryptozoology came into being in 1982, the terms "hominoid" and "hominology" have appeared on the pages of its cryptozoological journal and newsletters. So hominology has become part of cryptozoology, and that is quite natural and inevitable for the time being. Cryptozoology deals with cryptids, that is creatures unrecognized by zoology. And so does hominology: the Yeti, Sasquatch, Almas, Almasty, etc., are beings still unrecognized by zoology and anthropology.

But that is not all. There's much more to be said about the place of hominology within cryptozoology. As a matter of fact, an official scientific organization for activities, later termed hominology, was set up 20 years before the International Society of Cryptozoology. I mean The International Committee for the Study of the Humanlike Hairy Bipeds (Gini 1962), created, here in Rome by Dr. Corrado Gini, professor emeritus of the University of Rome. The committee included some 30 outstanding persons from different countries, among them Boris Porshnev, Bernard Heuvelmans, Ivan Sanderson, John Green, René Dahinden, Tom Slick, Roberto Dorion, Bob Titmus, Ralph Izzard, George Agogino, Charles Cordier, Osman Hill, Raymond Dart, Phillip Tobias, and Academician Rinchen (of Mongolia).

"Formation of the Committee," wrote Gini, "presupposes that the Snowman and other hairy bipeds present a subject worthy of a profound scientific study. The promoter of the Committee even thinks that this is a subject of the greatest importance for understanding the origin of man and the initial stages of human society." (Gini 1962, 4)

The journal *Genus*, published by Corrado Gini, printed many important articles by its committee members in the early 1960s. The Committee ceased to function because of the death of its promoter, not because it had failed as an enterprise. Had the institution continued to exist I am sure our situation today would be quite different.

As for hominology's theoretical basis, it was laid down by my teacher, Professor Boris Porshnev. In 1966, he published an article with a provocative title: "Is a Scientific Revolution in Primatology Possible Today?" (Porshnev 1962) The author answered the question in the affirmative. What's more, he predicted that such a revolution was not only possible but necessary and inevitable.

I have now to explain, very briefly due to lack of time, the essence of Porshnev's theory. Primatologists distinguish human and non-human primates. The latter are apes, monkeys, lemurs, tree shrews; the former are all of us humans. The two categories of primates are separated by a wide gap filled in with fossil forms. How can we know which of these fossil forms are human or non-human? What criteria should we use? Where should the line be drawn between the two categories?

When skeletal remains were found that looked more manlike than apelike, scholars, without much further thought, started labeling them "man." Thus such terms as Java man, Peking man, and Neanderthal man came into usage. Images of ourselves were projected into the unfathomed past and, once placed there, they began to be treated as facts of prehistory.

These fossil hominids made stone tools and used fire. Therefore they were certainly men, not animals, argued paleoanthropologists. Wrong! said Porshnev. The crucial difference between the animal world and the human world is not in the making and use of tools, but in the use of conceptual language, the capacity for speech. That is the Rubicon that separated human primate from non-human, biological evolution from social history. Language is the universal tool of the intellect, which made an increasingly complex technological and sociological development possible.

When did it happen? When was the language Rubicon crossed? Porshnev argued that it happened during the Upper Paleolithic and that the only bipedal primate that broke the language barrier, so to speak, was modern man, *Homo sapiens*.

This means that all primates, except *Homo sapiens*, are non-human primates. It also means that man did not descend from the ape. Directly, that is. Between man and ape there is an entire and most important stage of evolution—that of non-human bipedal primates, classified as hominids, but in Porshnev's classification re-named as the taxonomic Family of Troglodytes.

Undoubtedly, it's a most radical revision of the existing theory of hominid evolution and the origin of man. I have no time to substantiate and explain it in detail here, and refer those interested to our articles in the journal *Current Anthropology*. (Porshnev 1974; Bayanov and Bourtsev 1974, 1976).

What has Porshnev's theory to do with our subject?

The crux of the problem is that we are dealing with a creature having human shape but leading a non-human way of life. It's quite a puzzling object for modern science. Neither anthropologists nor zoologists have an idea or explanation for such a creature, but "Porshnevism" does. It not only explains the nature of this creature, but postulates its present-day existence as necessary and inevitable. Indeed, if the origin of *Homo sapiens* was so relatively recent and rapid, it follows that our non-human bipedal ancestors could not have completely died out. They simply had not enough time to do so. Initially, they greatly outnumbered humans; today humans greatly outnumber them. But fortunately the process of their extinction, like that of the great apes, hasn't run to its conclusion. These bipedal primate relicts are still with us on the planet; they still survive.

How do we relate them to the fossil record? Porshnev laid special stress on the survival of Neanderthals, because, according to his theory, they were the latest form of non-human primates preceding modern man, *Homo sapiens*. The evidence of "wildmen" of ancient and medieval Western Europe seems to give support to this idea.

But Porshnev's theory does not rule out the survival of other fossil forms either, such as the so-called *Homo erectus* and *Australopithecus*. On the contrary, their survival may be as natural as that of gorillas, chimpanzees, and orangutans.

How has the Porshnev theory been treated by the scientific community? Alas, in the worst possible way. In Russia, anthropologists met his innovative ideas with a conspiracy of silence, while zoologists attacked him personally in a vicious way in academic journals. When publication of Porshnev's book, *On the Beginning of Human History*, was blocked at the last moment, the father of hominology died of a heart attack.

In the West, Porshnev's theory was published, with my comments, in *Current Anthropology* 25 years ago, but again it is surrounded by silence. When researchers come to the same conclusions as Porshnev did, they never mention his name and give him priority. What's worse, a prominent scientist, the late George Simpson, giving a negative account of cryptozoology, singled out Porshnev's ideas for his most contemptuous remarks, calling them "a succession of one blunder after another" and "downright silliness." (Simpson 1984)

The scholar in the West who has understood and accepted at least one aspect of Porshnevism is anthropologist Grover Krantz. In the manuscript of his article, sent over to me for possible publication in Russia, he wrote in 1976:

> Something very fundamental happened between the time of the Neanderthals and the rise of modern man. There is no confusion between the skulls of the two types. ... I must agree with the late Boris Porshnev that no matter how intelligent and resourceful Neanderthals were as individu als, they were still on the other side of the critical line separating animals and man. If one grants this point, then it makes less difference than might be supposed just which fossil form, or forms, might be surviving today. All suggested candidates for this position are prehuman, bipedal primates which differ from each other merely in their degree of intelligence and tool-making skills.

The "something very fundamental" that happened with the rise of modern man was, according to Porshnev, human language, which separated the human and animal worlds.

Such, in brief, is the theoretical basis of hominology. But is the discipline really necessary? Isn't cryptozoology sufficient to cover all enigmatic animals, including relict hominids? No, even with the advent of cryptozoology, I presume that hominology is not only necessary but inevitable. The reason is this. As soon as a cryptid is recognized by zoology, it stops being a cryptid and falls under the auspices of one or another branch of zoology. Such was the case with the okapi, the giant squid, the gorilla, and numerous other ex-cryptids. But relict hominids have no discipline to get into, except hominology. Anthropology has never dealt with a primate having human shape but leading a non-human way of life. Zoology has never dealt with a beast in human form. Between zoology and anthropology there is thus a big gap in knowledge, and hominology is called upon to fill in this gap.

The discipline does exist, even if only *de facto*. The task ahead is to get its *de jure* recognition. But when this happens, hominology will say goodbye to cryptozoology and the two disciplines will separate. Thus, we can say that hominology, being a branch and sub-discipline of primatology, currently finds itself in a cryptozoological phase of development.

This marriage of hominology and cryptozoology is only temporary, but not less significant for that. Cryptozoologists are interested in all sorts of cryptids, from snakes to snowmen. As a hominologist, I may be biased and unfair when I say: all cryptids are equal, but some are more equal than others. Hominology is heralding a revolution in the natural sciences perhaps only second in importance to that of the Darwinian revolution, of which it is a continuation. The rumbles and repercussions of this revolution are bound to entail reforms and revisions in quite a number of humanitarian disciplines as well.

So I view hominology as the most explosive, most enlightening, most philosophical, and most glorious part of cryptozoology. My only regret is that I haven't so far been able to get this view across to cryptozoologists, and the tremendous potential of hominology has not yet been properly used for the good of both hominology and cryptozoology.

## Historic Evidence

As the saying goes, there is nothing so practical as a good theory. At the birth of hominology we had only vague Yeti stories. But as soon as Porshnev offered his theory, some very important conclusions followed from it regarding the practical collection of information about relict hominids. If the separation of human and non-human hominids was so relatively recent, if these relicts are still surviving in historic times and even today, then all or almost all peoples in the world must retain the knowledge or at least the memory of such creatures. The names and images of them must be present in mythology and folklore, along with the names and images of other living beings.

If non-human hominids were surviving in historic times, then ancient and medieval travelers and natural history authors must have mentioned them in their travelogues and natural history books.

If prehuman hominids are surviving today, then inevitably there must be sightings of them, finds of their footprints, and even photographs.

And that is the case. All these areas of knowledge have proved to be viable sources of information for hominology, and have been successfully tapped by us since the birth of Porshnevism.

I am going to illustrate this with examples from our collections. Let us begin with natural history.

## Natural History

A celebrated source here is Titus Lucretius Carus (1st century B.C.), who in Book 5 of his famous *De Rerum Natura* (Lucretius 1947) described a race of wildmen who had very strong bodies, covered with hair, who lived in woods and caves; who had no language, nor clothes, nor any industry; who hunted animals with sticks and stones, and who ate meat and other foods raw.

Lucretius said that such was the condition of man in prehistory, long before the rise of civilization. Modern scholars wonder how the ancient philosopher could have fathomed the appearance and life of prehistoric man.

Let us note that Lucretius also wrote that these wildmen were initially literally "born by the earth," and accordingly applies to them the epithet "terrigenas" (earthborn). So we see a marvelous piece of truth, on one hand, and a tall tale, on the other. How could it be?

Hominology gives the answer: the ancient philosophers used relict hominids as models for the portrayal of prehistoric man. It was not much more difficult than looking at some wild animals to guess the origin of their domestic relatives. The popular names of relict hominids in the Greco-Roman world were satyr, silenus, faun, pan, etc. Lucretius did not use these appellations in this context, probably not wishing to mix natural history with folklore and mythology.

But other ancient authors did use these names in a down-to-earth manner. Thus, Pliny the Elder said in his *Natural History* that "Satyris praeter figuram nihil moris humani" ("the Satyrs have nothing of ordinary humanity about them except human shape"). (Pliny, 5.8) The hominologist readily underwrites these words, fully understanding what they mean.

Geographer Pausanias (2nd century A.D.), in his *Description of Greece*, said that when satyrs grow old, they are called sileni. He also remarked that the Silenus race must be mortal, since their graves are known. Pausanias confided that he was very eager to learn something about the nature of satyrs and for this purpose would meet and interview many sailors returning from voyages. This shows that already in ancient Greece relict hominids had become enigmatic creatures.

This is confirmed by Plutarch, who told of an actual capture of a satyr by the soldiers of the Roman general Sulla in the territory of modern Albania, in the year 86 B.C. The satyr was brought to Sulla and "interrogated in many languages as to who he was, but he uttered nothing intelligible; his accent being harsh and inarticulate, something between the neighing of a horse and the bleating of a goat." The general "was shocked with his appearance and ordered him to be taken out of his presence." (Plutarch 1792, 349)

Greco-Roman naturalists also used the word "troglodyte" (cave-dweller) to denote bipeds different from humans. The term was usually applied to wildmen reported beyond Europe—in Africa, the Middle East, and the Caucasus. Among characteristics of troglodytes that were stressed were their great running speed, lack of speech, and strange vocalization.

From the Middle Ages an important piece of information comes from the Persian scholar Nizami al-Arudi (12th century A.D.). In his book, *Chahar maqala,* he said that the lowest animal is the worm and the highest is Nasnas, "a creature inhabiting the plains of Turkistan ... This, after mankind, is the highest animal, inasmuch as in several respects it resembles man: first in its erect stature, secondly in the breadth of its nails, and thirdly in the hair on its head."

In addition, Nizami said that Nasnas is curious about man, that it can kidnap a lonely human being, and is even said "to be able to conceive by it." Nizami al-Arudi is a precursor of Linnaeus and Darwin, for he used Nasnas to fill in the gap between mankind and the animal world.

Interesting information comes from medieval Arab travelers who, being well aware of the existence of monkeys, often compared wildmen not only to man, but also to monkeys. Thus, an Arab traveler who visited the Caucasus in the 10th century A.D. wrote that forests and jungles there "are inhabited by a sort of monkey having an erect stature and round face; they are exceedingly like men, but they are all covered with hair .... They are deprived of speech .... They express themselves by signs." (*Ali* 1841, 440)

This author, Abul Hassan Ali Masudi, also mentioned the existence of "monkeys that approach in appearance the figure of man" in the northern regions, forests, and jungles, "in the country of the Sclavonians [i.e., Slavs—D.B.] and other nations."

Johann Schiltberger, a native of Bavaria, was taken prisoner by the Turks in the 15th century and sold to the Khan of Siberia. After 30 years spent in Asia, Schiltberger returned home to Bavaria and in his book of travels described:

> ... savages, who are not like other people.... They are covered all over their body with hair, except the hands and face, and run about like other wild beasts in the mountains, and also eat leaves and grass, and any thing they can find. The lord of the country sent to Edigei [another ruler—D.B.] a man and a woman from among these savages, that had been taken in the mountain. (Schiltberger 1871, 35)

The wildman's presence in medieval Europe is well documented by Richard Bernheimer in his book, *Wild Men in the Middle Ages*. Having studied a great number of original medieval sources, the author came to the following conclusion:

> About the wild man's habitat and manner of life, medieval authorities are articulate and communicative. It was agreed that he shunned human contact, settling, if possible, in the most remote and inaccessible parts of the forest, and making his bed in crevices, caves, or the deep shadow of overhanging branches. In this remote and lonely sylvan home he eked out a living without benefit of metallurgy or even the simplest agricultural lore, reduced to the plain fare of berries and acorns or the raw flesh of animals. (Bernheimer 1952, 9)

Let us note how similar this is to what Lucretius wrote about "earth-born" men and what Schiltberger wrote about hair-covered savages found in Siberia.

Unlike scholars of ancient and medieval times, who knew full well that wildmen, even if called satyrs or fauns, were an integral part of Nature, educated Europeans of the enlightened 18th century usually took them for "feral man," i.e., humans who had gone wild as a result of some misfortune. Such misidentification of relict hominids is partly responsible for their remaining beyond the grasp of science for so long.

The best example in this category is the so-called boy of Kronstadt, captured and held in captivity in Rumania in the 1780s. A German witness gives a most detailed description of this individual, both his anatomy and behavior. He was covered with hair, was devoid of speech, and had Neanderthal morphological features. (Singh and Zingg 1942, 237–40)

It is a curious fact of anthropology that its basic term—*Homo sapiens*—owes its origin to the existence of troglodytes. It is generally believed that the term was coined to distinguish modern man from extinct forms known from the fossil record. That is not so. The term *Homo sapiens* was introduced by Linnaeus in the middle of the 18th century, a hundred years before Darwinian theory and the systematic study of hominid fossils. Linnaeus had information about the existence of primate bipeds, hairy, speechless, non-sapient, and for the sake of contrast with them he designated our own species with the rather wishful term "sapiens." Accordingly, in the 10th edition of his *Systema Naturae*, published in 1758, Linnaeus described two species of man: *Homo sapiens*, defined as "diurnus," and *Homo troglodytes*, defined as "nocturnus" and "sylvestris."

After Linnaeus, less knowledgeable scholars crossed out *Homo troglodytes* from the taxonomy of primates and closed the question for two hundred years. It is only in our time, especially thanks to Boris Porshnev, that Linnaeus has been found to be right in proclaiming that *Homo sapiens* is not the sole living hominid on our planet.

Such is my very brief account of our evidence derived from natural history.

## Mythology and Folklore

Relict hominids are different from all other cryptids not only in anatomy and behavior, but also in the place they hold in man's culture. I dare say there is no other living creature, except man himself, that figures so prominently in religion, mythology, folklore, and the arts.

And yet, in modern European culture, this creature is believed to be non-existent. How can this be? How to explain this paradox?

Cryptids are usually hidden in forests, mountains, lakes, and oceans. The object of our research is also hidden in natural forests and mountains, but above all it is hidden in the "forests of the mind." If not for these forests the problem would have been solved long ago.

Cryptozoologists know very well what to do with animals found in folklore and mythology. By special biological analysis they attempt to demythologize such animals and uncover their biological features. Why then have relict hominids not been demythologized long ago? First, because they have been so highly mythologized. Second, because before Darwin and Porshnev there was no adequate biological theory to perform such work. That is why the evidence of travelers and naturalists of the past has been ignored by modern biology. In order to discover relict hominids, biologists have first to discover Porshnevism. They will then get out of their dark mental forests into a sunlit clearing and see that the evidence of wildmen is all around us in our culture.

In the hoary past, when humans were a minority confronted by an awesome preponderance of non-human bipeds, they had no choice but to find a *modus vivendi* with the non-humans. The latter ruled supreme in the environment. So humans offered a part of their hunting trophies to non-humans in order to placate them and so be allowed to hunt and gather food in territories dominated by non-humans. As this process went on, non-human bipeds became viewed as lords of nature and eventually worshiped as heathen gods. Food offerings to placate them turned into religious sacrifices.

"Div" is an ancient word in the Indo-European languages to denote a wild hominid. Etymologically related to "div" are the words "Deus" and "Divus" in Latin, "Zeus" in Greek, and "divine" and "divinity" in English. Let us note that according to Greek mythology, Zeus was born in a cave.

With the advent of major religions, such as Judaism, Christianity, and Islam, the heathen gods and their hominid prototypes were condemned and relegated to the status of demons. They were condemned not only on account of their beastly appearance, but also and mainly because of their beastly behavior, especially their inability to abide by human standards in sexual relations. According to the ancient historian Diodorus Siculus, satyrs "shamelessly seek interbreeding" with humans. As a result, European languages have acquired such ancient medical terms as "satyriasis" and "nymphomania."

Christianity won the day initially in the cities, while in villages and the countryside heathen cults went on to be observed clandestinely throughout later history. In Catholic countries it was the business of the Inquisition to fight such heathen cults and heresies. In this connection of special interest is the work by Italian theologian, Luigi Maria Sinistrari (1632-1701), jurisconsult of the Inquisition's High Tribunal in Rome.

Sinistrari argued that it was necessary for the Inquisition to distinguish between culprits who associated with real demons, and people who fell victim to certain manlike animals, mistaken for demons. Accordingly, Sinistrari's work has a long and instructive title:

*On demonism and the animals, incubi and succubi, where it is proved that there are reasonable creatures on earth, apart from man, which have like man a body and a soul, which like man are born and die, which are redeemed by our Savior Jesus Christ and capable of salvation and damnation. (Sinistrari 1875)*

Sinistrari mentioned popular names of these animals, such as *folletto* in Italy, *follet* and *lutin* in France, and *duende* in Spain.

His main argument for why these beings are animals, not evil spirits, is this: they are immune to exorcism. It happens, he wrote, that they "meet exorcism with a grin," or "even beat up exorcists and tear up sacred clothes." Hence, it is clear they "are not evil spirits or angels, nor are they human beings, even though they are endowed with reason."

Further biological traits of these animals, pointed out by Sinistrari, are the following:

- they seek sexual intercourse with humans;
- from such intercourse children are born who, when grown up, become very tall, strong and daring;
- these animals' vocalization resembles whistling;
- these animals are attracted by horses and like to plait their manes;

- these animals throw stones and pile them up;
- it is very difficult to see these animals, they are seen either by chance or of their own volition;
- they are capable of feeling and suffering, but being very swift and nimble in avoiding danger, it is surprising that they get killed or injured at all. This can happen when they are asleep or in some other inadvertent way.

Thus we see that the biological nature of the creatures, regarded as evil spirits by some and as figments of the imagination by others, was apparent to an open-minded theologian.

As for folklore proper, it is a very abundant and fascinating source of information for hominology, to which I devoted one of my books. I learned from folklore certain things about relict hominids that I had never heard from eyewitnesses and took them to be tall tales. I mean, for example, wildmen's relationship with wild dogs. To my amazement, this point was later confirmed by eyewitness evidence in America. Humans have "best friends" among domestic dogs, wildmen have theirs among wild dogs.

We learn from folklore some original methods of capturing hairy bipeds. Profiting by their interest in horses, it is recommended to put some sticky stuff, such as tar for example, on the back of the horse that the wildman likes to ride: his hair will stick to the tar on the horse and the creature won't be able to escape.

Usually, though, folklore recommends intoxicating a wildman, by offering him wine or vodka, and when he gets tipsy and asleep, taking him prisoner. Incidentally, King Numa Pompilius is said to have captured two woodland creatures by this method on Aventine Hill, right in the territory of what today is Rome. King Numa baited them with wine and honey. One creature was called Picus ("woodpecker"—I suppose from the habit of knocking on trees, a common trait of relict hominids), the other was called Faunus. Both are said to have offered strong resistance when being captured.(Tolstoy 1966, 97–114)

There are interesting explanations in folklore of the origin of those whom we define as relict hominids. Most explanations are legendary and poetic, but some are quite pragmatic and realistic. It is also noteworthy that popular names for wildmen in folklore are often the same as for apes and monkeys, even though apes and monkeys are not native to a particular country or continent, such as, for example, Russia or North America.

We know from folklore and mythology that man-wildman relationship throughout millennia has been one of a love-hate kind. The lords of nature have been deified and condemned, offered sacrifices and hunted as valuable game, both for food and medical purposes. We also know that separate specimens have been captured, tamed, and exploited as warriors, hunters, and unskilled workers. Why then have relict hominids not been turned into slaves or a kind of most sophisticated domestic animal? Wolves have always been man's enemies, but transformed into dogs have become man's best friends. Why hasn't this happened with pre-human hominids?

I think the answer is obvious: genetically they are so close to humans that they tend to interbreed with our kind. But, unlike human slaves, they are unable to understand and obey the ban on such interbreeding.

Closing this brief discussion of mythology and folklore as a source of our information, I'd like to mention some more words and notions in European languages derived from the existence of non-human hominids, such as "giant," "genius," "Egypt," "fauna," "panic," and "Pan" (the generic name of the chimpanzee).

## Ancient and Medieval Art

The evidence of mythology and folklore is paralleled and supplemented by that of ancient and medieval art. A pioneer in collecting such evidence was Professor Corrado Gini, who, in 1962, published images of a hairy bipedal primate depicted on a bowl of Carthaginian or Phoenician origin, dated 7th century B.C., found among the treasures of a Roman villa in Palestrina, Italy. (Gini 1962, fig. 5)

Today, the collection includes scores of images from across Eurasia. It presents two kinds of portrayal: realistic and "ritualistic," i.e.. symbolic. The first is true to life and helps the hominologist to study the creatures' appearance and anatomy. They show hairy bipeds with certain typical features setting them apart from humans.

Symbolic portrayals may be a virtual caricature that shows not so much the real object as the artist's attitude to it. Images of grotesque monsters in ancient and medieval art have therefore led scientists and art specialists into believing that these monsters were pure figments of the imagination and that nothing real stood behind them. Hominology puts an end to this self-deception.

## Modern Evidence

Relict hominids are animals, or rather superanimals, whose distinctive characteristics are manlike form and animal way of life. These super-animals are very cautious, very cryptic, and more intelligent than any animals. In order to find food and survive, they have to be very mobile and distributed very sparsely over vast territories.

One of the oldest known symbolic portrayals of the hominoid from an illustraton of the Gilgamesh epic. Babylonian King Gilgamesh fighting a bull with the help of Enkidu, according to the epic, a tamed hairy wildman depicted here as half man, half beast, by using the attributes of horns, hoofs and a tail. The symbolic character of these appendages is suggested by the fact that none of them are referred to in the epic's text. Imprint of a Babylonian cylinder seal, 3rd millennium B. C.

Depiction of hairy hominids on a bowl of Carthaginian or Phoenician origin, dated 7th Century, B.C. A distinct difference between humans and hominids is clearly seen here.

A hirsute couple in the Kelermes mirror of gold and silver dug up from a Scythian tomb (VII-VI centuries B.C.) near the village of Kelermes in the North Caucasus. The presence of a female hominoid is significant. Devoid of clothes, both male and female sport a protective coat of hair, just like the other wild animals of the Caucasian fauna depicted by the artist in the treasure piece.

As nimbuses serve to identify divine persons in Christian art, so horns, hoofs and tails indicate heathen gods of hominoid origin in ancient art. Here is an image of the ancient Greek god Pan, patron of herdsmen, hunters, beekeepers and fishermen (circa 100 B.C.) The word "panic" originates from the great fear experienced by people stumbling on the 'god.' When heathen gods were turned indo demons, their distinguishing features were also transferred to the latter.

Sculptured portrayal of silenus found in excavations of Nymphaion, an ancient Greek colony town in the Crimea. The term "silenus" in ancient Greece meant "old satyr." The portrayal is remarkable for it combines certain features (low cranium, bulging brow ridges, prominent cheekbones, deeply sunk nose bridge) typical both of fossil man and relict hominoids as described by witnesses.

Fragment of an old Russian icon from the collection of the Museum of the Moscow Kremlin. Called *The Virgin Bogolyubskaya with scenes from the lives of Sts Zosima and Savvati*, it was painted in 1545 at the Solovetsky Monastery in the north of Russia. The scene shows the hermits Zosima and Savvati being tempted by the devil. The latter is portrayed in the image of a shaggy biped better known today by the name of 'snowman.' The depiction of such creatures in ancient and medieval art is ubiquitous and very instructive. Deified and worshiped as lords of nature in heathen times, they were subsequently condemned and turned into demons by the major religions—Zoroastrianism, Judaism, Christianity and Islam. The demonological and religious connections of snowmen, alias relict hominoids, have for ages camouflaged their true nature and prevented science from investigating the question in earnest.

A traditional Persian style illustration in the epic *Iskander Namah* by Nizami. The epic's hero, Iskander, lassoes and captures a div whose image is far removed from biology into devilry.

A drollery from Queen Mary's Psalter, 14th Century, British Museum, London. Wildmen have always had problems with dogs.

A wildman family at leisure. Drawing by Hans Dürer (1478-1538).

Sculpture of a wildman. Church in Bretagne, France, 14th Century.

Wildman. Wood sculpture, 15th Century, located at Saint-Tugdale Cathedral, Treguier, Bretagne, France.

Wild woman with a baby. Sculpture, 15th Century, located at the Church of Ambierie, Forez, France.

Knight vs. wildman; two illustrations in the medieval epic *Valentine and Orson*, Lyone edition, 1605.

*Fight in the Forest* by Hans Burgkmair (1473-1531). Judging by this picture, a burly European wildman could be a match for any American Bigfoot.

A wildman said to live in the mountains. This image in a textbook of Buddhist medicine, entitled *The Anatomical Dictionary*, was discovered in Mongolia in 1959. The accompanying text says that the creature's body resembles that of a man and he has enormous strength. His meat may be eaten to treat mental diseases and his gall cures jaundice.

One of the most picturesque homin portrayals of the symbolic series. It's a Percian div, named Akwan, a formidable foe of the king in the ancient epic *Shah Namah* by Firdausi. The artist did his best to depict the hominid as a very "bad guy", the incarnation of a beastly demon.

Under these conditions, it's quite natural and logical for the present-day evidence of relict hominids to appear first of all in the form of sightings, footprints, and photographs rather than in the form of hides, bones, and bodies. And that's exactly what we see in practice.

## Sighting Reports

In the short history of hominology they are the first and main source of information on the present-day existence of relict hominids. In Russia hominid sightings have been analyzed most thoroughly, first of all by Boris Porshnev. His approach to the study of such evidence can be seen from his following words:

> It is precisely the use of non-traditional methods, such as the comparative analysis of mutually independent evidence, that has made it possible to establish the existence of this relic species and to describe its morphology, biogeography, ecology, and behavior. In other words, fact-finding methods have been used in biology that are usually employed by historians, jurists, and sociologists. This indirect research into the problem of relic Paleanthropus is now considerably advanced. (Porshnev 1974, 450)

Hundreds of sighting reports in the Caucasus have been collected and analyzed by Marie-Jeanne Koffmann, who, on this basis, presented a paper to the Vancouver conference in 1978 entitled, "Brief Ecological Description of the Caucasus Relic Hominoid (Almasti)." (Kauffmann 1984, 76–85) She also published detailed articles, based mainly on sighting accounts, in the French journal *Archeologia*. (Kauffmann 1991, 24–43; 1992, 52–65)

I also devoted one part of our report at the Vancouver conference to the discussion of eyewitness evidence. (Bayanov, Bourtsev, and Dahinden 1984)

Jordi Magraner has collected sighting accounts and written a work on this basis, devoted to relict hominids reported in north-western Pakistan. (Magraner 1992)

In North America, John Green's books are based mainly on sighting evidence, which he has now analyzed and classified with the help of computer processing. (Green 1978)

The latest most useful contribution in this category is the recent book by Canadian wildlife biologist John Bindernagel. Regrettably, following John Green, the author calls the North American relict hominid an "ape" in the book's title. (Bindernagel 1998)

## Footprint Evidence

Footprints are the second most important kind of evidence of the present-day existence of relict hominids. Tracks are traditionally regarded as perfect zoological material, but zoologists haven't been in a hurry to investigate the tracks of this purported new species of animal. The reason is obviously the shape of the tracks, which the zoologist finds to be outside the limits of his profession. Indeed, photographs and plaster casts of footprints under study in Russia, China, and North America clearly indicate that we are dealing with fully bipedal creatures, i.e., representatives of the hominid, not pongid, family of the higher primates.

Footprints allow the researcher not only to classify these primate cryptids as hominids, but also to estimate the track-maker's size—and sometimes weight (by the depth of impressions).

Having in general the shape of human footprints, relict hominid tracks possess a number of specific features that differentiate them from humans'. These features have been studied and discussed in the works by Boris Porshnev, Marie-Jeanne Koffmann, Dmitri Bayanov, Igor Bourtsev, Zhou Guoxing, John Green, and some others. Anthropologist Grover Krantz has devoted a whole volume to this subject. (Krantz 1992)

Bigfoot tracks (each 37 cm long) photographed by John Green on a mountain road in Northern California (compare the size with boot prints). Compared to human, they reveal not only a different type of foot (flat and enormous), but also a different type of walk. A man normally walks with his feet pointing outward, while Bigfoot strides as if following an invisible line, his feet pointing straight ahead or even turned in a little.

Wildman tracks, with the figure of the track-maker, are depicted in petroglyphs, discovered in 1926 in Karelia (north-western Russia). They show typical features of relict hominid footprints. (Savvateyev 1967)

## Photography

The Patterson-Gimlin documentary is the third most important proof of the current existence of relict hominids. The film was taken in the mountains of northern California in October 1967 by Roger Patterson and Robert Gimlin. American scientists, ignorant of hominology, wrote it off as a hoax. In December 1971, René Dahinden, of Canada, brought the film to Moscow where it was studied in depth by a team of researchers, and found to be authentic as early as 1972.

This conclusion was published in René Dahinden's book *Sasquatch* in 1973 and repeated in our report at the Vancouver conference in 1978 (Bayanov, Bourtsev, and Dahinden 1984). Detailed information about these events is available in my book *America's Bigfoot: Fact, Not Fiction—U.S. Evidence Verified in Russia.* (Bayanov 1997)

In 1992, Grover Krantz published his analysis of the Patterson-Gimlin film, fully confirming our positive conclusion. (Krantz 1992) Thus, after thorough scientific study, the film has been found genuine on both sides of the Atlantic. Regrettably, this fact is still being ignored by the scientific community, while desperate attempts are being made to debunk the documentary by fake analysts whom I call "glory scavengers."

The Patterson-Gimlin film remains for the time being the most informative piece of photographic evidence in the whole field of cryptozoology.

---

In the past I used to say, half-jokingly, that "relict hominids are unknown to science because there is no science to know them." Today such a science does exist, and is well aware of their existence. The task now is to share this knowledge with the scientific community.

## References Cited

*Ali Masudi's historical encyclopaedia, entitled meadows of gold and mines of gems.* 1841. Vol. 1. London.

Bayanov, Dmitri. 1996. *In the footsteps of the Russian snowman.* Moscow: Crypto-Logos.

Bayanov, Dmitri. 1997. *America's Bigfoot: fact, not fiction: U.S. evidence verified in Russia*. Moscow: Crypto-Logos.

Bayanov, Dmitri, and Igor Bourtsev. 1974. Reply (to Comments). *Current Anthropology* 15(4): 452–456.

———. 1976. On Neanderthal vs. paranthropus. *Current Anthropology* 17(2): 312–318.

Bayanov, Dmitri, Igor Bourtsev, and René Dahinden. 1984. Eyewitness reports and footprints: An analysis of Sasquatch data. In *The Sasquatch and other unknown hominoids*, ed. Vladimir Markotic and Grover Krantz, 176–185. Calgary: Western Publishers.

Bernheimer, Richard. 1952. *Wild men in the Middle Ages*. Harvard University Press, Cambridge.

Bindernagel, John. 1998. *North America's great ape: The Sasquatch*. Courtenay, BC: Beachcomber Books.

Gini, Corrado. 1962. Vecchie e nuove testimonianze o pretese testimonianze sulla esistenza di ominidi o subominidi villosi. *Genus* 18(1–4).

Green, John. 1978. *Sasquatch: The apes among us*. Seattle: Hancock House.

Hunter, Don, with René Dahinden. 1973. *Sasquatch*. Toronto: McClelland and Stewart.

Koffmann, Marie-Jeanne. 1984. Brief Ecological Description of the Caucasus Relic Hominoid. In *The Sasquatch and other unknown hominoids*, ed. Vladimir Markotic and Grover Krantz, 76–85. Calgary: Western Publishers.

———. 1991. L'Almasty, yeti du Caucase. *Archeologia*, June.

———. 1992. L'Almasty du Caucase, mode de vie d'un hominide. *Archeologia*, February.

Krantz, Grover L. 1992. *Big footprints: A scientific inquiry into the reality of Sasquatch*. Boulder, CO: Johnson Books.

Lucretius, Titus Cams. 1947. *De rerum natura*. London: Oxford University Press.

Magraner, Jordi. 1992. *Les hominoides reliques d'Asie Centrale*. [Paris?] Editions Association Troglodytes.

Pliny the Elder. *Natural History*.

Plutarch. 1792. *Plutarch's lives*. London: Silly.

Porshnev, Boris. 1966. Is a scientific revolution in primatology possible today? (In Russian). *Voprosy filosofii* (Questions of philosophy), no. 3: 108–119.

———. 1974. The troglodytidae and the hominidae in the taxonomy and evolution of higher primates. *Current Anthropology,* 15(4): 449–450.

Savvateyev, Yu.A. 1967. *Risunki na skalakh* (in Russian) (Drawings on rocks). Petrozavodsk.

Schiltberger, J. 1879. *The bondage and travels of Johann Schiltberger, a native*

*of Bavaria, in Europe, Asia, and Africa 1396-1427*. London.
Simpson, G.G. 1984. Mammals and cryptozoology. *Proceedings of the American Philosophical Society* 128(1): 15.
Singh, J.A.L., and Robert M. Zingg. 1942. *Wolf-children and feral man*. New York and London: Harper.
Sinistrari, L.M. 1875. De la demonialité et des animaux incubes et succubes .... Translated from Latin into French by Isidore Liseux. Paris.
Tolstoy, I.I. 1966. *Stat'i o folklore* (in Russian) (Articles about folklore). Moscow and Leningrad: Nauka.

# Hominology in Russia

## Overview of Field Investigations

(Presentation at the Willow Creek, California, Bigfoot Symposium, September 2003)

Hominology was born in Russia in the middle of the 20th century, mainly thanks to two factors: the Anglo-American Himalayan expeditions after the Yeti, and the ideas and work of Professor Boris Porshnev. When the news of the Yeti expeditions reached all corners of the Soviet Union, the media and the Academy of Sciences began to receive letters from people saying they had seen wild, hairy man-like creatures, which received the adopted overall name of "snowmen" (*snezhny chelovek* in Russian).

On Porshnev's initiative, the Academy of Sciences formed a Commission to investigate the snowman question. As there were reports from the Pamir mountains in Tajikistan, in Central Asia, not too far from Tibet and the Himalayas, it was believed that the Yetis may wander over to Tajikistan or even inhabit it. So a state-funded expedition was sent there in 1958 to search for the snowmen. A scientist working in Tajikistan, Professor Staniukovich, a botanist by profession, was appointed leader of the expedition. Porshnev was his deputy and Marie-Jeanne Koffmann was the expedition's doctor.

The expedition consisted of two teams, one supposed to engage in field-work, and another to gather local eyewitness accounts and folklore. The members of the first team had a very vague idea what they were searching for, so they were also collecting plants for their botanist leader who was preparing botanical maps of the Pamirs flora. After several months of work, the first team came back empty-handed as regards the official goal of the expedition, the second team brought back a thick file of snowmen tales.

Reporting on the results of the expedition at the Academy session, Staniukovich alleged that as the very thorough search had revealed no trace of snowmen in the Pamirs, it was clear the creatures were absent there. Koffmann totally disagreed with that verdict, while Porshnev said the following: "We were obviously not prepared to interrogate nature prior to querying people who have lived for generations in that environment." He meant the fruitful results of the second team, one that collected sighting accounts and folklore, and he stressed the necessity of continuing such work.

Let me note that evidence from Tajikistan included accounts not only of the natives, but also of Russians and newcomers, such as general Topilsky, geologist Zdorik, hydrologist Pronin. The combination of evidence by native and non-native informants is a most important indication of reliable information.

The negative report by Staniukovich put an end to the Academy's interest in the snowman problem. Porshnev's numerous opponents among anthropologists and zoologists used the failure of the Pamirs expedition to declare the snowman problem pseudoscientific and abolish the Snowman Commission. However, in the three years of its existence, the Commission had done a good job of collecting modern and historical information.

After that, in 1960, the chief curator of the Darwin Museum in Moscow, Pyotr Smolin, organized a permanent seminar at the museum that became and remains the theoretical and organizational center of hominology in Russia.

The focus of fieldwork then shifted to the Caucasus, which is not as distant from Moscow as Tajikistan. At first hominologists could hardly believe that homins could survive in that region, much traveled by tourists, mountaineers and holidaymakers. Koffmann later wrote: "When I left for the Caucasus to verify the first reports that had reached us, I considered the possibility of wild men living there to be ridiculous .... It took a long time and hundreds of conversations before I reached the conclusion, and later the conviction, that I was dealing with realities ..."

Remarkably, we find an exact parallel between events and attitudes at that time in Russia and America. Ivan Sanderson, who set Roger Patterson on the trail, first read about Bigfoot in the Willow Creek and Bluff Creek area in 1958, in a press report about Jerry Crew. "But this one," Sanderson wrote, "I frankly refused to believe, mostly because I rather naturally assumed that the location as given (California) must have been a complete error or misquote. It is all very well to have abominable creatures pounding over snow- covered passes in Nepal and Tibet .... But a wild man with a 16-inch foot and a 50-inch stride tromping around California is a little too much to ask even Californians to accept." (Roger Patterson, *Do Abominable Snowmen of America Really Exist?* 1966.)

Author at the podium.

We are pretty lucky to have people like Koffmann and Sanderson who are ready to spend time checking information that seems at first sight ridiculous and incredible.

So for many years the Caucasus became the central area of our fieldwork. Evidence there comes from antiquity, the Middle Ages, and modern times. Marie-Jeanne Koffmann is a key figure in the hominology of the Caucasus. A report on some of her findings was published by her in the book, *The Sasquatch and Other Unknown Hominoids,* edited by Vladimir Markotic and Grover Krantz (1984).

Koffmann, like Porshnev and Smolin, is of the first generation of Russian hominologists; Igor Bourtsev, myself and many others, which can be called a second generation, served their apprenticeship in Koffmann's self-funded expeditions. Her contribution to our knowledge is great, and what is remarkable, it coincides in many things with the knowledge about Sasquatch obtained by John Green.

It's also noteworthy that in the decades of her searches in the Caucasus, Koffmann never succeeded in clearly sighting an almasty, although she once clearly sighted an Unidentified Flying Object.

Other remarkable topics about the Caucasus are Bourtsev's investigations of the riddle of braids in horses' manes, Porshnev's and Bourtsev's searches for Zana's grave, and the study of the skull of Zana's son, a skull exhibiting a combination of modern and ancient features. Biologist Gregory Panchenko'a encounter and observation of a young almasty in a barn (1991) should also be mentioned here.

As for me, I learned from my expeditions to the Caucasus two main things: One, that almasty is a reality. Two, that folklore, along with mythology and demonology, is the richest and most useful source of hominology.

When the fact of the Caucasian habitat, previously deemed impossible, was swallowed and digested, we thought, why not investigate other areas of the vast country and look for homins there? So brave Alexandra Bourtseva took a flight to the far-off Chukchi Peninsula, which faces Alaska, and brought back positive information. Homins are still reported there and their local names translate as "broad shoulders," "swift runner," and "pointed head." We have now positive information regarding the Russian Far East, from the Chinese border in the south to the Chukchi Peninsula in the north. That is the presumptive route of homins' migration to North America.

A memorable episode from that area is in a newspaper report by a local searcher who lectured local reindeer herders on the origin of man. The lectures were accompanied by slide showings of reconstructions of fossil hominids. When a reconstruction of *Homo erectus* was shown, the onlookers began whispering "that's pikilian, that's pikilian"—the local name for the wild man.

We have positive information from the eastern part of Siberia, including the area of Lake Baikal. A memorable report from there is one by a local hunter who, sitting at night in ambush for game, was confronted by a hairy giant. Being terrified, and though an atheist, the hunter started praying, saying "Jesus Christ and Son and Holy Spirit, please help me and take him away!" The giant didn't budge. The hunter then prayed: "Holy Allah, please help me and take him away!" The giant didn't budge. The hunter went on praying: "Holy Buddha, please take him away!" Christ, Allah and Buddha seemed in no hurry to help the hunter, but after a while the giant walked off without doing any harm to his human cousin.

We have lots of information from Western Siberia, the latest encounter was reported by the media last year. Maya Bykova's observation of the homin dubbed Mecheny (Marked) is one of the most remarkable (1987).

Tajikistan and Soviet Central Asia had a comeback as our important region of fieldwork during the 1980s thanks to the expeditions led by Igor Tatsl and Igor Bourtsev. Nina Grinyova's encounter with a homin they dubbed Gosha (1980) is a very memorable event. Recently Nina appeared with Bourtsev on TV and she again confirmed her experience very convincingly.

In the European part of Russia we have a very interesting account by Alexander Katayev (1974), who claims to have observed in the Urals a male and female homins who were in a merry mood. They laughed and seemed to converse, then swam across the river, climbed a steep rock wall, and disappeared.

On the bank of the Volga, apple orchard guards caught a homin (1989), bound him, bundled him into a car boot, kept him there for a couple of hours, and then released him out of fear for the safety of the car.

There came a time when we received sighting accounts from places near Moscow and Leningrad (St. Petersburg now). And we didn't reject them out of hand, because our view of what is possible and impossible in hominology has radically changed compared to our perceptions at the very beginning.

Of course, nobody has financed hominological investigations; all of them have been conducted with the meager means of enthusiasts. In Soviet times, during their summer vacations our activists used to go on self-funded expeditions to all remote and not so remote corners of the Soviet Union. Marie-Jeanne Koffman headed such expeditions in the North Caucasus, Igor Bourtsev in Azerbaijan, Abkhasia and Tajikistan, Vadim Makarov and Michael Trachtengerts in Central Asia and Euro-Russia, Maya Bykova in West Siberia and the Kola Peninsula. A lot of eyewitness accounts have been collected, homin tracks photographed and cast, homin hairs and feces found and analyzed. Thanks to the cooperation of the local people, Maya Bykova was able to observe a male specimen of the homin in West Siberia (1987); our colleague, biologist Gregory Panchenko, tipped by the locals, observed a homin in the Caucasus (1991).

Perestroika, the end of the Soviet Union, had both positive and negative consequences for hominologists. The negative result is in the fact that whole regions, where our expeditions determined the presence of homins—Tajikistan, the North Caucasus, the Transcaucasus—have become areas of social instability or even military actions, barring investigations there at present. Another snag was the country's economic situation, which endangered sometimes the survival of hominologists themselves.

The positive side of the situation is in the acquired freedom of the press, in the removal of the unofficial ideological taboo on hominological information. This made it possible the publication of my books in Russian and English, as well as books by my colleagues.

In 1997, despite economic difficulties, we held an international conference at the Darwin Museum, to mark the 30th anniversary of the Patterson-Gimlin documentary, and Igor Bourtsev's publishing firm Crypto-Logos put out my book *America's Bigfoot: Fact, Not Fiction—U.S. Evidence Verified in Russia*, telling the story of the first Bigfoot documentary film and its study and verification in my country.

In Soviet times I was prevented from going to Vancouver to speak at the first sasquatch conference in 1978. Today here I am in front of you. If any of you still doubt that Bigfoot is real I trust nobody doubts any longer that Bayanov is real, who doesn't doubt that Bigfoot is real.

Now the economic situation is slowly improving and a trickle of hominological information is beginning to reach us in Moscow. The most important and promising is information from the Kirov Region, seven hundred kilometers northeast of Moscow. An adult male and females with young homins have been repeatedly observed there recently by foresters, game wardens and huntsmen. Igor Bourtsev and Gleb Koval made trips there to investigate, came to positive conclusions, and enlisted the cooperation of local hunting and nature preservation authorities. We plan to focus our fieldwork there; thanks to relatively liberal times in Russia today, we are inviting possible North American sponsors and expedition participants to come and investigate in the Kirov Region.

Dear friends, I trust all the above-said sounded familiar and comprehensible to you. Now I'm going to say something that will probably cause surprise and objections, but I must say it because such are my thinking and conviction. I think that one of the great scientific results of the 20th century was the discovery of relict hominids (homins, for short), popularly known as *Abominable Snowman, Yeti, Yeren, Almas, Almasty, Bigfoot, Sasquatch*, etc. Actually, it was a re-discovery by hominologists of what had been known to western naturalists from antiquity to the middle of the 18th century, when wild bipedal primates were classified by Carl Linnaeus as *Homo troglodytes* (i.e., caveman) or *Homo sylvestris* (i.e. woodman, forestman). As for eastern scholars and rural populations in many parts of the world, they have always been aware of wild hairy bipeds, known under diverse popular names.

For science in the West, the re-discovery occurred thanks to the influence of two major factors: the Himalayan expeditions in search of the Yeti and the exceptional theories of Russian Professor Boris Porshnev, who, after a gap of 200 years, restored and validated the Linnaean idea of *Homo troglodytes*.

It is necessary to distinguish between a scientific discovery and its general recognition by scientists: the time gap between them may last from weeks and months to centuries and millennia (the idea of the Earth's flight in space took two thousand years to be generally recognized. It was first put forward by ancient Greeks in the 3rd century B.C.)

It is true that the existence of relict hominids is not yet officially recognized by the scientific community, which results in two kinds of illusion: most scientists believe that relict hominids do not exist, while most investigators who admit the creatures' existence work under the illusion that the discovery has not been made, and many dream to make it. Both opinions are illusory: wild hairy hominids do exist on earth today, and on the agenda is not their discovery but general recognition of their re-discovery in the 20th century.

Such recognition is expected to make a tremendous impact on science, affecting its overall strategy and methodology. But the event will not come about by itself; it has to be diligently worked for by widely disseminating the already existing knowledge and seeking new tangible evidence. We have tons of good evidence. What we need now is the straw that breaks the camel's back. I mean the back of the Establishment's resistance.

(First printed in *Bigfoot Co-op*, December 2003, pp. 6-10)

# Open Letter to *The Sunday Denver Post*

**(Not published)**

Dear Theo Stein,

I was a speaker at the International Bigfoot Symposium in Willow Creek, California, in September 2003, during my first, very pleasant and enjoyable visit to America. Among some fine gifts I received from Richard Noll was a copy of *The Sunday Denver Post*, January 5, 2003, with your article, "Bigfoot Believers." In this connection, I'd like to share some thoughts with you.

First of all, many thanks for the article—for its very positive and serious contents. Verbal support for Bigfoot researchers from some leading primatologists is refreshing news indeed. I only wish it wouldn't remain just sort of lip-service, but turn into concrete and tangible action. Back in 1973, I wrote the following to a leading primatologist, Dr. John Napier of the Smithsonian Institution: **"The living missing link is 'unknown' to science because there is no science to know it."** Today there is such a science, the science of living nonhuman hominids, called hominology. This discipline is a new branch of primatology (just as paleoanthropology was once a new branch of paleontology).

All researchers versed in this science do know that Bigfoot is a mammal, not myth, because of the females' conspicuous mammae. All know that Bigfoot is a primate because of the dermal ridges on its soles, a diagnostic characteristic of primates. All hominologists, respectful of logic and the current classification of primates, know that Bigfoot is a non-sapiens hominid because of its nonhuman way of life and bipedalism.

These conclusions are scientific knowledge contained in numerous books and articles. What remains still hypothetical is the exact relationship of the species with the fossil hominids, on the one hand, and *Homo sapiens*, on the other. So in regard to hominology there is knowledge and there is ignorance, and it is on the basis of ignorance and reluctance to know that the reality of Bigfoot is denied. There are also scientists who know the truth, but dare not admit it out of fear for their reputations, which is a shameful situation for science.

There is nothing uncommon for a newborn science to be in a sorry plight. The history of primatology itself is a telling example. (The Order of Primates was established by Carl Linnaeus in 1758, discarded after his death and re-established a hundred years later.) It was only in the 19th century that this part of zoology acquired a scientific basis. The very term "primatology" only began to be used as late as the 1940s. Or take paleoanthropology. Its birth was also long and difficult, and it was by no means immediately that the first fossils of Neanderthal, *Homo erectus*, and *Australopithecus* were recognized as such.

There are special ideological and methodological reasons why primatology, paleoanthropology, and now hominology, initially came up against strong opposition, but that's a long story to tell. What is relevant here is that in the cases of primatology and paleoanthropology the resistance of conservative circles in science was broken with the help of progressive and open-minded scientists of other disciplines. And this is what hominology needs today.

In this connection, I paid attention to these words in your article: "The key, Schaller said, will be finding dedicated amateurs willing to spend months or years in the field with cameras. So far, no one has done that." This reminded me of the two dedicated amateurs who had become the most fruitful contributors to the science of primates: Jane Goodall and Dian Fossey. Besides dedication, talent and courage there was another indispensable ingredient to their success—money. It was the famous paleoanthropologist Dr. Louis Leakey who obtained funds to launch the long-term field study of chimpanzees by Goodall and mountain gorillas by Fossey.

Now, there is no lack of dedicated hominologists willing to spend months or years in the field, but where is the money? Where is the famous primatologist or anthropologist who, like the late Dr. Leakey, would obtain funds for the dire needs of hominology?

I also noted with approbation the word "cameras" in your reference to George Schaller's advice. To prove the creature's existence with the help of a rifle is not worthy of *Homo sapiens*. To all aspiring to kill a Bigfoot I advise to read and remember the words of Dian Fossey in the Acknowledgments of her book *Gorillas in the Mist*: "Lastly, I wish to express my deepest gratitude to the gorillas of the mountains, for having permitted me to come to know them as the uniquely noble individuals that they are." Dian was killed by those who kill gorillas.

Lastly, may I draw your attention to the concluding part of my presentation at the International Bigfoot Symposium in Willow Creek.

> I think that one of the great scientific results of the 20th century was the discovery of relict hominids (homins, for short), popularly known as Abominable Snowman, Yeti, Yeren, Almas, Almasty, Bigfoot, Sasquatch, etc. Actually, it was a re-discovery by hominologists of what had been known to western naturalists from antiquity to the middle of the 18th century, when wild bipedal primates were classified by Carl Linnaeus as *Homo troglodytes* (i.e., caveman) or *Homo sylvestris* (i.e., woodman, forestman). As for eastern scholars and rural populations in many parts of the world, they have always been aware of wild hairy bipeds, known under diverse popular names.
>
> For science in the West the re-discovery occurred thanks to the influence of two major factors: The Himalayan expeditions in search of the Yeti and the exceptional theories of the Russian Professor Boris Porshnev, who, after a gap of 200 years, had restored and validated the Linnaean idea of *Homo troglodytes*.
>
> It is necessary to distinguish between a scientific discovery and its general recognition by scientists: the time gap between them may last from weeks and months to centuries and millennia. (The idea of the earth's flight in space took two thousand years to be generally recognized. It was first put forward by ancient Greeks in the 3rd century B.C.).
>
> It is true that the existence of relict hominids is not yet officially recognized by the scientific community, which results in two kinds of illusion: most scientists believe that relict hominids do not exist, while most investigators, who admit the creatures' existence, work under the illusion that the discovery has not been made, and many dream

to make it. Both opinions are illusory: wild hairy hominids do exist on earth today, and on the agenda is not their discovery but general recognition of their re-discovery in the 20th century.

Such recognition is expected to make a tremendous impact on science, affecting its overall strategy and methodology. But the event will not come about by itself; it has to be diligently worked for by widely disseminating the already existing knowledge and seeking new tangible evidence. We have tons of good evidence. What we need now is the straw that breaks the camel's back. I mean the back of the Establishment's resistance.

October 2003

**Dmitri Bayanov**
Chairman, Smolin Seminar on Questions of Hominology
Darwin Museum, Moscow, Russia

# The Ape Misnomer

I am glad it [the idea of killing a specimen] arouses protests now from North American researchers themselves. First from Gordon Strasenburgh (*Bigfoot Co-op*, August 1998, p. 5), now from Bobbie Short (August 2000, p. 5–6). The "villain of the piece" is John Green, long responsible for the confusion and still, defending it (October 2000, pp. 6–7).

John and I are good friends and colleagues on the question of sasquatch existence, but bitter opponents on the question of an appropriate name for the creature. Sasquatches are primates; primates are divided into human and non-human. All agree that sasquatches are non-human, but can we call them apes? Apes are quadrupedal and arboreal; sasquatches are bipedal and land-dwelling. Apes don't swim, sasquatches are great swimmers. Apes are diurnal, sasquatches are very active at night.

Let's take for comparison other related animals, as for example such rodents as squirrels and beavers. The former are arboreal, the latter land-dwelling, but both are quadrupedal. So there is more misnomer and confusion in calling the sasquatch an ape than in calling the beaver a squirrel!

Bobbie Short is quite right saying "neither-nor"—sasquatches are neither humans nor apes, although related to both, as, for example, wolves are related both to dogs and foxes. The intriguing question is: Who is a closer relative of sasquatches—humans or apes? Undoubtedly, humans (somewhat as wolves are closer relatives of dogs than foxes). I provided arguments for this view in a recent debate, by e-mail, which John is silent about. Participants in the debate, besides myself and Green, were Krantz, Meldrum, Greenwell, Coleman, Fahrenbach, Colarusso, Morant, and Loofs-Wissowa.

The scientific name for apes is "pongid," for bipedal primates—"hominid." When Krantz said in the debate "I call the sasquatch an ape, though it most probably is a hominid," I answered: "I presume there is as much science in it as in the appellation 'naked ape' or 'son of a bitch.' No doubt, a rose would smell as sweet if called by any other name. And a sasquatch would smell as strongly if called an ape or a pongid. It's only our thinking that stands to suffer by losing sharpness of perception and understanding."

Bobbie Short says the sasquatch/bigfoot is "perhaps astonishing to science." It sure is, Bobbie. Paleoanthropologists deemed fossil hominids (i.e., fossil bipedal primates) to be human, while extant pre-*sapiens* hominids turned out to be non-human. As paleoanthropologists have no explanation for this "astonishing" phenomenon, they keep mum about it.

The explanation is provided by hominology—a new branch of primatology. To put an end to confusion in terminology it provides a common name for all non-human bipedal primates—the homin. The ape misnomer is bad enough but easy to spot and refute. The opposite thing—the *Homo* (man) misnomer, plaguing paleoanthropology (*Homo habilis, Homo erectus, Homo neanderthalensis*) is much worse.
This one is a real roadblock for hominology. When it is removed the larger part of paleoanthropology will become paleo-hominology.

(Originally printed in *Bigfoot Co-op*, December 2000, pp. 4–5.)

SECTION 8

# The Evolution of Higher Primates

## Introduction

In the spring of 1972 Porshnev played host to Sol Tax, Editor of *Current Anthropology*, a world journal of the sciences of man, as the publication defines itself. Being an intellectual of the older generation, Porshnev was fluent in French and German but needed help in English which was provided by me in his talks with the editor. Porshnev proposed that the world's most prestigious anthropological journal publish a translation of his recent article carried by the Soviet academic publication *Doklady Akademii Nauk SSSR* (Reports of the USSR Academy of Sciences). Sol Tax agreed and soon after Porshnev's paper, in my translation, was posted to *Current Anthropology.*

By the time the article was ready for publication Porshnev was no longer with us. We received a letter from the Editor saying that publication could only take place if we replied to the comments made by several scholars who had read Porshnev's paper. The comments were sent over, I wrote a reply and had to send it to *Current Anthropology*. It was to be my first mailing to a foreign journal (the previous was done by Porshnev himself through the official channels at his Institute). Private citizens in the Soviet Union were supposed to engage only in private correspondence with counterparts abroad. All non-private correspondence—scientific, cultural, business, etc.,—had to be conducted through relevant official organizations. Thus, wishing to send an article of my own to a magazine abroad, I had to get a visa from the director of the Darwin Museum who would then be officially responsible for the contents and consequences of my writing. As the director was under pressure from the conservatives in science urging her to ban our seminar, she asked us to keep a low profile if we wished to use the museum as our sanctuary, and so I was reluctant to ask her for a visa on my paper.

Under the circumstances, not being sure that the paper would reach its destination overseas, and to increase its chances of doing so, I asked Igor Bourthsev to undersign the article along with me. His official standing with the authorities was better than mine, and he had done so much for hominology. After some hesitation he agreed. I mailed the article, it safely reached the Editor, and was published in *Current Anthropology*, December 1974, pp. 452-456, under the heading *Reply* by Dmitri Bayanov and Igor Bourtsev. I reveal now how our co-authorship came about in order to state that all possible mistakes in that *Reply* and the next article in *Current Anthropology* (June 1976) are my own responsibility.

# The Troglodytidae and the Hominidae in the Taxonomy and Evolution of Higher Primates*

*by B.F. Porshnev*
Moscow, U.S.S.R. 1 May 71

The present crisis in current ideas on the evolution of the higher primates (see Porshnev 1966) calls for revision of certain postulates and rehabilitation of the Haeckel and Vogt hypothesis (1866–68) of a "missing link" between apes and man. Haeckel and Vogt called this hypothetical form "ape-man" (or "man-ape"), but in accordance with the rules of taxonomy Linnaeus's term *Homo troglodytes* or simply troglodytes should have been used. *H. troglodytes*, according to Linnaeus, is a species characterized among other things by hairiness of the body and absence of speech. In the last hundred years, especially since Dubois's discovery of *Pithecanthropus* bones in Java in 1891–94, a fairly large amount of fossil material, representing many species of extinct bipedal higher primates, has been collected. The generally accepted evolutionary interpretation of this material seems to me incorrect: all these species (except *Australopithecus* and *Meganthropus*) are bracketed with man. Linnaeus and Haeckel and Vogt would have classified them as troglodytes, or ape-men, having affinity to man in morphology, including bipedal locomotion, while lacking the higher cerebral functions which make speech and reason possible. Under the pressure of philosophical concepts, this fruitful idea has been almost totally abandoned in this century. Darwinian evolutionism (and the corresponding view of Engels) has been weakened by the concept of man's direct descent from apes without an intermediary zoological link. Now we can restore this link by embracing in taxonomy all the extinct and living forms belonging to it.

The main criterion for placing fossil forms in the family Hominidae is in practice the presence of accompanying stone implements. Such practice contradicts the purely morphological principle of classification. The creature discovered in Africa that was first named *Pre-Zinjanthropus* and later *Homo habilis* made crude pebble tools but had the brain of an anthropoid. Nonetheless it is considered that the discovery of *Pre-Zinjanthropus* increases the antiquity of hominids ("humans") some 2,000,000 years. At the same time, the contemporary and subsequent morphologically similar Australopithecinae are set apart as a subfamily, for their tool-making is considered doubtful or rudimentary. The geologically contemporary *Meganthropus* and *Gigantopithecus* are not included in the Hominidae at all, because they undoubtedly made no tools.

Bunak (1966, 536) interprets Paleolithic stone implements as "exosomatic organs." Their production was to a high degree a stereotyped and automatic function: only slight changes of the prevailing pattern occur over a period of about 1,000 generations in the Lower Paleolithic and over a period of about 200 generations (taking a generation as 30 years) in the Middle Paleolithic; such ethological changes lie quite beyond the level of consciousness. The psychotechnical analysis of Paleolithic tools shows that the process of their production did not involve speech, but was sustained by automatic imitation within populations. Modern neurophysiology and neuropsychology have found it possible to locate in the brain the areas of speech control and its bearing on behavior in that destruction of certain fields and zones in the frontal, temporal, and

---

*This paper first appeared, in slightly different form and in Russian, in *Doklady Akademii Nauk SSSR* (Reports of the USSR Academy of Sciences) 188(1).

[The above idea was sent to the same scholars who were invited to comment on the papers of Butzer, Tuttle, Todd, and Blumenberg in this issue. The responses are printed after the text. B.F. Porshnev died while his article was in press; his colleagues Dmitri Bayanov and Igor Bourtsev of the Darwin Museum agreed to reply to the comments, and their remarks appear below–Editor (of *Current Anthropology*)] sincipital regions of the cortex renders the above-mentioned functions in one respect or another impossible; and these particular fields and zones arc developed only in *H. sapiens* (Shevchenko 1971). The comparison of the data of the morphological evolution of the brain of fossil hominids and the study of aphasia excludes the possibility of articulate speech in the pre-*H. sapiens* stages of evolution, and the concept of inarticulate speech is rejected by psycholinguistics as senseless.

Hence, it is advisable to abandon the current practice of including all bipedal higher primate fossils in the Hominidae. It is preferable to include in this family just one genus, *Homo*, represented by a single species, *H. sapiens* (subdivided into *H. sapiens fossilis* and *H. sapiens recens*). The main diagnostic distinction of the Hominidae is the presence of those formations in the structure of the brain which make speech possible and the correlative features in the organs of speech and in the face (Porshnev 1971). All the other bipedal higher primates should be embraced by the family Troglodytidae (or Pithecanthropidae), whether they made tools or not. Their main diagnostic distinction from the family Pongidae is bipedal (erect, orthograde) locomotion, with all the correlative features in the structure of the body, head, limbs, and internal organs. The Troglodytidae (or Pithecanthropidae) may be subdivided into the following genera: (1) *Australopithecus*, (2) *Meganthropus*, (3) *Pithecanthropus (Archanthropus)*, (4) *Troglodytes (Paleanthropus)*, subdivided into *Troglodytes fossilis* and *Troglodytes recens*. This fourth genus (commonly known as the Neanderthal) can in its turn be divided into the following species: (a) Southern (Rhodesian type), (b) Classical (La Chapelle type), (c) Presapient (Steinheim-Ehringsdorf type), (d) Transitory (Palestine type).

The family Pongidae branched off the Primate tree in the Miocene. Currently it is represented by four genera: the gibbons (sometimes separated as a distinct family), the orangutans, the gorillas, and the chimpanzees. The family Troglodytidae diverged from the anthropoid line in the Pliocene. At present it is represented by one genus, probably one species (*Troglodytes recens* L.sp.?), which can be described as "relic hominoid" (Porshnev 1963 & 1969). From the hominoid line (Troglodytidae) in the Upper Pleistocene there separated a family of hominids in which the tendency towards the formation of species did not prevail and which from the very start and up to the present has been represented by the species *H. sapiens*. The taxonomic rank of family for *H. sapiens* is justified by the great biological significance of such new formations as the organs and functions of speech, i.e., the second signal system. The unusually fast tempo of this evolutionary progress (naturally, on the basis of useful variations of ancestral forms, i.e., late *Paleanthropus*) indicates a mechanism of selection somewhat akin to artificial selection. In this comparatively speedy divergence the two species were juxtaposed in such a way that connecting links were washed away. This process was intimately connected with the genesis of the second signal system (Porshnev 1968a). The question is open now which species of *Paleanthropus (Troglodytes fossilis)* was the direct ancestor of *H. sapiens fossilis*. Perhaps we shall know this when study of the relic *Paleanthropus* yields serial morphological material, for there can be no doubt that the *Troglodytes recens L.* is a direct left-over of the divergence of the Troglodytidae and the Hominidae. Therefore study of these relics becomes a cardinal objective for the theory of anthropogenesis (Porshnev 1966b).

The thesis that all *Paleanthropus* forms died out or were assimilated almost immediately (not more than 3,000 years) after the appearance of *H. sapiens* is totally a priori and biologically absurd. Neanderthaloid skeletons found in the more recent strata of the earth, including some of the various periods of historic time, are looked upon as "pseudo-Neanderthal." The argument here is based on the absence of Mousterian tools, though this can be explained by the above-mentioned divergence resulting in the disappearance of the transitional forms and in the new ecology of relic *Paleanthropus*. Archeological and historical evidence shows the latter's coexistence with man, sometimes in symbiotic relationship, sometimes in parasitic relationship with various species of the Carnivora and the Herbivora, and, lastly, in more recent time, subsisting vicariously in the most deserted but ecologically varied biotopes.

The highest possible degree of negative adaptiveness of relic *Paleanthropus* to man, on the one hand, and its great outward likeness to man, on the other, jeopardize its study and explain why primatology and anthropology have fallen so far behind on this problem. It is precisely the use of nontraditional methods, such as the comparative analysis of mutually independent evidence, that has made it possible to establish the existence of this relic species and to describe its morphology, biogeography, ecology, and behavior. In other words, fact-finding methods have been used in biology that are usually employed by historians, jurists, and sociologists. This indirect research into the problem of relic *Paleanthropus* is now considerably advanced. Yet the road traveled so far can only be described as the initial (though perhaps the hardest) stage of research. The way is not yet clear for the next step—the planned acquisition of a specimen, living or dead. The state of research on the problem allows one to think it is high time certain postulates in primatology and anthropology were rewritten.

# Reply by Dmitri Bayanov and Igor Bourtsev

Moscow, U.S.S.R. 15 July 74

We are grateful to Sol Tax for acquainting CA readers with Porshnev's anthropological ideas and for giving us the opportunity to discuss them here. The vastness of the problems embraced by the Porshnev theory, its (in our opinion) truly revolutionary character, and the fact of its presentation for discussion in an overly summarized form make many queries on the part of the reader inevitable. Besides, as we see from the comments, Porshnev's works are not known to those who kindly agreed to take part in the discussion. Therefore we would like to provide an explanation of our late colleague's theory before answering concrete questions and comments offered by his critics.

There are two cardinal notions in anthropology on whose mutual relation the very essence of this science depends: man and animal. In pre-Darwinian times the relation between these notions was of one kind, in post-Darwinian times of another, and the changeover from one to the other signified an unprecedented revolution in man's thought and world outlook. Before Darwin, a supernatural schism divided animal and man; after Darwin, we accept a natural affinity and transition between one and the other. But the more science tries to solve the riddles of this transition, and the deeper in time it looks for minute details of it, the less distinct the notions of man and animal become, so that one is left with the question, "Transition from what to what?" To understand the origin of man, we have to know exactly what he is, and to know that we have to understand his origin.

Porshnev offered to break this vicious circle by restoring and reemphasizing the difference between the notions of man and animal, but this time on a scientific basis. In fact, his theory is a colossal attempt to stress and define the uniqueness of man in the light of modern science.

Science consists of facts and their interpretation. America is a fact of geography; Columbus's taking it for India is a famous example of interpretation. Let us state from the outset that Porshnev never quarreled with facts, but he was up against some very sacred interpretations.

How could science possibly go awry in interpreting facts of paleoanthropology? First of all, by uncritically using the ready-made, unscientific, pre-Darwinian, intuitive concept of man in the study of fossil material. When skeletal remains were found that looked more manlike than apelike, scholars, without much further thought, started labeling them "man." Thus such terms as Java man, Peking man, and Neanderthal man came into usage. Using a familiar name for an unknown thing, one inevitably imagines that unknown entity in terms of the makeup of the familiar one of the same name. In other words, images of ourselves were projected into the unfathomed past, and once placed there they began to be treated as facts of prehistory.

Another possible cause of misinterpretation in paleoanthropology is the fact that this science is manned by osteologists, who know everything about skulls and very little about their contents, while it is the latter and not the former that have anything to do with the life of all brainy creatures.

A third cause is the fact that modern evolutionary-anthropology was born in Western Europe, and the closest living animal relatives of man known to the European scientist were representatives of the Pongidae. The evolutionist's thought could have taken a somewhat different direction had he set his eyes on a *Troglodytes recens*.

The sacred interpretations challenged by Porshnev are:
(1) that primate bipedalism is sufficient for human status;
(2) that any of the pre-*sapiens* higher primates were big-game hunters;
(3) that certain primates' tool-making activity and use of fire are sufficient evidence of their human intellects; and
(4) that any of the pre-*sapiens* primates had speech and abstract thinking. All of this adds up to his denial that man descends directly from the ape.

Between ape and man Porshnev places a whole zoological family of higher bipedal primates: the Troglodytidae. In his view, instead of primitive man and developing man there was an extremely developed animal, an animal of the highest possible order, which at a certain point of evolution became man—*Homo sapiens*, the only species of man in existence. We don't know what will become of man in the future, but so far he is the only model of this type of "production."

To illustrate this phylogenetic point with an example from ontogenesis, let us note that there is no such thing as a primitive butterfly. It's either a butterfly, or a pupa, or a caterpillar, yet these vastly different things are intimately connected by their origin.

Borrowing a simile from a more topical realm of present-day reality, we could also liken the origin of man to a space shot. It was a multistage rocket of evolution that put humans into orbit, and the rocket went faster and faster, but no matter how high the stages got it was only those of our ancestors who were actually in orbit who can be called human beings, according to the Porshnev theory.

True, in the final phases of their steeply rising evolutionary curve the animals become very strange and unusual and deserve a place of their own in biology and philosophy. The old Aristotelian problems of the actual and potential of these borderline cases are somewhat similar to those confronting the biologist in some fungi which "behave" sometimes as animals and sometimes as plants, or in viruses which display characteristics of both animate and inanimate objects, or those facing the physicist studying "liquid crystals." Yet, according to Porshnev, on the basis of what we know at present, our unusual creatures in their usual state have to be classed beyond the pale of man. Compared with such common beings as, say, cats and dogs, anthropoids are very strange animals indeed, more manlike than doglike. And even compared with apes, Porshnev's troglodytes are very unusual animals, more manlike than apelike. But this still doesn't make them men.

Nobody ever raises an eyebrow over the fact that such different things as, say, the amoeba and the gorilla belong in the same world and are called by the same name, "animal." If the animal world encompasses things as different as this, how can we know where it should end? Why couldn't Nature have created animals even more developed than apes? Who has proved that the anthropoid is the last word of zoological evolution? Who can say to Nature, "Here and no more. This is the limit of thy power"?

In fact, there must be a limit to the animal kingdom and a boundary between man and beast, but is it not reasonable to assume that life moves on to a new stage of creativity only after it has fully displayed its talent in the old one?

What about tool making and the use of fire by our primate ancestors? Doesn't this prove beyond all doubt their human intelligence? Well, do the beaver's dams or the squirrel's storing of food for a "rainy day" signify their human intelligence? Extrapolation in biology from similar effects to similar causes is very risky. Similar functions may and do appear at very dissimilar levels of biological organization.

Still, persists the critic, there is no phylogenetic connection between the squirrel's or the beaver's activity, on the one hand, and man's activity, on the other, while there is every reason to believe that *H. sapiens* inherited tool making from his pre-*sapiens* ancestors. Doesn't this show that the squirrel-and-beaver argument is irrelevant here? Not quite. To make the point clearer, let us take a function man shares with animals and inherited directly from them, sexual reproduction. Can we infer from the obvious similarity of this function in man and animal their similar intelligence? Is it not more reasonable to assume that an animal engaged in propagation doesn't really know what it is doing? This example shows that even in phylogeny a function can first be devoid of sense and later acquire it.

We agree that Porshnev's theory sounds very strange at first hearing. How did he arrive at such unorthodox ideas, and is there more justification for them?

Boris Porshnev was a man of encyclopedic erudition and interests. Besides his main subjects of history and philosophy, he actively worked in and published papers on psychology, sociology, and archeology. Taking part in archeological and paleontological expeditions, he not only looked for facts but also searched out threads of logic to connect them. This is normal practice for the theoretician and has nothing to do with bias. The mere empiricist can't see the woods for the trees, whereas the creative theoretician soars on high and takes a bird's-eye view of the forest of facts below.

History and philosophy taught Porshnev to look for trends and tendencies in processes of historic dimensions. They also taught him to take account of the immense diversity of causes and effects and their interactions in evolution, thus whetting his interest in problems of ecology. Here he had a worthy forerunner, Academician Pyotr Sushkin (1868–1928), also a scholar of diverse interests and great erudition. In an article published in 1928, Sushkin stressed the necessity to take ecology into account in solving the problem of man's origin: "I ... strive to see emerging man not in isolation but as an element of certain fauna which is part of the environment and its changes."

Ecology combines the concreteness of the natural sciences with the broad outlook of philosophy; in fact, in its broad-mindedness ecology is second only to genuine philosophy, and therefore it was not by chance that Porshnev found an ecological approach to the problem of man's origin most appropriate.

To be exact, Porshnev applied the ecological approach not to the study of the origin of man per se (in his classification), but to the origin and development of that zoological stage of evolution which directly precedes man and paves the way for his emergence, i.e., the origin and development of the Troglodytidae. Let us briefly trace his train of thought, sometimes expanding upon what he left in parentheses and making explicit what he implied.

Fact: abundance of splintered animal bones found in association with hominid (Troglodytidae) fossils. Orthodox interpretation: hominids were hunters, killing various animals (including some very big ones), eating their flesh, and crushing their bones for marrow. Porshnev's interpretation: early Troglodytidae were "bone hunters," collecting the leavings of predators' feasts. As is known, carnivores with their stomachs full are no threat even to the meekest of animals. Besides, Troglodytidae stole bones in broad daylight, while predators are most active and dangerous at night.

When the anthropoid ape found himself on the ground and in the savanna as a result of ecological changes in the Tertiary period, he suffered a decrease in food supply from what he had enjoyed in the forest; hence his search for dietary substitutes. Because of his morphology, he could not consume grass the way herbivores do, nor could he feed on herbivores the way carnivores do. But he had hands formed in the forest, and it didn't take him long to put this biological pre-adaptation to good use. Abundant bones, especially skulls, of savanna-dwelling animals were like shells and nuts which the ape knew how to crush with stones. The only problem was to bring bones and stones together.

Thus bone-carrying and -crushing was the main factor of selection which made the anthropoid ape bipedal and marked the beginning of the Troglodytidae as such. In this respect, Porshnev's theory closely coincides with Hewes's (1961) food-transport hypothesis, the only difference being that the former suggests scavenged bones as the objects carried by would-be bipedal primates while the latter suggested scavenged meat. Writes Hewes (1961, 687): "DuBrul (1958, 90) notes that upright posture is essentially a 'reduction of the repetition of structures serving the same function,' with the forelimbs becoming 'as it were, accessory mandibles rather than locomotor devices,' leading to a 'new mode of feeding and feeding niche.'"

Indeed, the troglodyte's hands became mighty accessory mandibles, with ever replaceable teeth of stone, which could crush bones of such strength and in such numbers as were beyond the power of all other scavengers, including the hyenas. This bone-cracking, brain and marrow-eating stage in the evolution of the Troglodytidae, which we may call a stage of cerebro-and-myelophagia, must have lasted at least a couple of million years.

As a result of this million-year-long process, the ground-dwelling higher primates not only became bipedal, but also got the knack of using stones to provide for their livelihood. A million-year-long application of stones to skeletons taught the troglodytes that stones were good not only for cracking bones, but also for cutting and mincing meat that remained on some bones they collected. They also learned in the process that only sharp stones, appearing as a by-product of bone smashing, are good for meat cutting. Thus the next and most important phase in the process was their hunting for skeletal remains with ever more meat on the bones and eventually for whole carcasses, on one hand, and their systematic making of sharp stones, on the other. Such a reconstruction of events makes comprehensible how bipedal primates came to apply hard objects (stones) to soft material (meat), which otherwise seems a stroke of genius.

Another, and ultimately the most important, "by-product" of the process was the unusually swiftly growing brain of our bipedal scavengers. What were the causes of natural selection of the brainiest in this case? The answer is probably provided by realization that the troglodyte had not one but several rather demanding tasks on his mind during each feeding cycle:

(1) to watch the herbivores,
(2) to watch the carnivores,
(3) to look for results of their interaction,
(4) to be in the right place at the right time to find an adequate carrion supply,
(5) to outfox and outmaneuver carnivore enemies and competitors in getting away with it, and
(6) to solve the problem of consumption with the ever present handicap of inadequate teeth through finding and later fashioning "artificial teeth."

Thus the Troglodytidae became the brainiest creatures on earth prior to *H. sapiens*. For our theme, however, it is important to emphasize that in the broad context of evolution their intelligence was the result and not the cause of their way of life. And, according to Porshnev, their intelligence was still of an animal kind, still insufficient to classify them as humans.

What about fire? Isn't its use a clear and indisputable proof of the user's human status? No, it isn't, said Porshnev, the first scholar ever to utter such heresy. According to him, the use of fire was no invention by a pre-*sapiens* genius, but a natural and inevitable consequence of stone-tool production—a by-product again, if you wish. If bipedalism was the consequence of carrying and cracking bones, then the use of fire was the consequence of fashioning stones. Red-hot splinters produced by hammering one piece of flint with another were bound to make smoldering a common occurrence at the litter-strewn sites of our bipedal primate ancestors. Porshnev thought that for an unknown length of time troglodytes were a sort of firemen, extinguishing the nasty patches of smoldering with their broad hands and feet. By and by they got used to this nuisance and learned to turn it into flames and keep it going. If man can teach an anthropoid to smoke cigars and drive an automobile, then Nature, the greatest instructor of all, could have taught bipedal hominoids some tricky things too. Thus, according to Porshnev's logic, it seems not so much that bipedal primates adapted fire as that they became adapted to it.

To sum up, the Troglodytidae's making of tools and use of fire were more the result of their ecology than of their psychology, whereas with *H. sapiens* it was the other way around. This needs to be stated to show not only Porshnev's understanding of the events preceding the appearance of *H. sapiens*, but also his idea of the subsequent divergence of man and the Troglodytidae. Since the tool-making activity of the Troglodytidae was mainly stimulated by ecology, they were bound to lose it with a sufficient change in the environment. And, conversely, since such activity of *H. sapiens* was deeply rooted in his intelligence, he went on developing it despite the environment. Thus the troglodytes and *H. sapiens* headed in opposite directions: the first slipped back to the tool-less and fire-less life of other animals; the second marched on to ever new vistas of technological innovation.

Now we come to the crucial question of the whole theory: How and why did *H. sapiens* come into being? According to Porshnev, the appearance of *H. sapiens* is connected with the formation in the brain of the second signal system (Ivan Pavlov's term), which makes speech and conceptual thought possible. The second signal system emerged, Porshnev thought, not as a result of the primates' work with any inanimate objects (such as stone tools, for example), but as a result of their intergroup relations, of activities directed at each other. The suggested mechanism of such interaction is described in detail by him in a work which is due to be published posthumously in a few months. [The title is *On the Beginning of Human History (Problems of Paleopsychology)*, (in Russian), 488 pp, 1974, Moscow: Mysl. Now we expect the second, full (with all the dropped material included) edition.—D.B.]

Certainly, Porshnev was not the first thinker to believe that the power of speech is the true mark of man, but he was the first to think it appeared so suddenly and so late in anthropogenesis. The event can be compared to an atom bomb explosion. Just as a critical mass of uranium is needed to produce such an explosion, so a certain critical amount of brain of a certain complexity is required to make speech and abstract thinking possible. Therefore Porshnev denied the possibility of any rudimentary, inarticulate, and primitive speech prior to this postulated "verbal explosion."

To test this heretical theory, we have to find out whether Neanderthals had the power of speech or not. We'll say more on this issue below, but, assuming for the moment that Porshnev is right and all the pre-*sapiens* primates were truly devoid of language, what status—human or animal—are we going to grant to Neanderthals? For our part, we'd rather accept a species or genus of tool-making and fire-using animals than a species or genus of speechless humans.

Poirier asks what is meant by "the present crisis in current ideas on the evolution of the higher primates." As we understand it, the crisis is evident from the following:

1. The more facts are obtained (to wit, the Leakeys' discoveries), the less clear the overall picture of man's origin becomes from the viewpoint of the orthodox version.
2. The more fossil forms are found, the more insistent becomes the unspoken question of what made the whole stage of primate evolution between the apes and *H. sapiens* so promptly extinct. While paleontologists hotly debate the question "What did in the dinosaurs?", paleoprimatologists keep silent about the immeasurably more relevant question of higher-primate extinction.
3. Orthodox primatology has not recognized and, apparently, has no clues for analyzing the evidence of the continued existence on earth of higher primate forms distinct from both the Pongidae and *H. sapiens*. Such a turn of events is completely inconsistent with the orthodox version and therefore is quietly ignored.

What is the "mutually independent evidence that has made it possible to establish the existence of this relict species"? A detailed answer is provided in Porshnev's (1968b) work *Borba Za Troglodytov* (The Struggle for Troglodytes), which is now available in a French translation (Heuvelmans and Porshnev 1974). Here we list the categories of independent evidence as the matter stands at present:

1. Mention, description, and/or drawings of what Porshnev, following Linnaeus, calls troglodytes (or relict hominoids; i.e., higher bipedal primates different from *H. sapiens*) in accounts of ancient or medieval travelers, in natural-history books, medical books, etc.
2. Mention or description in ancient or medieval poetry, art, folklore, demonology.
3. Sightings by modern outdoorsmen.

4. Photographs and plaster casts of footprints.
5. The Patterson film, which at last makes the creature's photographic appearance and movements available to everybody's eyes.

As an example of the first category, we can cite Nizami al-Arudi, who says in his *Chahar maqala* (c. 1150–60, quoted in Bernheimer 1952, 190): "The highest animal is the Nasnas, a creature inhabiting the plains of Turkestan, of erect carriage and vertical stature, with wide, flat nails. . . . This, after mankind, is the highest of animals, inasmuch as in several respects it resembles man: first in its erect stature, secondly in the breadth of its nails, and third in the hair on its head." Also in this category is the fact, strangely overlooked, that modern anthropology bears in its very heart an indirect mark of the troglodytes. It is generally believed that the central term of modern anthropology—*H. sapiens*—was coined to distinguish modern man from the forms we know from the fossil record. Nothing of the sort. The term was introduced by Linnaeus in the 18th century, 100 years before the Darwin theory and systematic studies of hominid fossils. Linnaeus had information about the existence of another, kind of "man," hairy, mute, non-sapient, and for the sake of contrast with it he designated our species "*sapiens*."

Examples in categories 2 and 3 are legion. As for categories 4 and 5, we have studied the photographs and plaster casts of footprints ascribed to relict hominoids, on the one hand, and the Patterson film (made available to us by René Dahinden, to whom we express our gratitude), on the other. In the latter examination, biomechanist Dr. Dmitri Donskoy also took part, supplementing our analysis with his conclusions (Hunter with Dahinden 1973, 189–92). We have established five solid correlations between the footprints and the creature seen walking in the Patterson film, all five distinct from or totally nonexistent in *sapiens* characteristics. This leaves no doubt in our minds whatsoever that both the film and the footprints we studied are genuine.

Poirier wonders about the "planned acquisition of a specimen, living or dead," of the so-called relic *Paleanthropus*. According to the theory expounded here, man is a unique offspring of a unique family. One potent proof of man's unsurpassed originality is the fact that he decided and managed to reach the moon prior to meeting and officially recognizing his unique animal cousins on earth. As to the whys and hows of this fantastic situation, see Green (1968, 1970, 1973), Hunter with Dahinden (1973), Heuvelmans and Porshnev (1974), Krantz (1971, 1972), and Sanderson (1961).

What is "negative adaptiveness"? By this term Porshnev meant that after *H. sapiens* and the troglodytes had diverged and the former got the upper hand, the latter had to adapt themselves to the conditions and environments the former found negative. For example, *H. sapiens* prefers daylight; troglodytes had to be active at night (hence Linnaeus [1758] defines *H. sapiens* as "diurnus" and *H. troglodytes* as "nocturnus"). Again, *H. sapiens* prefers fertile plains; troglodytes had to settle in high mountains, deserts, dense forests, and swamps.

Malik argues that "the concept of automatic imitation of implements makes our ideas about cultural tradition and change absurd." This is *argumentum ad hominem*, and as such of no use in science. Many things in science first seemed right, then absurd, and vice versa. Porshnev objected to the application of the term "culture" or "cultural tradition" to pre-*sapiens* forms, but he never denied change in their tools or tool making. If these forms themselves changed morphologically, why shouldn't their "exosomatic organs" have changed? Porshnev also argued that these "ethological organs" could change somewhat faster than the morphology of their owners. From the viewpoint of Porshnev's theory, the right use of the term "culture" is seen from the following example: Dances of *H. sapiens* are an element of culture and are studied by ethnography; dances of the chimpanzee are an element of zoology and are studied by ethology.

In response to Raemsch: As is known, size alone cannot be the criterion of a brain's function: both size and structure should be taken into account. Though equal to the *sapiens* brain in size, the Neanderthal brain is different from it, especially in its underdeveloped frontal lobes. (This is apparent from a look at a Neanderthal frontal bone.)

Among other considerations, Porshnev based his belief that Neanderthals were speechless on the study of their morphology, on the one hand, and on the data of *sapiens* brain pathology resulting in aphasia, on the other. He also mentioned the following consideration: no drawings of any identifiable objects made by Neanderthals are known to science. As far as we know, such drawings appear only with the advent of *H. sapiens*. A drawing is a definite sign of abstraction, just as words of a language are. Therefore, the absence of Neanderthal fine art indicates indirectly an absence of language.

That the emergence of language in anthropogenesis was rather sudden seems probable from the following: Though man's physical tools at present include everything from stone axes to earth satellites, we don't find any comparable gradation in his mental tools, i.e., languages. "Nowhere in the world has there been discovered a language that, can validly and meaningfully be called 'primitive' " (Hockctt 1900:89).

Raemsch holds that "we cannot now examine the neuroanatomical structures for speech in Neanderthal man." Let us answer by quoting from a newspaper account sent to us by our Canadian colleague René Dahinden (Agnew 1971):

> The vocal tract of Neanderthal Man—who lived some 40,000 to 70,000 years ago—lacked most of the pharynx and was capable of producing only "inefficient and monkey-like" sounds, according to researchers from Yale and the University of Connecticut.
>
> They undertook studies of the vocal system of Neanderthal Man for the National Institute of Dental Research after noticing that some mongoloid children who do not talk have heads with an infantile shape. Internally, Neanderthal skulls have similar shapes, they found.
>
> The researchers also found that Neanderthal Man had a voice box high in the throat—a condition present in apes, monkeys and human infants—that made it possible for him to breathe and swallow simultaneously without choking.
>
> This capability is lost in the modern adult human when his vocal tract becomes a sophisticated structure linking larynx, pharynx and mouth with complex neurological controls.
>
> The researchers suggested that Neanderthal Man may have disappeared because of his speech deficiency. ...
>
> "We may speculate on the disappearance of Neanderthal Man, and we can note that his successors—for example, Cro-Magnon Man—had the skeletal structure that is typical of man's speech mechanism," they added.
>
> "Neanderthal Man's disappearance may have been a consequence of his linguistic—hence, intellectual—deficiencies. ... in short, we can conclude that man is human because he can say so."

We hasten to add that in Porshnev's opinion Neanderthal's muteness accounts for his disappearance from the tool-working record only; he never disappeared from life itself. If Porshnev is right, we should still have a chance to examine the neuroanatomical structures for speech (or lack of it) in Neanderthals *in vivo*.

We share Blumenberg's warm feelings for the chimpanzee, and we love other animals which vocalize even less than chimpanzees. What if the baboon could learn the equivalent of 50 or 25 words in the use of plastic objects? Should he be "sunk in the genus *Homo*" too?

We think that to compare man and animal in terms of their communication abilities we should first of all examine their natural communication systems and not such artificial things as Yerkish. There are many points on which man's speech and the communication systems of animals coincide, but there are others on which they are as far apart as heaven and earth. By the communication means at their disposal, animals can greet, warn, threaten, frighten, order, tease, invite, entice, deceive, ask for, beg, give consent, and show indifference, surprise, bewilderment, respect, contempt, contentment. A bee through her dances can indicate to her sisters the direction and distance to nectar-laden flowers, which the instructed bees don't fail to find. Thus both animals and humans do use symbols to influence their counterparts' behavior in their respective kingdoms. But what animals can't do, what is the sole prerogative of man, is to engage in a symbolic give-and-take which we happen to be performing right now and which is called discussion. Animals can "argue" with paws and claws, but not with symbols. To be fair to the chimpanzee, we must at least ask his opinion before plunging him into our excessively vocal genus. If Blumenberg can produce a chimp which can argue the point, be it in Yerkish and within 100 words, we will promptly capitulate.

In reply to Aguirre: Porshnev mostly referred to points and practices of taxonomy accepted by the majority of Soviet anthropologists at the time of the writing of his article. As for his estimate of the number of generations, he didn't mean that the whole of the Lower Paleolithic lasted 1,000 generations, but only that it took about 1,000 generations for stone tools to change slightly in the Lower Paleolithic and 200 generations for slight changes in the Middle Paleolithic.

Touching on the problem of continuity in evolutionary and historical processes, we can say that Porshnev proceeded from the thesis that in evolution and history slow processes of quantitative change alternate with sudden and stormy processes of qualitative change—in other words, that there is no evolution without revolution.

Aguirre says, "Let us look for a specimen, alive or dead, of an inframan, but let us not classify it before we find it." Well, you can't even start looking for something before you have some idea what you are going to look for. It was precisely the development of such ideas on the issue that led Porshnev to the taxonomy described in the article under discussion, which can be helpful both for the mounting and conduct of the search.

As for the possible racist connotations referred to by Aguirre, it was Porshnev's opinion that, on the contrary, current recognition of lower and higher forms of humanity, such as *H. erectus, H. neanderthalensis*, and *H. sapiens,* constitutes a potential basis for racism. Porshnev's insistence that there is and has always been just one species of humans—*H.*

*sapiens*—leaves no room for racism even in prehistory.

In conclusion, we want to thank all the participants in the discussion and hope that they will read Porshnev's article once again to see the depth and breadth of his theory.

## References Cited


Agnew, Patricia. 1971. Early man unable to talk properly. Toronto *Sun*, July 24.

Aguirre, E. 1966. "Las primeras huellas de lo humano," in *La evolutión*. Edited by M. Crusafont, B. Melendez, and E. Aguirre, pp. 617–75. Madrid: B.A.C. [EA]

Bernheimer, Richard. 1952. *Wild men in the Middle Ages*. Cambridge: Harvard University Press.

Bunak, V.V. 1966. "Speech and intellect, stages of their development in anthrolpogenesis" (in Russian), in *Iskopaemye Hominidy i Proiskhozhdenie Cheloveka,* Moscow: Nauka.

Day, M.H. 1969. Omo human skeletal remains. *Nature* 222:1135–38. [EA]

Green, John. 1968. *On the track of the Sasquatch.* Agassiz: Cheam.

____. 1970. *Year of the Sasquatch*. Agassiz: Cheam.

____. 1973. *The Sasquatch file.* Agassiz: Cheam.

Heuvelmans, Bernard, and Boris Porshnev. 1974. *L'homme de Néanderthal est toujours vivant*. Paris: Plon.

Hewes, Gordon W. 1961. Food transport and the origin of hominid bipedalism. *American Anthropologist* 63:687.

Hockett, Charles F. 1960. The origin of speech. *Scientific American,* September, p. 89.

Hunter, Don, with René Dahinden. 1973. *Sasquatch*. Toronto: McClelland and Stewart.

Kochetkova, V.I. 1970. On brain size and behaviour in early man. *Current Anthropology* 11:176. [EA]

Krantz, Grover S. 1971. Sasquatch handprints. *Northwest Anthropological Research Notes* 5:145–51.

————. 1972. Anatomy of the Sasquatch foot. *Northwest Anthropological Research Notes* 6:91–104.

Larsell, Olof. 1952. 2d edition. *Anatomy of the nervous system*. New York: Appleton-Century-Crofts. [BER]

Linnaei, Caroli. 1758. 10th edition. *Systema naturae*.

Porshnev, B.F. 1963. *The present state of the problem of relic hominoids* (in Russian). Moscow: Viniti.

————. 1966a, Do we face a revolution in primatology? (in Russian). *Voprosy Filosofii* 3:108–19.
————. 1966b. L'origine de l'homme et les Hominoides velus. *Revue Internationale de Sociologie*, ser. 2, 2(3): 76–83.
————. 1968a. "Les aspects anthropogénétiques de la physiologie de l'activité nerveuse supérieure et de la psychologie." *Proceedings VIIIth International Congress of Anthropological and Ethnological Sciences*, Tokyo, vol. 1.
————. 1968b. The struggle for troglodytes (in Russian). *Prostor* 4:98–112, 5:76–101, 6:108–21, 7:109–27.
————. 1969. The problem of relic Paleanthropus (in Russian). *Sovetskaya Etnografia* 2:115–30.
————. 1971. Second signal system as a diagnostic line of distinction between the Hominidae and the Troglodytidae (in Russian). *Doklady Akademii Nauk SSSR* (Reports of the USSR Academy of Sciences) 198(1): 228–31.
Premack, A. J., and D. Premack. 1972. Teaching language to an ape. *Scientific American* 227:92–99. [BB]
Rumbaugh, D.M., T.V. Gill, and E.C. Von Glasersfeld. 1973. Reading and sentence completion by a chimpanzee (*Pan*). *Science* 182:731–33. [BB]
Sanderson, Ivan T. 1961. *Abominable snowmen: Legend come to life.* Philadelphia: Chilton.
Shevchenko, Yu. G. 1971. *Evolution of the cortex of the brain in primates and man* (in Russian). Moscow: Moscow University Press.
Thoma, A. 1966. L'occipital de l'Homme Mindélien de Vértesszöllös. *L'Anthropologie* 70:495–534. [EA]

(Originally published in *Current Anthropology*, December 1974)

# On Neanderthal vs. Paranthropus

*by Dmitri Bayanov and Igor Bourtsev*
Darwin Museum, Moscow, 119 435, U.S.S.R. 30 July 75

We warmly welcome Strasenburgh's response (CA 16:486-87) to Porshnev's article and our comments. It is encouraging and refreshing to deal with a critic who does not "quietly ignore" the subject. Moreover, it is evident that Strasenburgh is one of us—a hominologist, one who recognizes and attempts to substantiate the existence of relic hominoids. His argument with us is not about whether the creatures exist or not, but about their exact identification. It is probably this "family bond" that explains the curtness of his comment. We don't mind the form, but we take exception to much of the substance of Strasenburgh's retort.

To begin with, it seems a contradiction to pose as the defender of both relic hominoids and orthodoxy in primatology. There is no such thing as a one-man (or few-men) orthodoxy—the notion takes greater numbers than that—and no matter how Strasenburgh defines his views, we doubt that "the majorities on both sides of the question" can take them as orthodox. The orthodox view in primatology is that *Homo sapiens* is the only living species of the family Hominidae. To prove the point, it is enough to quote Napier (1973, 204), who may be supposed to know what orthodoxy means in his science, and who says that if any of the Sasquatch footprints "is real then as scientists we have a lot to explain. Among other things we shall have to re-write the story of human evolution. We shall have to accept that *Homo sapiens* is not the one and only living product of the hominid line, and we shall have to admit that there are still major mysteries to be solved in a world we thought we knew so well."

Strasenburgh finds simply untrue our statement that "orthodox primatology . . . apparently . . . has no clues for analyzing the evidence of the continued existence on earth of higher primate forms distinct from both the Pongidae and *H. sapiens*." His objection apparently arises from his attaching a different meaning to the word "clues" from the one we intended. We meant conceptual clues, part of a theory to guide the researcher in analyzing factual material, not facts or evidential "tips" which help substantiate the existence of relic hominoids. Such tips abound, but they are ignored by orthodox primatologists because all their theorizing is done on a different wave-length.

As for Strasenburgh's unorthodox idea that *Paranthropus* is the cause of all our troubles, emotionally we have nothing against it: the hominologist's dream is that all the hominid forms known from the fossil record, and even those not known from it, will turn out to be alive. The dream, however, has to be checked against reality. The burden of proof as regards the relevance of *Paranthropus* to our problem is on Strasenburgh, and we regret that he leaves us guessing as to how he checked out his idea.

One tip seems to be his use of the adjectives "robust" and "gracile" to describe hominid fossils in Africa. Considering that relic hominoids can be safely described as "robust," and *H. sapiens*—though not so safely—as "gracile," the idea is probably that *A. robustus (Paranthropus)* must be the ancestor of relic hominoids, just as gracile *A. africanus* is considered ancestral to *H. sapiens*. If so, the argument seems to us simplistic and superficial. What about the Broken Hill skull? Doesn't it indicate a creature by no means less robust than any *Paranthropus*? And can classic Neanderthals by any stretch of the imagination be considered gracile?

The truth of the matter is that relic hominoids, or at least some of them, resemble *Paranthropus* about as much as we and Strasenburgh resemble, say, *H. habilis*. Strasenburgh is therefore called upon to explain how and why such changes occurred in relic hominoids. We find that relic hominoids more closely resemble fossil forms other than *Paranthropus* and shall explain and substantiate this view below. First, however, we want to dwell on our understanding of the Hominidae in general (in current classification) in the light of our present knowledge of relic hominoids, adopting Porshnev's ideas as a working hypothesis.

## The Uniqueness of Hominids

Porshnev's most striking and unorthodox thesis is that what at present are termed *H. neanderthalensis*, or even *H. sapiens neanderthalensis*, were actually animals, not men. Stunned by this thought, the reader should not overlook some other things we have said about these creatures, namely, that they were the highest animals possible: any further advance meant their turning into man. Thus, to understand the new theory it is necessary to remember that it not only assigns pre-*sapiens* hominids to the animal kingdom, but also recognizes their qualitative difference from all other animals. We have mentioned some aspects of hominid uniqueness earlier (CA: 15 452–56) and want to stress some others here.

When the anthropoid ape evolved into a hominid, it was a case of a warm-clime-forest-dweller becoming a creature capable of living in any landscape—forest, desert, rocky mountains, swamps—and eventually in any climatic zone. It was also, judging by the evidence of *H. sapiens* and relic hominoids, a matter of the appearance in the order Primates of able swimmers to whom rivers were no barriers. In itself a great evolutionary achievement, this ecological autonomy of the hominids, especially in their late stages, was mainly due to their biggish brains and free hands, which managed to eke out a hand-to-mouth existence under any conditions. Hence the hominid conquest of the earth.

Normally, in the animal kingdom such vast and varied geographic distribution of a taxon would have led to adaptive radiation into a great many species, but this apparently was not the case with the Hominidae. Paradoxically, they seem to be genetically a more closely knit family, especially in the late stages of their development, than the Pongidae. This is suggested by man's genetic and physiological closeness to the chimpanzee and the gorilla. Since pre-*sapiens* hominids in a sense occupy an intermediate position between *H. sapiens* and the apes, it is clear that they had even greater affinity to *H. sapiens* and to one another than exists between *H. sapiens* and the living anthropoids.

Morphologically, however, they varied considerably, as is known from the fossil record. How can we account for this? Can we imagine a systematic group whose morphological diversity has outstripped, so to speak, its taxonomic diversity? Such a situation is quite familiar among the domestic animals. The astounding variety of dogs, for example, all of whom belong to one species, far exceeds in morphological terms the difference that exists between species of the wild Canidae such as, say, the wolf and the jackal.

The canine breeds, however, are the product of man's activity. Who or what was responsible for the variety of the hominids? We believe that the near-human qualities of these creatures could have been at least partly responsible for the conditions of their evolution, though, of course, pre-*sapiens* hominids did not pursue or even perceive the process consciously. Simply put, a higher primate that had risen from all fours to a habitual upright position and had a big brain and lots of curiosity was more than any other animal liable to, figuratively speaking, behold the horizon and start wondering what was beyond it. Hominids' mobility took them into all sorts of environments in many a land, and their ethological ingenuity and vitality made it possible for them to colonize those newly discovered areas.

Undoubtedly, new habitats made new demands on the colonizing hominids and produced selective pressures of different kinds. According to Porshnev's theory, all early hominids passed through what we have termed a stage of brain-and-marrow eating. This made them bipedal, and at the same time it must have influenced their dentition. We know that peculiarities of the hominid dentition appear very early in the history of the family and continue to develop in the same direction—though not without setbacks and deviations—from the Australopithecinae to *H. sapiens*. It is the setbacks and deviations that interest us here.

Hominids were omnivorous, and, because the mode of feeding plays a major role in natural selection, it mattered a great deal which part of the diet predominated in the particular habitat. In habitats where vegetable ingredients in the hominid diet predominated over such soft animal foods as brain and marrow, the creatures must have developed or retained bigger molars, stronger chewing muscles, and rougher surfaces of the skull—in short, some of the features of specialized forms. Along with some other students, however, we assume that specialization in hominids was not nearly as intense as in other zoological families and that the tendency to speciate was accordingly reduced. As Mayr (1970, 394–95) has put it,

> In the animal kingdom the invasion of a new adaptive zone usually results in a burst of adaptive radiation into various subniches. This has not happened in the history of the family Hominidae. . . . [One] reason is that isolating mechanisms in hominids apparently develop only slowly. There have been many isolates in the polytypic species *Homo sapiens* and in the species ancestral to it, but isolation never lasted sufficiently long for isolating mechanisms to become perfected.

This means in practical terms that there was a lot of interbreeding among hominid populations as wave of invasion followed wave in the history of the family. It is as if dog owners were to let their pets loose to interbreed freely from time to time, with dog breeders continually starting their work anew from the motley material of sundry mongrels.

Yet there must have been a time and an event which set a limit to the hominid-family "incest" that follows from Porshnev's theory, and that momentous event was the emergence of *H. sapiens*. Apparently our sapiens ancestors were sapient enough to realize the uniqueness of their breed and take great pains to preserve it. We surmise this from the universal taboo surrounding the relic hominoid among all the indigenous populations today. We think that much of the mystery and deification or condemnation of the creature in historic times is due to the fact that he has been a potential, and sometimes actual, "diluter" of the human race. Thus, if no permanent natural barriers to hominid interbreeding existed—at least in the late hominids—then there appeared an artificial, social barrier with the advent of *H. sapiens*. Like any social barrier, it was not absolute, and there must have been a certain amount of interbreeding between *H. sapiens* and the pre-*sapiens* creatures. It is by gene flow resulting from this that certain racial traits common to *H. sapiens* and some pre-*sapiens* forms are to be explained, as is suggested by Roginsky (1969, 139)—to whose theory of "broad monocentrism" in the origin of *H. sapiens*, in opposition to the polycentrism expounded by Coon and others, we subscribe.

Such, in brief, is our understanding of the nature and history of the Hominidae in the light of our present knowledge. Much of the taxonomic relationships and status in the family remains unclear, and any hominid classification (including Porshnev's) has to be taken as a tentative one. Nonetheless, just as in translation a poor dictionary is better than none at all, so in zoology an inadequate classification is better than no classification. What we find valuable in Porshnev's classification of the higher primates is its central idea, the principle of drawing a line between man and animal, not necessarily its details of taxonomic names and ranks.

Any classification is the work of man, and its existence can ultimately be justified only by its correspondence to the work of Nature. Relic hominoids being flesh and blood, their existence does not depend on any classification, but the existence of any hominid classification is bound to depend on the nature of relic hominoids. We expect that when the creatures are finally discovered and recognized by science, the history of primatology and related sciences will be sharply divided into "before" and "after" this event. In short, to size up the creature we seek with the existing taxonomy is like measuring an object with a measure which is bound to be changed when the measurement is finished. Yet engage in this strange procedure we must, if only to show that our "wards" are no freak *sapiens* or visitors from outer space.

Before applying the hominid yardstick to the hominoid (we use this term in its etymological sense of "manlike" and *not* in its taxonomic meaning)—that is, trying to solve the mystery of the "snowman," "wild man," or whatever you choose to call it with the help of existing paleoanthropological knowledge—let us recall another mystery, one residing within this very body of knowledge. By this we mean the generally accepted view that *H. sapiens* is the only surviving species of the Hominidae. Isn't it mysterious, if not mystical, that we should be the sole-survivors of the whole family, while the nearest family, the Pongidae, boasts several surviving forms? And isn't it possible that by confronting one mystery with the other we shall be able, like detectives, to unravel them both?

The question then arises from which end of the Hominidae to start this confrontation—which form, early or late, to use as the yardstick? Hominologists in the West, including Strasenburgh, start with early forms, whereas Porshnev, relying on his revolutionary theory, used not simply late forms, but the latest, *H. neanderthalensis*. In retrospect, it is clear to us that methodologically, and even simply in terms of common sense, Porshnev was absolutely right. If an orphan were to discover that one of his relatives happened to be alive, whom would he think of as possible candidates? Probably sisters and brothers, then parents and grandparents, and then in the same order cousins and aunts and uncles. Indeed, since *H. sapiens* knows he exists, it is simple logic for him to wonder about the existence or non-existence of his nearest kin. As we well know, however, there has been, and still is, a formidable mental block preventing scientists from heeding this logic in the case of Neanderthalers. Since, thanks to Porshnev, we have no such impediment, we can unhesitatingly scrutinize Neanderthalers as candidates for being the ancestors of at least some of the relic hominoids.

## The Riddle of Neanderthal Disappearance

In the mystery of the Hominidae, the riddle of Neanderthal's disappearance ranks first and foremost. The creatures were the latest of the pre-*sapiens* hominids and, according to the fossil and archaeological record, more widespread on earth than any other hominids except modern *sapiens*. In fact, their traces have been found in all corners of the Old World. Whatever laws or patterns of evolution or social history we apply to the Neanderthalers, we know of no reason they should have disappeared from life in the relatively very short time that separates their recognized fossils or artifacts from our day (not to mention the so-called pseudo-Neanderthal remains, of which we shall write below). We can safely assume they did not have atom bombs, inflation, pollution, big cities, etc. They did have natural cataclysms, on no less a scale than Ice Ages, but they are known to have weathered them. Thus, of all the pre-*sapiens* creatures they were the most recent, the brainiest, the hardiest, and the most numerous. Yet, where are they?

The supposition that they were wiped out by *H. sapiens* is not convincing, because initially they must have outnumbered *H. sapiens*, and there has always been and still is room on earth to avoid the sight of our glorious species. That they might all have mixed with and been absorbed by *H. sapiens* is also implausible, because, as we've said, *H. sapiens* were not that numerous then; there must have been more danger of the latter's being diluted and even absorbed by the Neanderthalers. The absorption hypothesis ignores the existence of social barriers to intermarriage among *sapiens* races and nationalities; it is only reasonable to expect even stricter bans vis-à-vis hominids of different evolutionary status.

Porshnev's theory provides a neat and simple answer to the riddle: today's relic hominoids are yesterday's Neanderthals. An immediate obstacle to the general acceptance of this answer is the discrepancy between the animal ways of the relic hominoids and the human ways ascribed to Neanderthalers. Some of these ways (the making and use of stone tools and fire) are facts, while others (hairless bodies, clothing, religious rites) are interpretations. The interpretations are not much of an obstacle at present to the adherents of the new theory; as to the facts, there is nothing impossible in Porshnev's idea, which we share, that in the past all or some Neanderthalers made stone tools and used fire, while today none, or few, of their relic descendants do so.

Napier (1973, 191) admits the possibility of "pockets of Neanderthalers" surviving today "in geographically remote regions," but only on the condition that they have human ways: "Unless something very extraordinary took place in the evolution of Neanderthalers, any connection between the relatively sophisticated people of the Mousterian culture period and the lumbering hair-clad giants of the Cascade Mountains is highly improbable." Something very extraordinary *did* take place in the evolution of Neanderthalers: their giving rise to our own type (*H. sapiens*, in the nomenclature we prefer, or *H. sapiens sapiens* in the nomenclature of Napier's choosing). Knowing our kind as we do, we realize that nothing more extraordinary could have happened to them! Being by far the more able and eager tool maker, *H. sapiens* ousted the Neanderthaler from all the world's best stone-tool workshops and made him retreat into the wilds, where he had to rely for survival more and more on his animal powers. In fact, we don't know whether all Neanderthal populations were more or less equally advanced in their heyday with regard to tool making and the use of fire, or whether they were ethologically as diverse as *H. sapiens* is culturally diverse today.

Napier says that Neanderthalers "appear to have advanced sufficiently in the ability to conceptualize their thoughts to have conceived of an after-life" (p. 185). They may well appear so to Napier, but the question arises in what form they did their conceptualizing if their very linguistic ability is called into question (and not only by Porshnev; cf. Lieberman and Crelin 1971). Neanderthals are not known to have been able to draw an outline of an animal, which *sapiens* three-year-olds can do, and whether they made up for the deficiency with thoughts of an after-life will probably never be known. What we do know is that the late, or "classic," Neanderthalers have morphological features indicative of a so-called *retrogressive* evolution. Thus Porshnev's theory posits and supplements in ethological (or cultural) terms what is already accepted in morphology.

Besides, as Porshnev has pointed out (CA 15:450), at least some of the Neanderthaloid skeletons found in more recent strata and looked upon as "pseudo-Neanderthal" may be real Neanderthalers, among them the Neanderthaloid Podkumok (Caucasus) skullcap, which is of as recent origin as the Bronze Age, and the remains dealt with by Stolyhwo (1937). Having noted what look like rather late or recent remains of Neanderthalers in the ground, we can go on to search for traces of them in historic times on the ground. Obviously, we cannot expect the object of our interest to be referred to in historical sources in accordance with any nomenclature prevailing today. Accordingly, Porshnev's theory envisages a search for *H. neanderthalensis recens* under such names as pans, satyrs, fauns, sileni, sylvans, nymphs, and countless others. Indeed, the discovery of relic hominoids may be expected to bring about a revolution no less resolute and resounding in what Napier [following F.W. Holiday] calls the "Goblin Universe"—the study and understanding of mythology, and demonology in particular—than in primatology. Since we are preoccupied with biology here, let us start with the presentation of evidence from sources which pertain to natural history. Classic Neanderthalers are known to have lived in Europe, so both in a geographical and historical sense Europe is a testing ground for the Neanderthal-hominoid hypothesis. Our task, then, is, first, to show that Europe has been a habitat of hominoids in historic times and, second, to argue that these hominoids have been none other than relics of the Neanderthalers.

## Some ancient and medieval evidence

A celebrated source of information on hominoids is Titus Lucretius Carus (c. 99 – c. 55 B.C.), who in Book 5 of his *De rerum natura* (Lucretius 1947, 217–18) describes a race of "earth-born" men which

> ...was built up on larger and more solid bones within, fastened with strong sinews traversing the flesh; not easily to be harmed by heat or cold or strange food or any taint of the body. . . . Nor as yet did they know how to serve their purposes with fire, nor to use skins and clothe their body in the spoils of wild beasts, but dwelt in woods and the caves on mountains and forests, and amid brushwood would hide their rough limbs, when constrained to shun the shock of winds and the rain-showers. . . . And like bristly boars these woodland men would lay their limbs naked on the ground, when overtaken by night time, wrapping themselves up around with leaves and foliage.

Modern naturalists and historians of science have praised Lucretius for his foresight (or is it hindsight?) in portraying what one specialist called "for his time a surprisingly accurate picture of the appearance and life of prehistoric man." Yet, nobody has ever wondered out loud how Lucretius succeeded in fathoming things which science only learned some two millennia later, thanks to Darwin and modern archeology. That Lucretius did not rely on clairvoyance or on knowledge confided by Martians is evident from his fantastic description of the origin of these very woodland men. Let it also be noted that Lucretius's prehistoric man did not have the power of speech, did not make tools or use fire, and did not wear clothes or build houses.

For the hominologist there is only one answer to the secret of the ancient philosophers' insight in this matter: they used relic hominoids as models for their portraits of prehistoric man (fig. 1). From contemporary reality they were aware of the hairy hominoid, the skin-clad barbarian, and their own civilized selves, and on the basis of these three points in man's development they traced in their imaginations a curve of man's historic rise. It was not much more difficult than, looking at the "bristly boars," surmising the origin of domestic pigs and describing the life of their wild ancestors.

We do not know why Lucretius did not name his models, but we guess that he did not feel like mixing natural history with the names of satyrs, fauns, etc.—for such were the popular appellations of hominoids in the Greco-Roman world of his day. Yet ancient authors did use these names from time to time in a rather down-to-earth manner. According to Plutarch (1792, 349), when the Roman general Sulla (old spelling Sylla), having sacked Athens in 86 b.c., was returning with his army to Italy, he came to Dyrrachium (modern Durres in Albania):

> In that neighborhood stands Apollonia, near which is a remarkable spot of ground called Nymphaeum. The lawns and meadows are of incomparable verdure. ... In this place, we are told, a satyr was taken asleep, exactly such as statuaries and painters represent to us. He was brought to Sylla, and interrogated in many languages who he was; but he uttered nothing intelligible; his accent being harsh and inarticulate, something between the neighing of a horse and the bleating of a goat. Sylla was shocked with his appearance and ordered him to be taken out of his presence.

This gem of ancient evidence is corroborated in certain details by reports of modern sightings: we have two reports, for example—one from Central Asia, the other from the Caucasus—of people stumbling on a sleeping hominoid. Capture or killing of hominoids during *sapiens* military activities in various epochs is also quite well referenced in our files.

Geographer Pausanias (2d century a.d.), in his *Description of Greece*, says that the silenus race (fig. 2) must be mortal, since their graves are known. He also says that when satyrs grow old, they are called sileni. And when we read in Pliny the Elder's *Natural History* (5.8) that "the Satyrs have nothing of ordinary humanity about them except human shape," we know exactly what he means. As a matter of fact, there are both realistic and "surrealistic," or symbolic, representations of the hominoid in ancient art. Those students who, mindful of the satyrs' and others' traditional beastly attributes, such as hoofs, horns, and tails, are prone to think of them in purely mythological terms, seem to do no better than those of tender age who take the fairy-tale attributes of a Santa Claus too much to

FIGURE 1. Hominoid in attack, one of four hominoid figures depicted on a bowl of Carthaginian or Phoenician origin, dated 7th century B.C., found among the treasures of a Roman villa in Palestrina. Note the low cranium, prognathism, nose with deeply sunk bridge, and considerable knee bending in locomotion. (Reprinted from Gini 1962: fig. 5.)

FIGURE 2. Silenus; note lack of hair on hands, knees, and feet. (Reprinted from Reinach 1906, 414.)

heart and fail to see a biological reality behind his mask. The symbolic signs of the hominoid in art and folklore initially served the purpose of identifying the creature and distinguishing him from both humans and animals (fig. 3). Besides, symbolism, like euphemism, tends to sprout under the influence of emotion and mystery, and these have always been part and parcel of man's relationship with the hominoid. This is, however, another vast theme which for lack of space we dare not pursue.

Besides a fair number of mythological images, which have played not a small part in European culture, we owe to the "classic" hominoids of the Greco-Roman epoch such notions as "satyriasis" and "nymphomania," which reflect certain traits of hominoid ethnology and hint at the problems of man's relationship with creatures endowed with such patterns of behavior, and even one or two anatomical terms which seem to reflect certain peculiarities of hominoid physique. (This is not to forget, of course, the term "fauna" and the erroneous shifting of the names "Pan" and "Satyrus" to give zoological terms to apes.)

FIGURE 3. Hominoid and human from a Greek vase, suggesting that the ancients had ways of making friends with hominoids. (Similarly, Jane Goodall has shown that it is possible to make friendly contact with anthropoids in the wild.) Note the lack of hair on the hands and feet of the creature, called "Silenopappos" by art specialists, and the more than normal human knee-bending in locomotion. The presence of a tail is symbolic. Other scenes in Greco-Roman art show hominoids making love, drinking wine, sleeping, carrying loads, stealing fruit from orchards, and pulling thorns from one another's soles. (Reprinted from Reinach 1899, 19.)

The hominoid's presence in medieval Europe is amply documented in Bernheimer's *Wild Men in the Middle Ages* (1952). The evidence amassed therein is the more impressive when we consider that the author is overtly biased against the "wild man" and regards the hero of his book as fiction, not fact. Even this treasure-trove of hominology could not include all the wealth of the theme, however, and as an example of the omitted material, we should mention Albertus Magnus (1193–1280), who is characterized by encyclopedias as a philosopher deeply interested in natural science. In *De Animalibus* (2.1.4.49–50), he cites the recent capture in Saxony of two (male and female) forest-dwelling hairy monsters much resembling human beings in shape. The female died of blood poisoning caused by dog bites, while the male lived on in captivity and even learned the use, albeit very imperfectly, of a few words. The creature's lack of reason, concludes Albertus, is evidenced by, among other things, his ever trying to accost women and exhibit lustfulness.

## Eighteenth-century Evidence

In 18th-century Europe, hominoids became "the last of the Mohicans" in the West of the continent. Here a good source of information in English is *Wolf-Children and Feral Man* (Singh and Zingg 1942). As regards "feral man," the authors' credentials are above suspicion because, like Bernheimer, they didn't know what they were writing about. For example, they report (citing Tafel 1848, 123–24) the following case from Spain (p. 230):

> According to Le Roy, in the Pyrenees, shepherds who herded their flocks in the wood of Ivary, saw a wild man in the year 1774, who lived in clefts in the rocks. He appeared to be about thirty years of age, was very tall with hair like that of a bear, who could jump and run as quickly as a chamois. He appeared to be bright and happy and, according to appearances was not ungentle in character. He never had anything to do with anyone, and had no apparent interest in so doing. He often came near to the huts of the shepherds without making any attempt to take anything. Milk, bread, and cheese appeared to be unknown things to him, for he would not even take them when they were placed in his way. His greatest pleasure was to frighten the sheep, and break up the herds. When the shepherds put dogs on him, as they often did, he would disappear as quickly as an arrow shot from a bow, and he allowed no one to come near him. One morning he approached one of the huts, and as one of the people who belonged there approached him to catch him by the foot, he laughed and fled. . . . No one knows what became of him.

Turning to Eastern Europe, we learn (p. 219, citing Rauber 1888, 49–50, who cites Virey 1817) that in 1767 the inhabitants of Fraumark in lower Hungary, pursuing a bear in the mountains, came to a cave in the rocks in which a completely naked wild girl was found. She was tall, robust, and seemed to be about eighteen years old. Her skin was brown and she looked frightened. Her behavior was very crude. They had to use violence to make her leave the cave. But she did not cry and did not shed any tears. Finally they succeeded in bringing her to Karpfen, a small town of the county of Atlsohl, where she was locked up in an Asylum. She would only eat raw meat.

A very detailed description of a hominoid's morphology and his behavior in captivity (which contrasts dramatically with the creature's state in the wild) is given as follows (pp. 237–40, citing Rauber 1888, 49-55, who cites Wagner 1796; we have added emphasis to the description of those features which help identify the creature taxonomically):

Here you have information about the wild boy who was found a few years ago in the Siebenburgen-Wallachischen border [Rumania] and was brought to Kronstadt [now Brasov], where in 1784 he is still alive. How the poor boy was saved from the forests ... I cannot tell. However one must preserve the facts, as they are, in the sad gallery of pictures of this kind.

This unfortunate youth was of the male sex and was of medium size. He had an extremely wild glance. *His eyes lay deep in his head*, and rolled around in a wild fashion. *His forehead was strongly bent inwards*, and his hair of ash-gray color grew out short and rough. He had heavy brown *eyebrows, which projected out far over his eyes*, and a small *flat-pressed nose*. His *neck* appeared *puffy,* and at the windpipe he appeared goitrous. His *mouth stood somewhat out* when he held it half open as he generally did since he breathed through his mouth. His tongue was almost motionless, and his cheeks appeared more hollow than full, and, like his face, were covered with a dirty yellowish skin. On the first glance at this face, from which a wildness and a sort of animal-being shone forth, one felt that it belonged to no rational creature. . . . The other parts of the wild boy's body, especially the back and the chest were very hairy; *the muscles on his arms and legs were stronger and more visible than on ordinary people*. The hands were marked with calluses (which supposedly were caused by different uses), and the skin of the hands was dirty yellow and thick throughout, as his face was. On the fingers he had very long nails; and, on the elbows and knees, he had knobby hardenings. The toes were longer than ordinary. He *walked* erect, but *a little heavily.* It seemed as if he would *throw himself from one foot to the other. He carried his head and chest forward.....* He walked bare-footed and did not like shoes on his feet. He was completely lacking in speech, even in the slightest articulations of sounds. The sounds which he uttered were ununderstandable murmuring, which he would give when his guard drove him ahead of him. This murmuring was increased to a howling when he saw woods or even a tree. He seemed to express the wish for his accustomed abode; for once when he was in my room from which a mountain could be seen, the sight of the trees caused him to howl wretchedly....When I saw him the first time, he had no sense of possession. Probably it was his complete unfamiliarity with his new condition, and the longing for his earlier life in the wilds, which he displayed when he saw a garden or a wood. Similarly I explain why, at the beginning, he showed not the slightest emotion at the sight of women. When I saw him again after three years this apathy and disrespect had disappeared. As soon as he saw a woman, he broke out into violent cries of joy, and tried to express his awakened desires also through gestures. . . . Yet he showed anger and unwillingness when he was hungry and thirsty; and in that case would have very much liked to attack man, though on other occasions he would do no harm to men or animals. Aside from the original *human body* which usually causes a pitiful impression in this state of wildness, and aside from walking erect, one missed in him all the characteristic traits through which human beings are distinguished from the animals; it was rather a much more pitiful sight to see how this helpless creature would *waddle around* in front of his keeper growling and glaring wildly, and longing for the presence of animals of prey, insensible to everything which appeared before him. In order to control this wild urge, as soon as he came near to the gates of the city, and approached the gardens and woods, they used to tie him up in the beginning. He had to be accompanied by several persons, because he would have forced himself free and would have run away to his former dwelling. In the beginning his food consisted only of all kinds of tree leaves, grass, roots, and raw meat. Only very slowly did he accustom himself to cooked food; and, according to the saying of the person who took care of him, a whole year passed before he learned to eat cooked food when very obviously his animal wildness diminished. I am unable to say how old he was. Outwardly he could have been from twenty-three to twenty-five years old. Probably he will never learn how to speak. When I saw him again after three years, I still found him speechless, though changed very obviously in many other respects. His face still expressed something animal-like but had become softer. . . . The desire for food, of which he now liked all kinds (particularly legumes) he would show by intelligible sounds. He showed his visible contentment when one brought him something to eat, and sometimes he would use a spoon. He had gotten used to wear shoes and other clothes; but he was careless about how much they were torn. Slowly he was able to find a way to his house without a leader; the only work for which he could be used consisted of giving him a water jug which he would fill at the well and bring to the house. This was the only service which he could perform for his guardian. He also knew how to provide himself with food by diligently visiting the houses where people had given him food. The instinct of imitation was shown on many occasions; but nothing made a permanent impression on him. Even if he imitated a thing several times, he soon forgot it again, except the custom which had to do with his natural needs, such as eating, drinking, sleeping,

> etc., and everything which had connection with these. He found his home in the evening, and at noon, the house where he expected food, led only by his habits. He never learned to know the value of money. He did accept it but only with the intention of playing with it, and did not care when he lost it again. Chiefly he was in every respect like a child whose capacities had begun to develop, only with this difference that he was unable to speak and could not make any progress in that regard. He showed his likeness with a child in the fact that he would gape at everything which one showed to him; but, with the same lack of concentration, he would change his glance from the old objects to new ones. If one showed him a mirror he would look behind it for the image before him. But he was completely indifferent when he did not find it, and would allow the mirror to get out of his range of vision. The tunes from musical instruments seemed to interest him a little, but it was a very slight interest which did not leave any impression. When I led him in front of the piano in my room, he listened to the tunes with an apparent pleasure, but did not dare to touch the keys. He showed great fear when I tried to force him to do so. Since 1784, the year he left Kronstadt, I never had a chance to receive any more reports about him.

If only all observers of unusual phenomena recorded them as thoroughly as the author of this report! We hold that the evidence speaks for itself and requires little comment, despite our urge to dwell at length on these marvels.

It is clear from the above that there is no need to travel to "geographically remote regions" to find evidence of the hominoids. With the right theory in mind, one need only visit libraries and museums to discover ample proof of their presence in historic times in the center of Europe. It is also clear that the creatures of the cases we have cited are by no means abnormal *sapiens*, but morphologically resemble modern man to the same extent as Neanderthalers do. Furthermore, the features of the "wild-boy of Kronstadt" perfectly match Neanderthal characteristics, both in morphology and locomotion (the latter as deduced from fossil remains). Examining the artistic representation of the hominoid, we also see unmistakably Neanderthaloid features in all those snub-nosed pans, fauns, satyrs, etc. The best portrayal of the hominoid, side by side with *H. sapiens* so that the former's Neanderthal traits become absolutely clear by comparison, is the sculpture of a wild man on the north portal of Notre-Dame, Semur-en-Auxois, Burgundy (fig. 4). The low cranial vault, the size of the facial skull and some of its features, the "seat" of the head all bespeak a very good Neanderthaler.

FIGURE 4. Wild man and peasant, north portal, Notre-Dame, Semur-en-Auxois, Burgundy, France, 13th century. (Reprinted from Bernheimer 1952: fig. 7 by permission of the President and Fellows of Harvard College; original in Archives Photographiques, Paris.)

Let us sum up: Neanderthal fossils (or at least fossils that look Neanderthal) in Europe are known from prehistoric and historic times. The causes of Neanderthal disappearance are unknown. Europe in historic times is known to have been the habitat of hominoids that looked Neanderthal. Ergo, these hominoids are Neanderthalers.

Now, where does *Paranthropus* fit into this picture? We think nowhere. If Strasenburgh were to claim the cases we have cited for his hypothesis, he would have to explain how *Paranthropus* had evolved to resemble Neanderthal so much. Considering the different evolutionary paths implied for these types of higher primates, in the case of *Paranthropus* for as long as millions of years, such parallelism in morphology needs some explanation. If Strasenburgh were to put forth his own cases, he would have to accept three types of bipedal primates in Europe in historic times: *H. sapiens,* Neanderthal, and *Paranthropus*. As Ockham used to say, "Plurality is not to be assumed without necessity." It's up to Strasenburgh to show such a necessity in this case. In the meantime, *Paranthropus* in Europe is the odd man out.

## Evidence from the Caucasus and Central Asia

Even if Strasenburgh doesn't want to engage us in Europe and seeks battle elsewhere, it is clear that what is true in the west of Eurasia cannot be ignored in its other parts, especially since Neanderthal fossils and/or artifacts have been found in all corners of it. Let us move east now to the Caucasus, lying on Europe's border with Asia, and still a habitat of relic hominoids. That the creatures there are of Neanderthal origin is likely from, among other things, the Neanderthaloid Podkumok skullcap, dated to historic times, and V.S. Karapetian's evidence pertaining to a definitely non-*sapiens* creature so much resembling *sapiens* that it was suspected of being an enemy saboteur in disguise (Hunter with Dahinden 1973, 179–80).

Farther east, in Central Asia, we have the Teshik-Tash Neanderthal find, the Nizami al-Arudi evidence (CA 15:454), the "wild man" in the *Anatomical Dictionary* (Vlcek 1959), and the modern eyewitness evidence of Major General M. S. Topilsky. Topilsky's account, which deals with events of 1923 in the western Pamirs, runs in part as follows (Zerchaninov 1964):

> We recovered the body all right. It had three bullet wounds. . . . At first glance I thought the body was that of an ape: it was covered with hair all over. But I knew there were no apes in the Pamirs. Also, the body itself looked very much like that of a man. We tried pulling the hair, to see if it was just a hide used for disguise, but found that it was the creature's own natural hair. We turned the body over several times on its back and its front, and measured it. Our doctor (who was killed later the same year) made a long and thorough inspection of the body, and it was clear that it was not a human being.
>
> The body belonged to a male creature 165–170 cm. tall, elderly or even old, judging by the grayish color of the hair in several places. The chest was covered with brownish hair and the belly with grayish hair. The hair was longer but sparser on the chest and close-cropped and thick on the belly. In general the hair was very thick, without any underfur. There was least hair on the buttocks, from which fact our doctor deduced that the creature sat like a human being. There was most hair on the hips. The knees were completely bare of hair and had callous growths on them. The whole foot including the sole was quite hairless, and was covered by hard brown skin. The hair got thinner near the hand, and the palms had none at all, but only callous skin.
>
> The color of the face was dark, and the creature had neither beard nor moustache. The temples were bald and the back of the head was covered by thick, matted hair. The dead creature lay with its eyes open and its teeth bared. The eyes were dark, and the teeth were large and even and shaped like human teeth. The forehead was slanting and the eyebrows were very powerful. The protruding jawbones made the face resemble the Mongol type of face. The nose was flat, with a deeply sunk bridge. The ears were hairless and looked a little more pointed than a human being's with a longer lobe. The lower jaw was very massive.
>
> The creature had a very powerful chest and well developed muscles. We didn't find any important anatomical differences between it and

man. The genitalia were like man's. The arms were of normal length, the hands were slightly wider and the feet much wider and shorter than man's.

Says Strasenburgh: "The data which have been amassed on the unknown hominid my Russian colleagues refer to as a 'relic hominoid' attest to the similarity between it and *Paranthropus* in every particular which can be compared. Those under the impression that the supposed extinction of *Paranthropus* has any valid theoretical or evidential basis would do well to reexamine the question." We readily endorse this statement, but with a slight modification: *Paranthropus* should be replaced by Neanderthal.

## A Quick Look at the American Hominoids

We deliberately leave aside the Himalayan Yeti for the time being, because we do not yet have the same quality of material on it as we have on the hominoids in Europe and in this country. Judging by the available data, the American hominoids look more "archaic" than their European counterparts. Whether this "archaism" is due to the specific environmental conditions and geographical isolation of the hominoids in America or to the fact that they represent an earlier stage of hominid evolution than the Neanderthal is not clear at the moment, but some preliminary considerations of a general nature are in order. It is recognized that the higher primates initially entered America from Asia via the Bering land bridge, which was situated, let it be stressed, in northern latitudes. Not all higher primates living at the time in Asia could get to America. (The orangutan, for example, couldn't, because it lives in the south.) That *H. sapiens* could and did is a matter of fact, and it is only logical to suppose that his immediate evolutionary predecessors, who are known to have lived in northern lands, were also able to cross from Asia to America. Thus any proponent of *Paranthropus* or *Gigantopithecus* or any other early form of higher primates in America must first show that his chosen form could live in the north and then solve the problem of its coexistence with later non-*sapiens* forms that must have entered the continent just as *H. sapiens* did.

## Conclusion

We must apologize to the reader for the unusual number and length of the quotations in this paper. We hope it is realized that these literary references are as valuable and revealing for anthropology in this case as fossil relics are in other cases. Without such a broad-minded approach, this research could not even have been begun,

We hold that the immediate task of the hominologist is to take care of the Neanderthalers, assuming in the meantime that they may have taken care of *Pithecanthropus*, and the latter of *Paranthropus*—or, perhaps, *Paranthropus* took care of himself. In other words, in the case of early hominids there may have been enough time for such factors as evolution, absorption, or extinction to account for the absence of these forms today, whereas with the late forms such factors don't seem to make a convincing case. Yet we realize that in matters of natural history man proposes and Nature disposes, and if, despite our arguments, all relic hominoids turn out to be none other than descendants of *Paranthropus*, we will sincerely congratulate our colleague Strasenburgh.

## References Cited


Bernheimer, Richard. 1952. *Wild men in the Middle Ages*. Cambridge: Harvard University Press.

Gini, Corrado. 1962. Vecchie e nuove testimonianze o pretese testimonianze sulla esistenza di ominidi o subominidi villosi. *Genus* 18(1–4).

Hunter, Don, with René Dahinden. 1973. *Sasquatch*. Toronto: McClelland and Stewart.

Lieberman, Philip, and E.S. Crelin. 1971. On the speech of Neanderthal man. *Linguistic Inquiry* 2:202–22.

Lucretius, Titus Carus. 1947. *De rerum natura*. Translated by Cyril Bailey. Oxford: Oxford University Press.

Mayr, Ernst. 1970. *Populations, species, and evolution*. Cambridge: Harvard University Press.

Napier, John. 1973. *Bigfoot*. New York: Dutton.

Plutarch. 1792. *Plutarch's lives*. Translated from the Greek by John Langhorne and William Langhorne. London: Dilly.

Rauber, August. 1888. *Homo sapiens ferus, oder Die Zustände der Verwilderten und ihre Badeutung für Wisserschaft Politik und Schule*. 2nd ed. Leipzig.

Reinach, Salomon. 1899. *Repértoire des vases peints grecs et étrusques*. Vol. 1. Paris: Ernest Leroux.

———. 1906. Deuxieme edition. *Repértoire de la statuaire grecque et romaine*. Vol. 1. Paris: Ernest Leroux.

Roginsky, Yakov. 1969. *Problems of anthropogenesis* (in Russian). Moscow: Vysshaya Shkola.

Singh, J.A.L., and Robert M. Zingg. 1942. *Wolf-children and feral man*. New York and London: Harper.

Stolyhwo, K. 1937. *Les praenéanderthaloides et les postnéanderthaloides et leur rapport avec la race du Néanderthal*. Ljubljaná.

Tafel, J.F.I. 1848. *Die Fundamentalphilosophie in genetischer Entwickelung mit besonderer Rucksicht auf die Geschichte jedes einzelnen Problems*. Tubingen: Expedition.

Virey, J.J. 1817. *Nouvelle dictionnaire d'histoire naturelle*. Paris.

Vlcek, Emanuel. 1959. Old literary evidence for the existence of the "snow man" in Tibet and Mongolia. *Man* (London), August, pp. 133–34.

Wagner, Michael. 1796. *Beiträge zur philosophischen Anthropologie und den damit Verwandten Wissenschaften*. 2 vols. Vienna: Joseph Stahel.

Zerchaninov, Yuri. 1964. Is Neanderthal man extinct? *Moscow News*, February 22.

(Originally published in *Current Anthropology*, June 1976)

SECTION 9

# Hominology and Linguistics

## Some Thoughts on the Origin of Speech


### Abstract

The paper presents a theoretical and historical basis of hominology—the science of extant non-human hominids, termed homins. The origin of speech is the key problem in this new discipline. Presented are views of Charles Darwin on the nature and origin of speech, his mention of sound imitation as a first step in the process; as well as the views of Darwin's opponent, St. George Mivart, who called language the "rubicon of mind." In this connection the question of continuity and discontinuity, i.e., change in kind, in evolution is discussed. Scavenged bone-carrying and crushing is suggested as the main factor of selection responsible for bipedalism and the emergence of hominids from the pongids. Possible causes of encephalization in hominids are suggested.

Charles Darwin encountered evidence of "wild men" in Tierra del Fuego. Linnaeus introduced the term *Homo sapiens* because he had information on non-sapient *Homo troglodytes*. The article explains why paleoanthropologists discarded and forgot *Homo troglodytes*. The author equates homins with purported "wild men" and cites historical and modern data on their existence. The creatures' vocalizations are discussed and their sound imitating ability is specially noted. In the author's opinion, the homins are devoid of speech but have come close to obtaining it. Boris Porshnev's paradigm and contribution to man's self-knowledge and hominology are discussed, with the conclusion that language is not only a means of human communication but also of reproduction.

The subject of the article is at the root of such problems as the origin and nature of man, the origin and nature of homins, and the origin and nature of academic resistance to hominology. The author discusses only a limited number of aspects of this vast and profound theme, namely, those aspects with relevance to the present state of hominology.

## The Second Signaling System

Speech can be defined as verbal communication through air vibration. Non-verbal communication through air vibration is not speech, neither is verbal communication without air vibration (telepathy, for example). Speech is an exclusively human faculty and function. "Non-human speech" makes as much sense as "non-human poetry." Speech cannot be inarticulate. "Inarticulate speech" is as absurd as "inarticulate eloquence." There is no speech without words and it is words that make speech articulate. Appropriate names for inarticulate vocalizations are mumbling, gabbling, babbling, and gibbering.

Speech has evolved from non-verbal animal communication, termed by Ivan Pavlov the first signaling system, while speech he called the second signaling system. Because the second emerged from the first, the two systems have much in common, a fact I stressed in *Current Anthropology* (December 1974, p. 455):

> There are many points on which man's speech and the communication systems of animals coincide, but there are others on which they are as far apart as heaven and earth. By the communication means at their disposal animals can greet, warn, threaten, frighten, order, tease, invite, entice, deceive, ask for, beg, give consent, and show indifference, surprise, bewilderment, respect, contempt, contentment. A bee through her dance can indicate to her sisters the direction and distance to nectar-laden flowers, which the instructed bees don't fail to find. Thus both animals and humans do use symbols to influence their counterparts' behavior in their respective kingdoms. But what animals can't do, what is the sole prerogative of man, is to engage in a symbolic give-and-take which we happen to be performing right now and which is called discussion. Animals can 'argue' with paws and claws, but not with symbols.

The difference between the first signaling system and the second is not just a matter of degree, but of kind. The secret of speech is in the secrets of the word, its meaning, formation, and combination. Isn't it wondrous that a mere alteration in the order of sounds turns "dog" into "god"?

Words have the capacity to evoke mental images. Saying "dog" or "cat," one evokes a corresponding mental image in the mind. The word "dog" has nothing canine, the word "cat" nothing feline; so most words are arbitrary sound symbols. But, it may not have been always like that. If I say "bow-wows growl, meow-meows hiss," you probably understand what I mean. The linguistic term for sound-imitating words is "onomatopoeic" (Greek for "name making").

Many words in various languages are onomatopoeic; for example, in English the words hoot, zoom, buzz, whisper, whistle, bang, plop, rumble.... Such names as "mumbling," "gabbling," and "babbling" are also onomatopoeic. From ancient times to the present, scholars have shared the view that sound imitation was indispensable in the origin of language. Onomatopoeia is perhaps the "most fruitful mother of language." One of the most detailed works on the subject that the author has come across is by A.M. Gazov-Ginzberg, published by the Soviet Academy of Sciences in 1965, entitled *Is Language Imitative by Origin*? The scholar comes to an affirmative conclusion.

There is a parallel between the development of spoken language and written language in the emergence of abstract and arbitrary symbols from the initially imitative (iconic) and non-arbitrary. As one reference encyclopedia says: "Early Man drew rough sketches in order to convey his ideas to another. This method was succeeded in the course of time by a system of hieroglyphics." Hieroglyphics in turn gave rise to alphabetic writing, with its abstract and arbitrary signs and sounds in the form of letters. To be able to write, man had first to learn to draw. To be able to speak, he first had to be able to imitate sounds.

Now, let us note that of all mammals, presently recognized by science, man is the only one capable of sound imitation. Curiously enough, man shares this faculty, as well as bipedalism, not with primates, but with evolutionarily distant birds. The lesson of the parrot is, first, that sound imitation is in the nature of biological things, and, second, the faculty is not necessarily a precursor of speech or eloquence.

Our primate ancestors acquired hands not in order to enable us to clutch a steering wheel or hold a cell phone. And they became bipedal not in order to waltz or skate. So, is it not possible that they developed their vocal and sound imitative abilities in advance of speech and not for that purpose at all?

An infant prattles before talking or speaking. A pre-human primate must have mumbled and gabbled before turning human. Thus, the antiquity of speech—the temporal aspect of the problem—is of special importance both for hominology and anthropology.

## Darwin's Thoughts about Language

According to Boris Porshnev, hominology manifests itself as "the third stage of the Darwinian revolution in science." Therefore it is relevant to review Charles Darwin's thoughts on the subject of speech and its origins. The following quotes are from *The Descent of Man and Selection in Relation to Sex*. First published in 1871, this book, in Darwin's words, has passed through "the fiery ordeal." Darwin's main purpose was to substantiate man's evolutionary origin; as such he paid more attention to the characters that unite man with the apes and other animals than those characteristics that set him apart. Some of the terms used by Darwin are not valid today. Quotations are taken from the 1874 edition:

**Chapter III. Comparison of the Mental Powers of Man and the Lower Animals**

Language—This faculty has justly been considered as one of the chief distinctions between man and the lower animals.... The habitual use of articulate languages is, however, peculiar to man; but he uses, in common with the lower animals, inarticulate cries to express his meaning, aided by gestures and the movement of the muscles of the face. This especially holds good with the more simple and vivid feelings, which are but little connected with our higher intelligence. Our cries of pain, fear, surprise, anger, together with their appropriate actions, and the murmur of a mother to her beloved child, are more expressive than any words.

The lower animals differ from man solely in his almost infinitely larger power of associating together the most diversified sounds and ideas; and this obviously depends on the high development of his mental powers.

With respect to the origin of articulate language, after having read on the one side the highly interesting works of Mr. Hensleigh Wedgwood, the Rev. F. Farrar, and Prof. Schleicher, and the celebrated lecturers of Prof. Max Muller on the other side, I cannot doubt that language owes its origins to the imitation and modification of various natural sounds, the voices of other animals, and man's own instinctive cries, aided by signs and gestures.

Since monkeys certainly understand much that is said to them by man and when wild, utter signal-cries of danger to their fellows; and since fowls give distinct warnings for danger on the ground, or in the sky from hawks (both, as well as a third cry, intelligible to dogs), may not some unusually wise ape-like animal have imitated the growl of a beast of prey, and thus told his fellow-monkeys the nature of the expected danger? This would have been a first step in the formation of a language.

A complex train of thought can no more be carried on without the aid of words, whether spoken or silent, than a long calculation without the use of figures or algebra. It appears, also, that even an ordinary train of thought almost requires, or is greatly facilitated by some form of language, for the dumb, deaf, and blind girl, Laura Bridgman, was observed to use her fingers whilst dreaming.

We have, also, seen that animals are able to reason to a certain extent, manifestly without the aid of language. The intimate connection between the brain, as it is now developed in us, and the faculty of speech, is well shown by those curious cases of brain-disease in which speech is specially affected, as when the power to remember substantives is lost, whilst other words can be correctly used or where substantives of a certain class, or all except the initial letters of substantives and proper names are forgotten.

From these few and imperfect remarks I conclude that the extremely complex and regular construction of many barbarous languages is no proof that they owe their origin to a special act of creation. Nor, as we have seen, does the faculty of articulate speech in itself offer any insuperable objection to the belief that man has been developed from some lower form.

## Darwin's Opponent

If in Darwin's opinion the faculty of speech offered no insuperable objection to the belief of man's origin from some lower form, that was not the view of Darwin's staunch critic, St. George Mivart, author of *Genesis of Species* (1871), *Contemporary Evolution* (1876), *Man and Apes* (1877), and *The Origin of Human Reason* (1889).

In *The Origin of Human Reason*, St. Mivart took to task Darwin's follower, G.J. Romanes, author of *Mental Evolution in Man, Origin of Human Faculty* (1888). Romanes speculated that *Homo sapiens* must have been preceded by a form of speechless man, *Homo alalus*. He states, "I believe this most interesting creature probably lived for an inconceivably long time before his faculty of articulate sign-making had developed sufficiently."

St. Mivart found fault as well with Darwin's hypothesis of "the spontaneous vocal imitation by a monkey of some other animals' voice as a sign to denote its presence." St. Mivart further states that he was more than ever confident "that between the intellect of man and the highest psychological power of any and every brute there is an essential difference of kind, also involving, of course, a difference of origin."

According to St. Mivart, there have been on Earth three major transformations of nature, these being:

1) the first introduction of life;
2) the first introduction of sensitivity … the mode in which a creature that did not possess the faculty of feeling, could have been endowed with that wonderful and unprecedented power…. At least two breaches of continuity have certainly occurred, and two novel natures (the living and the sensitive) essentially different in kind, have somehow come to be.
3) Under these circumstances a third breach of continuity was possible and probable.

St. Mivart argues that these breaches of continuity could not have occurred naturally and speaks of:

> ... the formation of a new creature, or the infusion of a new nature …a new nature infused into one which already existed. [Explaining,] the impossibility of the evolution of intellect from sense, [he wrote:] It is always the same kind of fallacy which besets these speculations: sensitive phenomena are supposed to be divided and subdivided till they are imagined to be subdivided enough for the entrance of a grain of conceptual power into them. Such a grain having once been smuggled in unnoticed, there is then really no difficulty in seeing how it may augment till it attains the level of the intellect of a Scotus. But phenomena are not really to be explained by merely being subdivided or even pulverized. Of course, Mr. Romanes himself thus slips in intellect, without saying so, although not with any personal disingenuousness, but with an entirely innocent unconsciousness of what he is doing.... Thus is intellect silently 'slipped in,' and when once it has been so smuggled in unnoticed, it is, of course, easy enough to explain any subsequent progress by it.
>
> With a mind thus freed from the mists of imaginary prejudice, let the reader next consider the arguments in favour of a difference of kind between man and brute–the presence in the former and the ab-

sence in the latter of intellect, as manifested by language, and, above all, by language expressing moral judgments and asserting merit and demerit. We are strongly persuaded that he will then clearly see that language is the 'rubicon of mind'.

## The Rubicon of Mind

Language, the "rubicon of mind." Hear! Hear! It is from St. George Mivart that I first heard of language thus defined.

On the one side, Darwin, with "*natura non facit saltum*" (nature makes no leap), on the other, St. Mivart with "breaks in nature," "breaches of continuity," and the "rubicon of mind." Who is right? Both are! For "Touching on the problem of continuity in evolutionary and historical processes, we can say that Porshnev proceeded from the thesis that in evolution and history slow processes of quantitative change alternate with sudden and stormy processes of qualitative change—in other words, that there is no evolution without revolution." (*Current Anthropology*, December 1974, p. 456.) The revolution that broke out with the origin of speech is known as The Human Revolution. The following are relevant thoughts by scientists, aside from Porshnev.

**Stephen J. Gould:**

Philosophies of change and progress formulating such laws as the 'transformation of quantity to quality' of the Hegelian dialectic: the addition of quantitative steps will lead eventually to a qualitative leap.

*(Ontogeny and Phylogeny)*

Perhaps the most amazing thing of all is a general property of complex systems, our brains prominent among them—their capacity to translate merely quantitative changes in structure into wondrously different qualities of function.

*(The Panda's Thumb)*

**Ernst Mayr:**

Progress is by trial and error. One organ may run far ahead, the others lag behind; periods of stagnation may alternate with periods of explosive advance.

(Accident or Design, The Paradox of Evolution, in *The Evolution of Living Organisms*)

Transformation of quantity to quality is ubiquitous, from physics to chemistry to biology to sociology to cosmology. The melting of ice and the boiling of water are the most familiar examples. Quantitative changes in temperature lead to qualitative changes in the state of matter. Water can be boiled or non-boiled. It can't be half-boiled. Puberty is an example from biology. A growing organism can be either fertile, or infertile, it cannot be partially fertile. Puberty hence is the "rubicon of fertility."

St. Mivart and, independently, Porshnev took speech to be the rubicon of mind, thus denying the possibility of the "speechless man"—*Homo alalus*—theorized by G. J. Romanes. A species of man with "inarticulate speech" is impossible too. This view implies that speech emerged in evolution fully grown-up, like Athena born from the head of Zeus. The difference between St. Mivart and Porshnev, however, is that the latter was a firm Darwinist and evolutionist whilst the former believed that the power of speech was "infused" into our ancestors by "the great Author of nature."

The question arises: Quantitative changes in what have resulted in the qualitatively unprecedented faculty and function called "speech"? Essentially in two biologically interconnected structures, the vocal system and the nervous system, especially the brain.

The vocal capacity that humans possess has been evolving gradually since the amphibious stage. Among primates, gibbons are very good singers, but apes and monkeys are not capable of imitating various sounds. Therefore this vocal ability must have developed at the hominid stage of evolution.

As for the nervous system and the brain, the latter's growth in size and complexity along the evolutionary way leading to man is well known and called "encephalization" (from the Greek for brain, "encephalos"). At the hominid stage, the brain was growing ever more rapidly from the pongid size and eventually doubled the highest pongid quantity. At the same time, as we have concluded, the vocal system attained the ability of sound imitation.

Are there any imaginable reasons for these developments?

## Bipedal Bone-Crackers

Proceeding from Porshnev's ideas, I wrote in 1974:

> When the anthropoid ape found himself on the ground and in the savanna as a result of ecological changes in the Tertiary period, he suffered a decrease in food supply from what he had enjoyed in the

forest; hence his search for dietary substitutes. Because of his morphology, he could not consume grass the way herbivores do, nor could he feed on herbivores the way carnivores do. But he had the hands formed in the forest, and it didn't take him long to put this biological preadaptation to good use. Abundant bones, especially skulls, of savanna-dwelling animals were like shells and nuts which the ape knew how to crush with stones. The only problem was to bring bones and stones together.

Thus bone carrying and crushing was the main factor of selection which made the anthropoid ape bipedal and marked the beginning of the Troglodytes as such. In this respect, Porshnev's theory closely coincides with Hewes's (1961) food-transport hypothesis, the only difference being that the former suggests scavenged bones as the objects carried by would-be bipedal primates, while the latter suggested scavenged meat. Writes Hewes (1961: 687): "Dubrul (1958:90) notes that upright posture is essentially a 'reduction of the repetition of structures serving the same function,' with the forelimbs becoming 'as it were, accessory mandibles rather than locomotor devices,' leading to a 'new mode of feeding and feeding niche.'"

Indeed, the Troglodyte's hands became mighty accessory mandibles, with ever replaceable teeth of stone, which could crush bones of such strength and in such numbers as were beyond the power of all other scavengers, including the hyenas. This bone-cracking, brain-and-marrow-eating stage in the evolution of the Troglodytidae, which we may call a stage of cerebro-and-myelophagia, must have lasted at least a couple of million years.

As a result of this million-year-long process, the ground-dwelling higher primates not only became bipedal, but also got the knack of using stones to provide for their livelihood. A million-year-long application of stones to skeletons taught the Troglodytes that stones were good not only for cracking bones, but also for cutting and mincing meat that remained on some bones they collected. They also learned in the process that only sharp stones, appearing as a by-product of bone smashing, are good for meat cutting. Thus, the next and most important phase in the process was their hunting for skeletal remains with ever more meat on the bones and eventually for the whole carcass, on one hand, and their systematic making of sharp stones, on the other. Such a reconstruction of events makes comprehensible how bipedal primates came to apply hard objects (stones) to soft material (meat), which otherwise seems to be a stroke of genius.

Another, and ultimately the most important, 'by-product' of the

> process was the unusually swift-growing brain of our bipedal scavengers. What were the causes of natural selection of the brainiest in this case? The answer is probably provided by realization that the Troglodyte had not one but several rather demanding tasks on his mind during each feeding cycle: 1) to watch the herbivores, 2) to watch the carnivores, 3) to look for results of their interactions, 4) to be in the right place at the right time to find an adequate carrion supply, 5) to out-fox and out-maneuver carnivore enemies and competitors in getting away with it, and 6) to solve the problem of consumption with the ever present handicap of inadequate teeth through finding and later fashioning 'artificial teeth.'
>
> Thus the Troglodytidae became the brainiest creatures on earth prior to *H. sapiens*. (*Current Anthropology*, December 1974, p. 453.)

The Troglodytidae is the term applied by Boris Porshnev to all hominids except *Homo sapiens*; i.e., he meant "non-human hominids," which I call "homins."

I did not mention sound imitation then, but the phenomenon fits well into the picture already described. Hominids must have interacted with wildlife to a much higher degree than pongids usually do, and having a bigger and more developed brain, they must have started to imitate the sounds of other animals, just as Darwin assumed.

## Looking for "Living Fossils"

What hominid crossed the "rubicon of mind"? All agree and none objects that *Homo sapiens* did. The difference between Porshnev and his countless opponents was his strong insistence that only *Homo sapiens* did the crossing, which actually turned a non-human biped into a human being. During the whole hominid stage (the "Troglodyte" stage) there was a step-by-step evolution by degree, culminating with a change in kind—the birthing of a talking animal, called man.

Porshnev's opponents, on the other hand, believe that *Homo sapiens* was not the first talking animal; the faculty was also present to a lesser degree in *Homo neanderthalensis* and even *Homo erectus*. If they are right, there was no "rubicon crossing," or the "rubicon" must have been as wide as the Gulf Stream, or there were several rubicons—which doesn't seem plausible either.

Is there a way to find out the truth? Charles F. Hockett says:

> A century ago there were still many corners of the world that had not been visited by European travelers. It was reasonable for the European

scholar to suspect that beyond the farthest frontiers there might lurk half-men or man-apes who would be 'living fossils' attesting to the earlier stages of human evolution. The speech (or quasi-speech) of these men (or quasi-men) might then similarly attest to earlier stages in the evolution of language. The search was vain. Nowhere in the world has there been discovered a language that can validly and meaningfully be called 'primitive.' Edward Sapir wrote in 1921: 'There is no more striking general fact about language than its universality. One may argue as to whether a particular tribe engages in activities that are worthy of the name of religion or of art, but we know of no people that is not possessed of a fully developed language. The lowliest South African Bushman speaks in the form of a rich symbolic system that is in essence perfectly comparable to the speech of the cultivated Frenchman."

(The Origin of Speech, *Scientific American*, September 1960).

No, the search was not vain, at least at the time when Hockett published that article. As for a century earlier, the budding scholar who came closest to discovering the "living fossil" of a man-ape was none other than Charles Darwin. In Chapter X of *A Naturalist's Voyage Round the World*, he mentions a story by a Fuegian, named York Minster, educated in England and serving as a guide and interpreter during the expedition's visit to Tierra del Fuego.

> In a wild and excited manner he also related that his brother, one day whilst returning to pick up some dead birds which he had left on the coast, observed some feathers blown by the wind. His brother said (York imitating his manner), 'What that?' and crawling onwards, he peeped over the cliff, and saw 'wild man' picking his birds; he crawled a little nearer, and then hurled down a great stone and killed him... What the 'bad wild men' were has always appeared to me most mysterious; from what York said, when he found the place like the form of a hare, where a single man had slept the night before, I should have thought that they were thieves who had been driven from their tribes, but other obscure speeches made me doubt this; I have sometimes imagined that the most probable explanation was that they were insane.

A 19th-century giant of Russian literature, novelist Ivan Turgenev, also took a mysterious gorilla-like creature for an insane human. The creature accosted him in the river and nearly made him drown. Today we know better. As for Tierra del Fuego, Charles Darwin is not alone in reporting the presence of "wild men" in that remote corner of the world. Without any reference to him, bipedal Fuegian "apes," locally called "Yoshil," have been reported in the 20th century by Dr. Manuel J. Molina, Professor en la Universidad San Juan Bosco de Comodoro Rivadavia, in his article "El Yoshil o Mono Fueguino" (*Karukinka*, 1971).

Commenting on Darwin's information, I wrote in 1984:

> I am much inclined to think that the creatures described as 'wild men' by the savages of Tierra del Fuego were not *Homo sapiens* but *Troglodytes recens* ubiquitous. Realizing that Darwin himself may have been close to a live object of our long and torturous research, undertaken in the light of his great and revolutionary theory, I can't help feeling sort of elation mixed with wonder. It is intriguing to conjecture what course anthropology might have taken had Darwin happened to see the 'bad wild man' whose sleeping place he was shown.
>
> (In *The Sasquatch and Other Unknown Hominoids*, edited by Vladimir Markotic and Grover Krantz)

Now, we have information on "wild men" that precedes Darwin not by centuries, but millennia. It has two main aspects and is reaching us in two main currents: mythological (demonological, folkloristic) and realistic (natural, historical). The first is predominant and ubiquitous (beginning with the Babylonian mythology, the Bible, etc.), the second is sporadic and surreptitious, for the following reasons. In a pre-Darwinian world the naturalistic and biological views on the living non-human hominids could not seriously resist and challenge religious and demonological views and explanations.

Examples of natural science's views on the subject are given in:

- the descriptions of the wild, "earthborn," woodland men (troglodytes, i.e., cavemen) by ancient Greek naturalists and the Roman philosopher Titus Lucretious Carus (1st century B.C.);
- the description of the nasnas by the 12th century Persian scholar Nizami al-Arudi ("This, after mankind, is the highest of animals, inasmuch as in several respects it resembles man; first in its erect stature, secondly in the breadth of its nails, and thirdly in the hair on its head.");
- mention in *De Animalibus* by Albertus Magnus (1193-1280) of two (male and female) forest-dwelling hairy "monsters" captured in Saxony;

- (the most salient fact) the Linnaean classification of primates in which *Homo troglodytes* plays a crucial part.

## Carl Linnaeus

A cardinal question of human life is how to class man. If, "when the saints go marching in," humans want "to build a city where all people can march in," they have to understand human nature, the nature of those who will build the city and those who will march in to live there. The City of God is God's business, the City of Man is Man's business. Therefore self-knowledge is indispensable. That was clear to the ancient Greeks, with their motto of "Know thyself," but apparently not to Marx and Lenin. The grandiose worldwide experiment in realizing their ideas floundered and flopped, as predicted, on account of human nature, or rather ignorance of it. That is why some Russian thinkers, such as Boris Porshnev, feel a keen interest in anthropology. And it was for this reason that I associated with him.

The point of departure for Porshnev was the Linnaean classification. This gives me reason and occasion to discuss Linnaeus and his contribution to our subject. Carl Linnaeus (1707-1778, originally named Karl von Linné) was as great a natural science celebrity in the 18th century as Charles Darwin was in the 19th. He established the binominal system of designation of plants and animals. "To him later naturalists owe the definition of genera and species and the uniform use of specific names. His style is a model of brevity and precision, with no possibility of ambiguous meaning." (Reference encyclopedia entry.) It was said at the time, "God created things. Linnaeus put them in order."

And this is how Linnaeus is viewed by a foremost Soviet primatologist:

> The first efforts in creating the modern basis of taxonomy in primatology are connected with the name of the great classifier of nature, Carl Linnaeus. The son of his century, a sincere religious man, fearful of collisions with the official dogmas, Linnaeus, as a primatologist, stepped nonetheless far beyond the limits of his time. In primatology he is immortalized as one of the most progressive and major builders of this science. It was Linnaeus who, in a century still risky for 'free thinkers,' courageously united man and ape in one zoological order, and named this order Primates, the term hitherto used by the Church.
>
> (Eman Friedman, 1979, *Primaty*, in Russian, Bayanov's translation).

What is less known is that it was Linnaeus who introduced in science the central term of anthropology–*Homo sapiens*–and did so a century before the discovery and study of fossil hominids. And nobody wonders why man was given such an incongruous scientific name. What is also generally unknown, even among members of our "club," is that Linnaeus stands out as the patriarch of hominology and cryptozoology, and that precisely because of his interests and pursuits in this sphere he coined "Homo sapiens" and included man, along with monkeys, in the Primate order.

It was with awe that one day in 1966 I opened and copied relevant pages in Latin from the original 10th edition of Caroli Linnaei *Systema Naturae* (1758), in the library of the Moscow Zoological Museum. This edition launched the Linnaean nomenclature. One of its salient features is that it presents two living species of man: *Homo sapiens* (man the wise) and *Homo troglodytes* (caveman). The first is described as "diurnus, varians cultura, loco," the second as "nocturnus" and "sylvestris." *Homo sapiens* is subdivided into races, and includes *Homo ferus*, which designated, in the opinion of Linnaeus, *Homo sapiens* gone wild (children captured and reared by animals), but actually embraced cases, as it is apparent now, of real "wild men" (i.e., relict hominids) reported at the time in Europe. Right after the term *Homo sapiens* Linnaeus put in the words to address mankind, "Nosce te ipsum" (know thyself).

As for *Homo troglodytes* and *Homo ferus*, Linnaeus made no secret of the fact that these terms and categories were established on the basis of circumstantial evidence supplied by ancient and modern authors and travelers. It was his interest in and use of anecdotal evidence, presented in his most serious scientific work, that makes Linnaeus a perfect hominologist and cryptozoologist and our eminent colleague in the 18th century.

*Homo ferus* and *Homo troglodyte*s evidently filled in for Linnaeus the gap between ape and man and prompted him to establish a single order of primates. On the one hand, there were *Homo sapiens* children reared by animals and turned into beasts; on the other hand, stood *Homo troglodytes* that seemed to be more manlike than apelike, especially on account of bipedalism and the dental system devoid of diastemata, the characteristic of apes and monkeys. (His information included this important detail.) So there is no doubt that man owes his undeserved name of *Homo sapiens* to the presence of non-sapient *Homo troglodytes* in the Linnaean classification.

Still, his information on the subject was so patchy, fragmentary, and contradictory that the great classifier, with his passion for order and exactness, must have been tormented by the lack of precise knowledge in the matter. This is seen from the dissertation *Anthropomorpha* (1760), which he dictated to his St. Petersburg student Christian Hoppius, saying in part the following:

> Is it not amazing that man, endowed by nature with curiosity, has left the Troglodytes in the dark and did not want to investigate the creatures that resemble him to such a high degree? A lot of mortals spend their days in feasts and banquets, and all they care for is how to prosper by honest and dishonest means. No better is the behavior of most navigators who sail to the Indies and who alone happen to see the troglodytes. Driven by greed, they despise the tasks of natural science, such as investigation of the way of life of troglodytes. Just imagine what wondrous objects of diversion for a monarch in his palace such animals could be, for one would never tire of marveling at them. Or is it really difficult for a monarch to get such animals, knowing that people vie with each other to fulfill his orders? And it would be of no small benefit for a philosopher to spend several days in the company of such an animal in order to investigate how much superior human reason is and thus discover the difference between those endowed with speech and those devoid of it. And should I mention what light could be shed for natural science from a detailed description of these animals. As for me, I remain in doubt what specific characteristic distinguishes the Troglodyte from man within the scope of natural history.
>
> (Bayanov's translation from a Russian translation from the Latin, published in St. Petersburg in 1777)

The fervent call of the great naturalist fell on deaf ears. Not only that, but his whole classification of primates, along with the latter term, was condemned and done away with by the scientific establishment of the century, whose creationist faith revolted against Linnaeus's innovation. The job was done by Johann Blumenbach (1752-1840), who in his *Manual of Natural History* (1775) established the order Bimanus for man and the order Quadrumanus for apes and monkeys. As for *Homo troglodytes*, Blumenbach discarded the species altogether as "an unintelligible mixture of pathological cases and the orangutan." He moved the term "troglodytes" to Simia and established "Simia troglodytes or Chimpansi," which implied that chimps were cave-dwellers.

According to S.J. Gould, "Historical changes in classification are the fossilized indicators of conceptual revolutions." Blumenbach's monumental change in the Linnaean classification was then a conceptual counter-revolution, which lasted nearly a hundred years, until resisted and reversed by Darwin's "bulldog," Thomas Huxley (1825-1895), who with *Man's Place in Nature* (1863) restored the single order Primates, as well as the term itself. But *Homo troglodytes* stayed in limbo for another hundred years, until resurrected and vindicated by Boris Porshnev (1905-1972), who proclaimed yet another conceptual revolution.

The struggle for primates is little known in the West today, as I conclude from the fact, among others, that encyclopedias are usually silent about the fundamental role of Linnaeus in primatology, and present him just as "a botanist," not as a philosopher and great naturalist. And, as to the struggle for troglodytes and its Linnaean connection, it's a complete secret to the scientific community. Says a modern specialist:

> Besides *Homo sapiens*, with various subspecies, he [Linnaeus] used the name *Homo troglodytes*, with a description based upon confused travelers' accounts, for a non-existent type of man, and this name was soon dropped. (Dobzhansky states /1962:162/ that Linnaeus included the anthropoid apes in the genus Homo, but a look at the original publication shows that the description of *Homo troglodytes* was not intended for any ape.)
>
> (Anna K. Winner, "Taxonomic Nomenclature in Paleoanthropology," *Current Anthropology,* April 1964)

## A Hominid Fossil in the Hand Is Worth Two Homins in the Bush

Let us imagine that living apes and monkeys were unknown to science, with only their fossils known. Then one day information started coming in of sightings, of footprints, and of a documentary film, testifying to the existence of live creatures corresponding to reconstructions from the primate fossils known to paleontologists. "Rubbish!" would say the paleontologists. "We have the fossils. This means apes and monkeys are extinct. All 'evidence' submitted by monkey fans is nothing but hoaxes and misidentifications."

Replace apes and monkeys by hominids, and paleontologists by paleoanthropologists, and you have the unbelievable-yet-real situation in primatology and anthropology today. Historians of science will doubtlessly dwell at length on how and why science was first blindly fixed on the fossils of hominids and ignored the ever-present indications of living creatures. A proverb comes to mind, "A bird in the hand is worth two in the bush," or paraphrased for the occasion, "A hominid fossil in the hand is worth two homins in the bush."

On a more serious note, it can be said that science looked not where it had lost *Homo troglodytes*, but under the streetlight. The troglodyte was lost in the mind and *on* the ground, while light had begun to be shed on paleontological digs *in* the ground. Thus paleoanthropology was born, and the science of anthropogenesis (origin of man) swerved to digs in the ground rather than to long and difficult searches on the ground. And the longer paleoanthropologists continued their fixation on the fossils, the more they were liable to go wrong in interpreting them. Their very thinking became "fossilized."

Explaining Boris Porshnev's views on this score, I wrote in 1974 the following:

> How could science possibly go awry in interpreting facts of paleoanthropology? First of all, by uncritically using the ready-made, unscientific, pre-Darwinian, intuitive concept of man in the study of fossil material. When skeletal remains were found that looked more manlike than apelike, scholars, without much further thought, started labeling them 'man.'
>
> Thus such terms as Java man, Peking man, and Neanderthal man came into usage. Using a familiar name for an unknown thing, one inevitably imagines that unknown entity in terms of the makeup of the familiar one of the same name. In other words, images of ourselves were projected into the unfathomed past, and once placed there they began to be treated as facts of prehistory.
>
> Another possible cause of misinterpretation in paleoanthropology is the fact that this science is manned by osteologists, who know everything about skulls and very little about their contents, while it is the latter and not the former that have anything to do with the life of all brainy creatures.
>
> A third cause is the fact that modern evolutionary anthropology was born in Western Europe, and the closest living animal relatives of man known to the European scientists were representatives of the Pongidae. The evolutionist's thoughts could have taken a somewhat different direction had he set his eyes on a *Troglodytes recens*.
>
> (*Current Anthropology,* December 1974, p. 452.)

According to G.J. Romanes, *Homo sapiens* was preceded by *Homo alalus*, which "probably lived for an inconceivably long time." However, Porshnev maintained that Homo cannot be "alalus" (speechless); but he fully approved of the German Darwinist Ernst Haeckel's idea of *Pithecanthropus alalus* (speechless apeman), proclaimed in 1868. Inspired by Haeckel, the Dutch physician Eugene Dubois went to Java and, after a considerable amount of excavation work, discovered in 1891 the hominid fossils that he named *Pithecanthropus erectus* (upright apeman). Note that "alalus" was dropped. As the fossil apeman, being upright, looked more manlike than apelike, it soon became known as "Java Man," and served as the type specimen of *Homo erectus* (upright man). Has anyone heard of a species of non-upright man?

The doubly inappropriate term *Homo erectus* served to bury for a long time the term and question of "alalus," so that today we hear Professor Philip Lieberman say that "speech and language must have already been present in *Homo erectus* and in Neanderthals." (Lieberman, 2000, p. 141) As for Neanderthals, who could doubt the presence of speech in the hominid called Neanderthal man, or *Homo sapiens neanderthalensis*? I can name at least three scientists who could—Boris Porshnev, Grover Krantz, and Ian Tattersall. Ian Tattersall wrote in his book *Becoming Human* that:

> Neanderthals, for example, had brains as large as ours, and I shall argue later that they probably did not have language (...) in any event, if we combine this with the absence of any substantive archaeological evidence of symbolic behavior, it seems reasonable to conclude that the Neanderthals did not communicate as we do.
>
> (Tattersall, 1998, pp. 74, 172)

Grover Krantz noted "the lack of full speech capability in Neandertals" and added, "Of course we will never hear the vocal utterances of Neandertals, but neither will we ever see a pterodactyl fly." (Sapienization and Speech, *Current Anthropology*, December 1980, pp. 775, 778.)

Boris Porshnev, on the contrary, believed that sooner or later scientists shall hear the vocal utterances of Neanderthals, as he took *Homo troglodytes* for the relics of Neanderthals.

Defending my teacher's concepts, I wrote in 1974:

> In Porshnev's opinion Neanderthal's muteness accounts for his disappearance from the toolworking records only; he never disappeared from life itself. If Porshnev is right, we should still have a chance to examine the neuroanatomical structures for speech (or lack of it) in Neanderthals *in vivo*.

Later, in 1976 I continued:

> Neanderthal fossils (or at least fossils that look Neanderthal) in Europe are known from prehistoric and historic times. The causes of Neanderthal disappearance are unknown. Europe in historic times is known to have been the habitat of hominoids that looked Neanderthal. Ergo, these hominoids are Neanderthalers.
>
> (*Current Anthropology*, June 1976, p. 317)

Bernard Heuvelmans, in his book *L'homme de Néanderthal est toujours vivant*, reproduced 16 photographs of Neanderthal heads, reconstructed by paleoanthropologists from fossil skulls. All the heads appeared quite different, in keeping with the specialists' personal tastes and preferences. It never occurred to the paleoanthropologists to consult the pictures and sculptures of "wild men" performed by ancient and medieval artists who worked not from fossils, but occasionally from life itself, i.e., living Neanderthal relics.

Why have anthropologists overlooked this anthropological treasure-trove? Simply because they have a wrong image of Neanderthals and never equated them with "wild men." A fair image of the latter is given by Richard Bernheimer in his *Wild Men in the Middle Ages* and reads as:

> About the wild man's habitat and manner of life, medieval authorities are articulate and communicative. It was agreed that he shunned human contact, settling, if possible, in the most remote and inaccessible parts of the forest, and making his bed in crevices, caves, or the deep shadow of overhanging branches. In this remote and lonely sylvan home he eked out a living without benefit of metallurgy or even the simplest agricultural lore, reduced to the plain fare of berries and acorns or the raw flesh of animals.

That is a fair image of "wild men," i.e., homins, not only in Europe, but around the world. Does it mean they are all relic Neanderthals? Not necessarily or even likely. It only means that they are all relics of the pre-*sapiens* stage of hominid evolution. According to Grover Krantz in *Sapienization and Speech*, Neanderthals had more traits in common with *Homo erectus* than with *Homo sapiens*, so that they "could all be classed with *erectus*."

I therefore conclude that homins in Eurasia, Australia, and the Americas are relics of the *Homo erectus* – Neanderthal stage of evolution.

## Historical and Modern Data

Now let us examine more closely the historical and modern data on the homins, especially their ability in regards to speech, which is the particular concern of this paper.

According to Pliny the Elder, in *Natural History*, "the Satyrs have nothing of ordinary humanity about them except human shape." Plutarch mentions a satyr being captured by soldiers of the Roman general Sulla (old spelling Sylla):

> In this place [in Albania], we are told, a satyr was taken asleep, exactly such as statuaries and painters represent to us. He was brought to Sylla, and interrogated in many languages who he was; but he uttered nothing intelligible; his accent being harsh and inarticulate, something between the neighing of a horse and the bleating of a goat. Sylla was shocked with his appearance and ordered him to be taken out of his presence. (Plutarch, 1792, p. 349)

In the Middle Ages, according to Richard Bernheimer, a wild man,

> ... described as tailless and hairy but bald, is supposed to have been caught in 1161 in the sea near Orford on the English coast, and to have been dumped back again, when it turned out that nobody could make him talk.
>
> Several authors inform us that the wild man did not enjoy the benefit of human speech .... Orson, the wild man, and Shakespeare's Caliban are both originally afflicted with aphasia. And Spenser relates of his wild man that:
>
> *... other language had he none, nor speech,*
> *But a soft murmurs and confused sound*
> *Of senseless words, which nature did him teach*
> *T'expresse his passions, which his reason did impeach.*
>
> His wild man is thus reduced to showing 'faire semblance ... by signes, by looks, and all his other gests.'
>
> This intellectual deficiency is paralleled and aggravated by a spiritual one. For the wild man is devoid – perhaps incapable – of any knowledge of God, and thus suffers from a defect which a religious age could not but regard as a decisive obstacle against brotherhood with civilized man. He could not even be looked upon as an infidel.... The wild man did not worship idolatrously because he did not worship at all. [...]
>
> Medieval writers are fond of the story which tells how hunters, venturing father than usual into unknown parts of the forest, would

> chance upon the wild man's den and stir him up; and how, astounded at the human semblance of the beast, they would exert themselves to capture it, and would drag it to the local castle as a curiosity... The wild man's own reaction to the sudden encounter with his civilized counterpart varies according to type and temperament. While some wild men, like one humorously depicted in the fourteenth-century Psalter of Queen Mary, are seized with panic and attempt to escape in headlong flight, others, like Orson, offer dogged resistance and are overcome only after a struggle in which they may defend themselves literally tooth and nail. But whether they be elusive or combative, the result of the encounter is the same: the wild man is dragged out of his habitat and brought to the castle, there confined, and immediately exposed to the efforts of his captors to return him to full-fledged human status. Only if all endeavor fails, and the hairy man remains morose and speechless in spite of blandishment or torture, can he hope to be released again.

*Wild Men in the Middle Ages* was published in 1952 by Harvard University Press, and its subtitle is *A Study in Art, Sentiment and Demonology*. Should I add that author Richard Bernheimer, along with corresponding scholars all over the world, takes the hairy hero of this monumental volume for a figment of the mind.

In contrast to Bernheimer and his peers, Albertus Magnus (1193-1280), a philosopher deeply interested in natural science, takes for a fact of life, in his work *De Animalibus*, the recent capture in Saxony of two (male and female) "forest-dwelling hairy monsters much resembling human beings in shape." The female died of blood poisoning caused by dog bites, while the male lived on in captivity and even "learned the use, albeit very imperfectly, of a few words." That is the only reference known to me, and if correct very significant, to a homin that was able to learn and use a few words.

One of the most detailed and trustworthy accounts of a homin in captivity was published in Vienna in 1796 by Michael Wagner in his scholarly *Beitrage zur philosophischen Anthropologie*. It dealt with a hairy wild man of perfect Neanderthal anatomy, captured in Rumania and held in captivity in the city of Kronstadt (now Brasov) in the second half of the 18th century. His vocal performance is described as follows:

> He was completely lacking in speech, even in the slightest articulation of sounds. The sounds which he uttered were ununderstandable murmuring, which he would give when his guard drove him ahead of

> him. This murmuring was increased to a howling when he saw woods or even a tree. He seemed to express the wish for his accustomed abode; for once when he was in my room from which a mountain could be seen, the sight of the trees caused him to howl wretchedly.... As soon as he saw a woman, he broke out into violent cries of joy, and tried to express his awakened desires also through gestures.... Chiefly he was in every respect like a child whose capacities had begun to develop, only with this difference that he was unable to speak and could not make any progress in that regard.

Going farther east, we reach the Caucasus, visited in the 10th century A.D. by the famous Arab traveler and historian Abul Hassan Ali Masudi, who in his historical narrative *Meadows of Gold and Mines of Gems* (translated from the Arabic by Aloys Sprenger) wrote of the Caucasian "forests and jungles which are inhabited by a sort of monkey having an erect stature and round face; they are exceedingly like men, but they are all covered with hair. Sometimes it happens that they are caught. They show very great intelligence and docility: but they are deprived of speech by which they could express themselves, although they understand what is spoken. But they express themselves by signs."

In 1888 the Caucasus was traveled by Douglas William Freshfield, president of the English Alpine Club, who published his impressions in the book *The Explorations of the Caucasus*, writing in part:

> I had for years been possessed by a strong desire to penetrate the 'No Man's Land' west of Svanetia.... All we heard in our travels had added to the mystery of the Great Forest. Russian officials gravely repeated strange tales of a race of wild men, who had no villages or language, but appeared naked and gibbering in the depths of the woods, who lived on berries and were without fire-arms.

In the second half of the 20th century, the Caucasus became the main area of fieldwork for Russian hominologists, the largest part undertaken by Marie-Jeanne Koffmann. Accounts of the local people who happened to observe the Caucasian homins and hear their vocalizations fully support the notion of the homins' verbal deficiency. It is also confirmed by the Zana case, a "wild woman" held in captivity in the 19th century. According to witnesses, "she could not speak; over decades that she lived with people Zana did not learn a single Abkhaz word; she only reacted to her name, carried out commands given by her master and was scared when he shouted at her."

North of the Caucasus, in the central part of European Russia, a good example is the already-mentioned case of the 19th-century novelist Ivan Turgenev, who encountered, while hunting in the forest, "the fearful human beast," "like a female gorilla." The sounds she made are described as "little cacklings of delight," "growling," and "howling."

The foremost Russian lexicographer, Vladimir Dahl, included in his dictionary a popular saying, "Leshy nem, no golosist" (Leshy is mute, but vociferous.) "Leshy" is derived from "les" (wood) and translates as "wood-goblin." A poet called this saying "a charming and eloquent contradiction." The poet was wrong, for there is no contradiction in the saying—Leshy is mute, being unable to speak, and he is vociferous, having a powerful voice.

Earlier I cited Charles Hockett as saying that the European scholar at one time suspected that beyond the farthest frontiers there might lurk half-men or man-apes who would be "living fossils" attesting to earlier stages of human evolution. "The search was vain," concluded Hockett. Actually, to be successful the scholar had to go not beyond the farthest frontiers but to the nearest public library and a museum of ancient and medieval art to read in the former descriptions, and see in the latter pictures and sculptures, of "living fossils" attesting to earlier stages of human evolution. In addition, the European scholar had a choice of several places in Europe where he could meet and interview people who had seen "half-men or half-apes" and/or heard their vocalizations. The only reason the scholar did not do that was his being a stranger to the science of hominology. Philip Lieberman states in *Human Language and Our Reptilian Brain* that, "Real science relates phenomena that previously appeared to be unrelated and explains those relationships." (Lieberman 2000, p.167). Being real science, hominology relates the phenomena of paleoanthropology, history, ethnography, art, folklore, and demonology that previously appeared to be unrelated, and comes to the conclusion, among others, that relic Neanderthals, and moreover earlier hominids, are devoid of speech.

## Gray Areas of Science

The picture real science presents is never neatly black and white: gray areas are always there, along with spots of different hues. In our case it means that the conclusion of the homins' total speech deficiency is not without uncertainties.

Surprisingly, the first objection came from North America. Surprisingly, because the homins of that continent are believed by many to be descendants of *Gigantopithecus*; removed, therefore, a lot farther from

*Homo sapiens* than relic Neanderthals and *Homo erectus*. My old friend and colleague John Green, a foremost authority on the subject, wrote me in 1973:

> Couldn't you and I agree that the creatures in Russia are likely to be Neanderthal man and shouldn't be shot, but the thing over here is a monstrous ape and a fit specimen for dissection." He also said that "when live Sasquatches are caught they will end up in the zoo alongside the gorillas. Or maybe alongside the grizzly bears, since the apes are usually indoors and well heated.

And yet it was John Green who included in his voluminous book, *Sasquatch: The Apes Among Us*, the story of Albert Ostman, which is in stark contradiction with Green's own views on the North American homins, called there Bigfoot and Sasquatch. Ostman claimed to have been kidnapped, in 1924, by a male Sasquatch, brought to a secluded campsite in the mountains of British Columbia, where he spent a whole week with a Sasquatch family before escaping and returning to civilization. There is nothing unusual about the abduction of humans by homins; this topic is well in place in hominology, but usually it is human females that are abducted by male homins. I know only one case (in Mongolia) of a man kidnapped by two female almases. In the Ostman case, a man was grabbed and carried by a male Sasquatch, along with his hunting gun and other possessions. Why?

I have it from René Dahinden, who heard it from Ostman himself, that the old Sasquatch kidnapped Ostman and brought him to his family as a future bridegroom for his young daughter! No doubt, that does set Sasquatches apart from bears and gorillas. Yet, it's hardly a credible supposition. Still, the question remains: why was Ostman seized by a male Sasquatch and carried on his back for three hours over rugged terrain before being dumped in front of the "missus" and children in their mountain abode?

The answer is missing, and we wouldn't need it if we could dismiss the whole story as a wild yarn. The problem is we cannot. By "we" I mean, at least, John Green and myself. The late René Dahinden, a most skeptical and distrustful investigator, who, like Green, personally interviewed Albert Ostman, could not dismiss the story either. Ostman may have made wrong interpretations of what he experienced and drawn unjustified conclusions, but his testimony can't be discarded and has to be taken into account.

So, here are Ostman's own words bearing on the theme of this paper as extracted from John Green's voluminous work.

> Finally he stopped and let me down...Then I heard chatter—some kind of talk I did not understand....
>
> It was still dark, I could not see what my captors looked like... They were standing around me, and continuously chattering.... I asked, 'What you fellows want with me?' Only some more chatter....
>
> The old lady did not seem too pleased about what the old man dragged home. But the old man was waving his arms and telling them all what he had in mind ...
>
> I wanted to go out. But he stood there pushing towards me—and said something that sounded like 'Soka, soka'....
>
> He [the young fellow] tasted it, then went to the old man—he licked it with his tongue. They had a long chat...
>
> He picked it up and looked at it then he went to the old man and showed it to him. They had a long chatter. Then he came to me, pointed at the dipper then at his sister. I could see that he wanted one for her too....
>
> The young fellow pointed to the old man, said something that sounded like 'Ook'. I got the idea that the old man liked snuff, and the young fellow wanted a box for the old man.

To believe Ostman, the Sasquatch communicated in a way no different from people speaking a foreign language. Could it be really so, or was it only his grossly exaggerated impression under conditions of the psychological stress he must have been experiencing at the time?

Arthur Buckley was not under any stress when he wrote about Sasquatches as follows:

> They communicate orally. On two separate occasions with colleagues, we have surprised a small group in their base camp—who upon a hurry retreat have resorted to a jargon that has the phonetics of a language when we got close to them.(Buckley 1981).

Finally, the famous Sierra vocalizations case, presenting the best tape-recorded sounds for the general public to hear and for scientific study. Warren Johnson, who with his brother Louis was the first to be involved in those events, wrote to Ivan Sanderson in 1971:

> They (the creatures) seem to be able to communicate with each other in some manner as there seemed to be a pattern to their calls at

times and they seemed to be able to let each other know each time we started out of the cabin.... During the times they were present they would be extremely vociferous for several minutes and then just mill around for a short time and then start their tirade all over again.... This time we discovered a new sound that they seem to use as some sort of signal. The sound is rather like the call of our mountain quail...

(Warren C. Johnson, *Our First Meeting with Bigfoot)*

Alan Berry, invited by Warren Johnson to witness the Sierra Sasquatch events, heard the homins' vocalizations and made the best tape recordings in existence so far. In his book *Bigfoot,* co-authored with B. Ann Slate, he describes the sounds with the following words and phrases:

"weird snarling and snortings," "screeches and cries," "the things jabbered a lot," "the snarls and growls," "jowls shaking," "lip-smacking and teeth-popping," "the chest beating," "spells and whining," "a gibberish that came in spurts punctuated by snorts and occasional long, drawn out nasal snarl," "some of it had the ring of monkey's chatter," "some of it seemed almost articulate and human," "nearly all of it was loud and raucous," "the creature-things responded ... sometimes with a snort or a high-speed burst of chatter," "the creatures began whining like kids in a kitchen waiting to be fed," "a vigorous and violent-sounding altercation between them ensued," "Warren whistled to them, and was answered in whistles in an exchange that lasted several minutes," "a deep, resonant, nasal voice mumbling incoherently," "a series of sonorous, almost singsong, 'umm-oh/oh-ahhs.' It had a primitive texture. It was like a chant."

Vocalizations thus described have no resemblance to the voices of people speaking a foreign tongue. Alan Berry has kindly provided me with his recordings, and I can confirm that his description of the sounds is accurate. I am confident that they are natural and do not belong to humans or any known animals. At the same time, I do not believe the creatures emitting such beastly sounds are capable of verbal communication. And they reminded me of an account of almasty vocalizations, received by Marie-Jeanne Koffmann in the Caucasus in 1962 from a 65-year-old shepherd named Daniel Khakonov.

He told her that one evening, in October 1947, when he and his mates returned from the pasture to their cabin, they saw that the caldron with boiled meat was missing. They immediately realized it had been stolen by the almasty. The episode is recorded in Russian in *Information Materials No. 7* as follows:

> They lived in an old dwelling, some 15 meters from our cabin. How many, don't know. A whole family, maybe 6 or 7. We heard their noises every evening. Horsing around, quarreling, playing. Very noisy crowd: screeching, howling, whining. They have no human language. Their talking is like drum beating: boom, boom, boom.
>
> None of us entered that dwelling. I offered a ram to any one who would. Nobody volunteered. They would come to our cabin and take food left-overs. Once they brought home a rain pipe and, beating it, played the whole night with it, giving us no sleep. We were 5 men, armed with hunting guns, but nobody dared go and look at them.
>
> We found the caldron in the morning, empty, of course, some place between their dwelling and the stream.
>
> Our dogs got used to them, they barked but did not bite them. But if the dogs surrounded one, the almasty would cry out very loudly. (Translated by Bayanov).

So much for "Neanderthal Man" in Russia and "a monstrous ape" in North America, as was suggested by John Green. It turns out that in vocalizations and behavior the two are very similar. Besides, we have an eyewitness account in Russia echoing Albert Ostman's impressions of Sasquatch vocalizations. It's a story by Alexander Katayev, who claims to have seen two human-like creatures one night in August 1974 in the Urals.

> One was male, the other female, both covered all over with gray hair. The male was over two meters tall, the female shorter. He had very long and hefty arms, she had very big breasts and an outstanding belly indicating pregnancy.... He said something to her, she answered and they began to eat from the box.... They talked again and she laughed.

I asked Katayev to explain what he meant by saying that the creatures "talked," and he wrote back:

> The sounds they made were very strange, resembling those of humans but dumb humans: Kh-Kh-Kh, M-M-M, No-No. The female made both high and low sounds, and he too, but she also laughed like a girl, only with a metallic sound. They talked in turn. After saying

something one would be silent for maybe a minute, then would speak again, waving the hands. The sound phrases were short.

(Bayanov 1996)

So what can we make, at this stage of research, of the vocal exchanges by homins? Do they exchange "notions," or "emotions," or both? With the insufficient evidence so far, there can be no absolute certainty one way or another, of course, but my feeling is they have only a semblance of "talk" and no verbal communication, i.e., speech as such. Let us recall Darwin's words, indicating what we may have in common in this respect with our hairy cousins:

> Man uses, in common with the lower animals, inarticulate cries to express his meaning, aided by gestures and the movements of the muscles of the face. This specially holds good with the more simple and vivid feeling... Our cries of pain, fear, surprise, anger, together with their appropriate actions, and the murmur of a mother to her beloved child are more expressive than any words.

If homins have not crossed the "rubicon of mind," there is little doubt they have come close to it, and some may even have stepped into it. This follows from evidence in different habitats (central Russia, the Caucasus, Tajikistan, China, North America) of their sound-imitating ability, which is *condition sine qua non* in the origin of speech. In Russia, for example, the leshy is said to be able to imitate the voices of human males, females, and babies, he can neigh like a horse, squeal like a pig, bark like a dog, meow like a cat, and cry like a cock and hen. Sasquatches, according to Arthur Buckley,

> ... are excellent mimics. In order to communicate and conceal their presence, they will at times employ the mimicking of other animals, birds and natural sounds. We have heard them bark like dogs and coyotes, whistle and pound rocks. I have even had them repeat my voice.

According to Warren Johnson, Sasquatches used some sort of signal sound, "rather like the call of our mountain quail." Let us also recall that the satyr brought to Sulla and interrogated in many languages, uttered "something between the neighing of a horse and the bleating of a goat."

So, it is reasonable to conclude, first that the sound-imitating ability in hominids had developed long before their linguistic ability, and not for that purpose. Second, there is little doubt that a homin's brain can form the image of the animal whose cry is imitated, thus making a connection between object and sound, between sight and hearing, which is a necessary condition of speech and "a first step in the formation of a language," as predicted by Darwin.

And let us not forget that if not all, then at least some, homins have been reported to be capable of utterances resembling words: "confused sounds of senseless words, which nature did him teach to express his passions" (Edmund Spenser); "imperfect use of a few words" (Albertus Magnus); "Soka, soka," "Ook" (Albert Ostman). No matter how primitive these utterances may sound to the ear of *Homo sapiens*, no ape is capable of vocalizations resembling words.

Speaking of apes, let's recall that young specimens reared by humans and taught sign language become able to communicate with man at the level of two-year-old children. The question is: What level of communication and humanness will be reached by a young homin reared by humans and taught sign or sound language? I guess the specimen will acquire what Pavlov called the second signaling system, i.e., language *per se*, which is the hallmark of a human being. But then the "rubicon crossing" will be made in a human-induced, i.e., artificial and cultural, not a natural evolutionary, way.

## Porshnev's Paradigm

The ancients have left us the behest "Know thyself." Linnaeus repeated it in the 18th century and made a major step in man's self-study by uniting *Homo* with apes and monkeys in the Primate order. And he left to the future generations the task of discovering the difference between primate beings "endowed with speech and those devoid of it."

We owe our next great advance in self-knowledge to Charles Darwin, who substantiated with a host of convincing arguments the idea of evolution and maintained that numerous common features in man and ape are due to their common evolutionary origin. Boris Porshnev called this the first stage of the Darwinian revolution in anthropogenesis. The second, according to him, was the discovery of man's fossil ancestors. And he called the third stage of the Darwinian revolution the discovery in the 20th century of relics of hominids believed by him to be ancestors of *Homo sapiens*. So the latest advance in man's self-knowledge is due to the philosopher and historian Boris Porshnev.

We are used to the terms "human" and "non-human" (e.g., "non-human primates"), but for these terms to be really meaningful we have to know the exact difference between human and non-human beings and the line of division between them in the process of evolution. Porshnev saw "the main diagnostic distinction" of human beings in "the presence of those formations in the structure of the brain which make speech possible and the correlative features in the organs of speech and in the face." He convincingly argued that this condition was attained in evolution only by *Homo sapiens*. Thus only *Homo sapiens* is human, the rest of hominids are non-human. To speak of *humans-without-speech* is a nonsense-concept, a contradiction in terms. In man's phylogeny, speech emerged as a completely novel and ready-made ability, somewhat akin to the way fertility appears as a novel and ready-made function in ontogeny. All changes in the evolution of pre-*sapiens* hominids were those of degree; the appearance of *Homo sapiens* was a change in kind. The Porshnev paradigm moves humans far away from the ape, but puts them cheek by jowl with wildlife because *Homo sapiens* is the first human on Earth, still retaining in body and mind a great many non-human features inherited from his very recent non-human ancestors.

The emergence of speech split our evolutionary past into history and prehistory. Even more, it signified the appearance of an absolutely novel world, not only distinct from the animal kingdom but in some ways opposite to it. This transformation is so significant that it is comparable only to the origin of life.

Proceeding from Porshnev's ideas, we conclude that language is not only a means of human communication, but also of reproduction. Human bodies ("hardware") are reproduced and multiplied by way of sexual relations; human minds ("software") are reproduced and multiplied by way of linguistic interaction. In the first case, parental genes are reproduced in children; in the second, certain neural patterns and structures of parental brains are reproduced in the brains of children learning the mother tongue. In genetics, the result is achieved by way of molecular replication; in linguistics, through sound imitation (a kind of replication as well); and both processes, biological and psychological, are connected with the world of physics by the universal phenomenon I call structural induction. The latter is responsible for transference and transformation of qualities, being distinct from kinetic induction, responsible for transference and transformation of motion. Thus, the throwing of a stone or launching of a missile is kinetic induction; a steel bar magnetized by a magnet is an example of structural induction. As for speech, structural induction is involved not only in its origin and reproduction, but also in every act of verbal communication.

Language is not only the main factor in the making of humans, both phylogenetically and ontogenetically, but also of nations, for every nation is determined, united, and sustained by a national language. What's more, language is the *sine qua non* of so-called globalization, turning mankind into a single whole, whatever the positive and negative sides of the process. In distinction from the animal systems of communication, human languages are translatable. The "bee language" cannot be translated into the "ant language," that is why globalization is not on the agenda of insects. If human languages were like that, international trade, science, and technology, and the United Nations itself, would be impossible. Thus, at the individual, national and global levels, language is the basis of human existence.

## A Quantum Leap

The primatologist and paleoanthropologist John Napier will be long remembered for the following revelation in his 1973 book *Bigfoot*: If any of the sasquatch footprints "is real then as scientists we have a lot to explain. Among other things we shall have to re-write the story of human evolution. We shall have to accept that *Homo sapiens* is not the one and only living product of the hominid line, and we shall have to admit that there are still major mysteries to be solved in a world we thought we knew so well."

Sasquatch footprints are real, as well as homins themselves on all the continents, except Antarctica. But Napier was wrong in saying "we shall have to re-write the story of human evolution," implying *paleoanthropologists*. The task is dealt with by specialists of another discipline—*hominology*, while paleoanthropologists are following, as Porshnev said, the example of ostriches, hiding their heads in the sand. Even the most perspicacious, such as Ian Tattersall, who doubts the presence of speech in Neanderthals, does not dare draw the right conclusion from his views. According to him, language "does appear to represent a quantum leap away from any other system of communication we can observe in the living world." (Tattersall 1998, p. 68.) A "quantum leap" means a leap in kind, and yet Tattersall blithely uses the term "human" for all and any hominids, before and after the "quantum leap." By taking non-human hominids for humans, anthropologists and paleoanthropologists, followed by the rest of the scientific community, have monumentally failed to see homins as a present day reality.

Shepherds, hunters, and lumberjacks around the world know that hairy "wild men" are real and part of the wildlife—academics around the world don't know this. How can it be? Is the paradox real? Can the scientific establishment be so divorced from life? Yes, it can. A similar situation occurred in the 18th century, when the French Academy (then the world's most eminent) refused to recognize the reality of meteorites, because "stones can't fall from the sky." Hence all evidence to the contrary was taken to be mythology and popular superstition.

Still, the difference between heavenly stones and earthly homins is not of degree, but of kind, and the consequences of one discovery and the other are incommensurate. Cosmologists maintain the Universe came into being with a Big Bang. I am not a cosmologist and don't know whether the claim is true, but, being a hominologist, I expect a Big Bang in science when it finally faces the truth of hominology.

(The article was written in 2002 and offered for publication to *Current Anthropology*. In a polite letter of July 24, 2002 the journal's Editor, Benjamin S. Orlove, wrote me: "I appreciate your interest in the journal. However, your manuscript does not fit the scope of the journal, therefore I am unable to accept it for publication.")

## Bibliography


Bayanov, Dmitri. 1974. Reply (to Comments). *Current Anthropology* 15(4): 449–50.

———. 1976. On Neanderthal vs. paranthropus. *Current Anthropology* 17(2): 312–18.

———. 1984. In *The Sasquatch and other unknown hominoids,* ed. Vladimir Markotic and Grover Krantz. Calgary, Western Publishers.

———. 1996. *In the footsteps of the Russian snowman*. Moscow, Crypto-Logos.

Bernheimer, Richard. 1952. *Wild men in the Middle Ages*. Cambridge, Harvard University Press.

Buckley, Arthur. 1981. *Bigfoot Co-op* (entry in), October.

Darwin, Charles. 1874. *The descent of man and selection in relation to sex*. New York/London, Merrill and Baker.

Freshfield, Douglas William. c. 1890. *The explorations of the Caucasus*.

Friedman, Eman. 1973. *Primaty* (Primates) (in Russian). Moscow: Nauka.

Gazov-Ginzberg, A.M. 1965. *Is language imitative by origin?* (in Russian). Moscow: Nauka.

Gould, Stephen J. 1977. *Ontogeny and phylogeny.* Cambridge, Harvard University Press.

———. *The panda's thumb*. 1980. New York, W.W. Norton.

Green, John. 1978. *Sasquatch: The apes among us.* Seattle: Hancock House.
Hockett, Charles F. 1960. The origin of speech. *Scientific American*, December.
Johnson, Warren C. 1971. *Our first meeting with Bigfoot.* Letter sent to Ivan Sanderson.
Krantz, Grover S. 1980. Sapienization and speech. *Current Anthropology* 21(6).
Lieberman, Philip. 2000. *Human language and our reptilian brain.* Cambridge/ London: Harvard University Press.
Linnaeus, Carolis. 1758. *Systema Naturae* (in Latin). Lipsiae.
Magnus, Albertus. *De Animalibus*.
Masudi, Abul Hassan Ali. 1841. *Ali Masudi's historical encyclopedia*, translated from Arabic by Aloys. London: Sprenger.
Mayr, Ernst. Accident or design: the paradox of evolution. In *The evolution of living organisms*, Proceedings of the Darwin Centenary Symposium, Royal Society of Victoria, 1-14. Melbourne, Melbourne University Press.
Mivart, St. George. 1889. *The origin of human reason.* London: Kegan Paul.
Molina, Manuel J. 1971. El Yoshil o Mono Fueguino. *Karukinka.*
Napier, John. 1973. *Bigfoot.* New York: Dutton and Co.
Plutarch. 1792. *Plutarch's Lives*. London: Dilly.
Porshnev, Boris. 1974. The troglodytidae and the hominidae in the taxonomy and evolution of higher primates. *Current Anthropology* 15(4): 449–50.
Romanes, G.J. 1888. *Mental evolution in man: Origin of human faculty.*
Slate, B. Ann, and Alan Berry. 1976. *Bigfoot.* New York, Bantam Books.
Tattersall, Ian. 1998. *Becoming human.* New York: Harcourt Brace.
Wagner, Michael. 1796. *Beitrage zur philosophischen Anthropologie.* Vienna.
Winner, Anna K. 1964. Taxonomic nomenclature in paleoanthropology. *Current Anthropology, April.*

# SECTION 10

# The Animal-Man Dichotomy

## Is a Manimal more Man than Animal?

Back in the 1960s, Jim McClarin dubbed sasquatch/bigfoot with the word "manimal." The neologism is at odds with the name applied to the creature by some leading investigators, as seen in the very titles of their books: *Sasaquatch. The Apes Among Us, North America's Great Ape: the Sasquatch*, and *Bigfoot! The True Story of Apes in America*. Grover Krantz, in his fundamental *Big Footprints: A Scientific Inquiry Into The Reality Of Sasquatch*, states categorically: "... the sasquatch is not an intermediate form at all. All available evidence points clearly to an animal status for this species in terms of its behavior and mental abilities. It walks bipedally, but so do chickens. It is highly intelligent, but dolphins are more so" (p.12). "It is not human, nor even semihuman, and its legal status would be that of an animal if and when a specimen is taken" (p. 173).

Accepting the theory of Boris Porshnev, I also believed all relict hominids to be less than human, but more than any known animals, which status I described as "superanimal." As for bigfoot, I wrote in *Current Anthropology* in 1976: "Judging by the available data, the American hominoids look more 'archaic' than their European counterparts." Since then available data concerning American homins has considerably increased, and my opinion has changed accordingly. Today I see big relict hominids of North America, Eurasia, and Australia to be probably at the same level of evolutionary development. I exclude the orang pendek of Indonesia for which information is still rather scanty.

Sasquatch the Manimal can be taken to mean "semihuman," a status rejected by Grover Krantz, who explains his opinion as follows: "On a more serious level the status of sasquatch can be tested against the three most basic traits that distinguish humans from animals—tools, society, and speech. ... Unless the sasquatch carefully conceals its tools, society, and speech, we must assume that they are absent." (pp. 171–72). In agreement with Porshnev, I accept only speech, not tools and society, as the most basic trait that distinguishes humans from animals. Let's note that humans are also manimals of sorts, because in all their biological structures and functions humans are totally animal. It is our intelligence that is not animal, or, let's say, not totally animal. A newborn human baby is in fact human only potentially, having intelligence at zero level. But a normal three-year-old child is human all right, being endowed with human intelligence. How do we know it? Just by speaking to the kid. If it can answer our questions and can put questions to us, we conclude the little one has human intelligence. Thus, language is the "Rubicon of mind" (St. George Mivart).

Krantz again: "Humans spend a great deal of time mumbling softly to one another with coded symbols that convey meanings. Again, nothing like this human speech has been reported for sasquatch" (p. 171). This is not exact, of course. At least two cases of talking sasquatches are registered in John Green's very books. One is in Green's reprint of two pages published by J. W. Burns in 1929 and titled "Introducing B.C.'s Hairy Giants, A collection of strange tales about British Columbia's wild men as told by those who say they have seen them." One strange tale mentions a wild woman who spoke "in the Douglas tongue" to an Indian hunter (John Green, *The Sasquatch File,* 1973, p. 11).

The other case is the famed Albert Ostman story, which deserved only a little paragraph in Grover Krantz's book: "In 1957 a Canadian man, Albert Ostman, recounted a story of being captured by a sasquatch some thirty-three years previously. He told of being held with a family of four of them for six days before he managed to escape and return to civilization. His description of them agrees with that of other observers, but some points of behavior, particularly the capture itself, seem incongruous" (p. 13).

Krantz placed this paragraph under the rubric Special Cases, and it is fair to say that of all special cases this one is the world's most special for the unique opportunity that the witness had to observe sasquatches right in their mountain home. Before his adventure really began, Ostman asked his Indian guide what kind of an animal he called a sasquatch, and the guide said: "They have hair all over their bodies, but they are not animals. They are people. Big people living in the mountains." The Ostman story is unique and incongruous because, if it is correct, then the Indian guide was absolutely right: Sasquatches are not animals. They **are people**, big **people** living in the mountains.

This conclusion is inescapable if the story is taken literally. The way Ostman was kidnapped and treated by the sasquatches is **not** the animal way. We don't know **why** he was kidnapped. Dahinden was told by Ostman that he "was taken for a mate for the daughter." If so, the aim was at least much nobler than the aims of kidnappings by modern terrorists. As remarked by Don Hunter, "Ostman wasn't with them long enough to find out whether his theory had any foundation."

The family communicated by means of a language, and Ostman even remembered two of their words: "soka" and "ook." This means that Ostman's captors were definitely on the human side of the "Rubicon of mind." There are two hints, though, that their minds and intelligence were less sophisticated than those of *Homo sapiens*, or at least of so-called civilized *Homo sapiens*. The first hint is that they did not bother to disarm the captive. The second is the ease with which the captive prevailed in the end over the captors by means of a ruse and made his escape. Incidentally, the motif of man getting the upper hand by ruse in confrontation with devils, wood goblins, etc., is well known in folkloristics and demonology. The Ostman case is also instructive regarding the supposed paranormal abilities of sasquatches, such as their alleged power of mind-reading, telepathy, etc. Either not all of them possess such abilities, or these powers are active only under certain conditions that were lacking in Ostman's case.

The first crucial question is this: Is the story believable—at least in general? John Green: "Albert Ostman is dead now, but I enjoyed his friendship for more than a dozen years, and he gave me no reason to consider him a liar. I have had him cross-examined by a magistrate, a zoologist, a physical anthropologist and a veterinarian, the latter two being specialists in primates. In addition to that all sorts of skeptical newsmen have grilled him. Those people didn't necessarily end up believing him, but none was able to trap him or discredit his story as a result of their questioning, although the magistrate in particular tried very hard to give him a rough time" (*Sasquatch*, 1978, p.110). To my mind, it's a good sign of Ostman's sincerity that he countered his doubters with the words: "I don't care a damn what you think" (Don Hunter with Rene Dahinden, *Sasquatch*, 1973, p.62).

So what did we think? Did we care a damn about the implications? What part has the Ostman case played in the development of hominology? Alas, virtually none at all. It seemed incongruous not only in North America, but also in Russia and around the world. John Green, René Dahinden and Grover Krantz continued to insist on the animal version, while in Russia, true to Porshnev's ideas, we believed that only *Homo sapiens* can have the power of speech. So Ostman's sasquatches remained in limbo, or as the current phrase goes, on the back burner. I remember trying to explain away talking sasquatches by supposing that Ostman, while in captivity, was under severe stress and thus his perception of the situation and his captors was not objective enough.

This does not mean that I was in full agreement with Green, Dahinden, and Krantz regarding the nature of the beings we investigated and the methods to be used to prove their reality. In the 1970s I initiated a debate on the "kill or film" question, later described in my booklet "Bigfoot: To Kill or To Film? The Problem of Proof." My true ally then was the late George Haas, of the Bay Area Group in California. Today his words ring as cogent and relevant as they did 30 years ago. George said:

> Most of us in our Bay Area Group feel that we are dealing with a creature that is more than a 'mere animal.'... What we must not forget or overlook is that in Bigfoot (and in other forms of relict hominoids) we now have a totally unique opportunity to do something worthwhile before it is too late: to demonstrate our integrity and to save and protect all the individuals of what we all agree is undoubtedly a rare and unique form of life.
>
> ... It seems to me to be a little reckless to advocate and encourage others to shoot something before we really know what it is. In this connection, let me quote the little Himalayan folk tale from Odette Tchernine's book, *The Snowman and Company*, page 158: "One day as I was walking on the mountainside, I saw at a distance what I thought to be a beast. As I came closer, I saw it was a man. As I came closer still I found it was my brother."

Unfortunately, the voices of those who advocated a killing were much louder than the voice of George Haas. At the turn of the century, I learned of other supporters of the non-violent method and opponents of the ape misnomer. Quite determined among them is Bobbie Short, who had a sasquatch sighting of her own. Just the other day she declared worldwide: "I've been saying all along that sasquatches weren't apes...." Another most determined proponent of the hominid version is Will Duncan, who substantiates this idea in two important articles,"What is Living in the Woods and Why it isn't *Gigantopithecus*" and "Predictability of Homin Behavior," published by Craig Heinselman in *Hominology Special Number I*, 2001, and *Hominology Special Number II*, 2002.

So I thought Will Duncan to be just the right man to investigate the Carter Farm habituation case in Tennessee when the relevant news reached us here in Moscow. The human-version implications of the case struck this time with a vengeance. No matter how much prepared I had been by previous experience for the idea of "superanimals" and how persistently I advocated the method of habituation, there was no end to surprise and bewilderment that overwhelmed me with the news gradually coming from Tennessee. On the whole, the Carter Farm habituation case is a hominological irony and paradox of global proportions.

Robert I. Carter discovered and befriended a young bigfoot on his property, named him Fox, and started to teach him English back in the 1940s. Then followed half a century of "co-existence" with a family of bigfoots. This means that the Carter Farm bigfoot adventure was concurrent with the world's snowman adventure, involving such countries as Nepal, Russia, China, Australia, and America itself. Members of numerous expeditions in far-off corners of the world had no inkling that the objects of their dreams were comfortably idling away on a farm in Tennessee, USA. Can you imagine what could have happened had Robert Carter Sr. invited Tom Slick to visit the farm and introduced Fox to the millionaire? The science of primatology and anthropology would be different today.

But Carter did nothing of the sort, and not only because he was indifferent to science. His involvement with bigfoots was in fact contrary to science. He believed that bigfoots "are from God like we are and the true Edomites," "descendants from Esau of the Bible." Janice says that her grandfather "never called the bigfoot by the name 'bigfoot,' he always called them The People of the Wandering Spirit" (*50 Years with Bigfoot*, p. 171). So the hospitality accorded by Carter to big wild fellows on his property was not for him an experiment in habituation but a kind of religious service, nay, a feat of faith, considering the problems the family always experienced and big material losses suffered as a result of friendship with hairy "Edomites."

That was the answer to my first bewilderment upon learning of the Carter Farm case. "50 Years with Bigfoot" and not a single recognizable photograph of the creatures! Is that possible? Yes, since religion cares for icons, not photographs. Yes, if the People of the Wandering Spirit, while having a good time on the farm, did not feel the least inclination to be caught and fixed by photography.

My next bewilderment, which stayed long with me, was Janice's description of how the bigfoots buried their baby that was born dead. I had heard in the Caucasus a local say that almastys bury their dead, but took it for just an opinion. According to Janice, the bigfoots dug a deep hole "mostly with their hands at first, then with pointed sticks they had chewed on." The unbelievable happened later: "They would take food to the grave of the little one they buried for a long time, laying it on top of the grave. ... Sheba [the mother of the baby—D.B.] sat on the grave and threatened the others not to come near for a while thereafter too" (p. 149).

Some relief came when I recalled seeing similar information elsewhere: "When he was working with Roger Patterson and headquartered at Yakima, Dennis Jenson saw a letter from a man who swore that he had watched three Bigfeet burying a fourth. They dug a deep hole, using only their hands as tools. After placing the body in the hole and covering it with earth they rolled huge boulders, each weighing many hundreds of pounds, onto the grave" (Peter Byrne, *The Search for Big Foot*, 1976, p. 109).

But the hardest stumbling block, which I painfully stumble against even today, is the unbelievable linguistic prowess of the Carter Farm bigfoots. The two sasquatch words of unknown meaning, remembered and brought from the wilderness to civilization by Albert Ostman, could easily be ignored and forgotten by hominologists, but how can you ignore and forget the published vocabulary of 223 bigfoot words and phrases presented by Janice Carter Coy, each word and phrase dutifully translated into English?

How did she manage to obtain and learn such a vocabulary? Listen to her answer: "I went daily with my Grandfather Carter to visit and feed them where they would say something, and either my Grandfather, Fox, Sheba or one of them would have to translate the words into English for me. I took notes in a little note pad of the words I would hear them say out in the woods or fields and brought them to my Grandfather Carter when I got the chance. I would ask him what they meant. ... This is the way I learned from Fox and his family how to speak in their bigfoot language. It is a practical skill, one might say. It is also very hard for a human to speak in bigfoot" (p. 196). "The sounds of some of the words are carried out, yet other words are chattered so fast that it is hard to catch what is being said. ... It took years for me to halfway understand them" (p. 205).

And here is some light on the way the bigfoots talk to each other and to their hosts: "Fox and his family can communicate with each other in a language of their own" (p. 196). Fox and Sheba "were chirping and chattering back and forth to each other. I don't know what they were arguing over" (p. 92). "They mostly talked in old Indian and used chirps and whistles and grunts and growls and such when they talked to each other. They would alternate this with English when they spoke to us. ... they would substitute an Indian word or a gesture or a grunt for some of the words when they talked to us" (p. 157).

Here's how language lessons began on the Carter Farm: As reported by Janice's grandmother, "The two, bigfoot and man, were said to spend hours together at the barn or in the field learning each others' language" (p. 57). Janice: "I always thought that my Grandfather Carter had taught Fox how to speak English, and that between Fox and Grandfather they had taught the rest of Fox's family how to speak English" (p. 196).

So how did they speak English? Fox "could say words in English that Papaw [that's how Janice calls her grandfather.—D.B.] taught him but not like human speech, as we know it. The sounds they make when saying words in English are not like the way humans speak" (p. 15). "While all of the bigfoot here on our place could speak their own language **fluently**, they can only speak a mixture of broken English language. Sheba struggled with it a lot. Her English was very limited... Fox could speak much more clearly and used longer sentences than any of the rest of his family when speaking in English" (p. 196).

An example given by Janice of Fox's utterance in English is his saying "Thank you" when she "scooped him up a pail of dog food" (p. 65). As to his speaking in Bigfoot, she offers, almost in the beginning of the book, an example which left me gaping. Once, when Janice, her four-year-old sister Lila, and another girl were playing in the woods, they were scared by the sudden appearance of Fox. Robert Carter intervened and told Fox off for scaring the girls. Janice writes that "he sure didn't mean to scare us, at least I gathered that much of it. Fox was not talking in English to Papaw, and Papaw was talking what I call bigfoot words to him. Fox looked right at Lila and said: "Yoohhobt Papi Icantewaste Mitanksi ... Posa ... Ka Taikay Kataikay Tohobt Wabittub". Translation by Janice: "Yellow Hair, be happy little sister. I naughty. Don't cry Blue Eyes." And her explanation: "Lila's eyes were blue when she was little and her hair was a yellow-red..." (p. 24). In other words, Fox was apologetic, tried to console little Lila and used her traits in naming and addressing her. All that in a few touching words. Call him what you like, bipedal ape, *Australopithecus robustus*, *Gigantopithecus blacki*, for me such an utterance, if it really happened, is the sure sign of a human being.

Some bigfoot words and phrases from the vocabulary compiled by Janice:

31. *Nenepi* = The malevolent little people (in reference to all humans, their word for human men).
80. *Ella Cona* = The Fire Rods of the White Men or Humans (Guns).
84. *Hah-Ich-Ka Po-Mea?* = Where is she going? (Asked twice to my Papaw about where I was walking off to.) Papaw told me this one's meaning.
96. *Me-Pe Mahtaoyo* = Poor little one or little baby (Refers to what Sheba kept chanting over and over the time her baby died and they buried it and she was sitting out there on its grave crying and chanting this.)
99. *Ob-Be-Mah-E-Yah* = Get out of here; get out of the way, leave from here. (This is what Sheba said when she knocked my horse over with me on it, along with telling me to leave and get out of here in English.)
129. *Nanpi yuze Sni Yo* = Take your hands off of me. (Sheba said this to the strange male bigfoot when he grabbed her.)
130. *Napi* = God, The Lord God (it is also Sioux for Great Spirit).
132. *Nicinca Tonape He?* = Do you have children? (Fox asked me this and I asked him to repeat the question in English, as I didn't know what he asked me.) In 1990 I was 25 and this is when he asked me this question. [Janice then did not live on the farm.—D.B.]
146. *Waste Ce Dake* = I love you (Papaw and Fox said this to each other when Papaw was in the road that time right before he died. Papaw said it means I love you in Bigfoot).
197. *Siyuhk* = Soul (Bigfoot).

Well, it took me three years to get somewhat "habituated" to the idea that homins can be as eloquent as that. Still it feels like a miracle. Jonathan Swift's speaking horses, the Houyhnhnms, are not a miracle because I know they are fictional. So why should speaking bigfoots be stranger than fiction? First, because this overturns my previous thinking. Second, because it confronts me with the incongruity between the bigfoots' human intelligence and their animal way of life. If they are so **clever**, why are they so **wild**? THAT IS THE QUESTION! Half a century of contact with civilization on the Carter Farm has not changed their animal way of life a bit. That is the conclusion I draw from *50 Years with Bigfoot*.

Janice confided to me by email three instances of her telepathic communication with the bigfoots. (She was advised not to include that in the book and it wasn't. But in her vocabulary you find under number 25: "*Mookwarruh* = Spirit Talker (What they call telepathic communications to each other and to people)". How about that? Believe me, I take this easier than their verbal skills, because telepathy doesn't clash with their wildness (who knows, maybe even supports it!), but language does. At least, in my present state of knowledge. (Thank goodness, she hasn't observed any Bigfoot-UFO connection).

I'll be grateful to anyone who can convince me that the Bigfoot language is fiction. Shall not be obliged then to revise the Porshnev theory that I accepted and spread for 40 years. I am in close contact with three people who have had long direct talks with Janice and investigated the matter on the spot. They are Mary Green, Will Duncan, and Igor Bourtsev. All three believe the case is genuine.

Dr. Henner Fahrenbach has not been to the Carter farm, but is in contact with Janice and examined some hair samples collected at the farm. He is also in touch with a woman in California who claims long-time observations of sasquatches on her property and being engaged in habituation attempts. Fahrenbach finds some observations by this informant and those by Janice to be "astoundingly identical," and this "has added immeasurably" to his confidence in the testimonies of both women, "because the coincidences exceed chance."

In May 2004, Dr.Fahrenbach analyzed hairs that Janice claims to have pulled from Fox's wrist, and the scientist's conclusion was this: "The morphology of this hair is clearly primate in character, all standard mammals of N. America are ruled out, and the remaining confounding variable—human hair—is not similar to this hair at all, in that the density of pigmentation far exceeds that of the blackest human hair. These observations provide a legitimizing underpinning to the factual details reported by Jan Coy (Carter) [as co-author] in the book by Mary Green, deviant interpretations thereof notwithstanding."

I haven't been to the Carter farm either, and want now to share with hominologists my opinion of the testimony by Janice Carter Coy, as published in the book *50 Years with Bigfoot*. If she could have made all that up, she would sure be an illustrious winner of grand prizes in literature. It is my conviction that no genius of belles-lettres or science fiction could compose what Janice has told and written. With my experience in hominology, I see that she knows what I know and also much, much more. It is just the excess of her knowledge, especially its linguistic part, that is so surprising and disturbing.

More than once I discussed the issue with Will Duncan, and this is what he wrote me in part in November 2004: "I have been investigating the Carter Farm for almost three years." "Janice's story is not consistent with models of what bigfoot is, as developed by many people over decades of investigation. However, it is consistent with both Native accounts and with the Ostman story. I can only conclude that either 1) portions of Jan's story (and the Natives', and Ostman's) are exaggerated, or 2) the prevailing ape-like model preferred by many long-time investigators is based on very limited observation of sasquatches in remote settings."

Remote settings... This reminds me of the little Himalayan folk tale, quoted by George Haas. Yes, most of our knowledge comes from observations at a distance, and it is Albert Ostman and Janice who observed the creatures at close quarters. As regards Janice, such closeness lasted not hours and days, but years and decades, and this is the only logical explanation of her superior knowledge.

As for the thought of "exaggeration," it was my line of "defense" when I began and continued to receive information from Tennessee. But this "fortification" was getting weaker and weaker as I continued to note instances of sasquatch "linguistics" in reports coming my way from North America. As, for example, this one: "The sounds were all jumbled together and it sounded like whatever it was, was trying to put words of sorts together, like it was trying to communicate with us. ... The individual sounds themselves sounded a bit like the sounds made by Tahltan Indians I used to hear long ago, but it wasn't any of the Native or white languages I have ever heard" (J. Robert Alley, *Raincoast Sasquatch*, 2003, p. 197).

Or this item I received recently from Chris Murphy, quoting *The Daily Colonist,* September 24, 1972, by T. W. Patterson: An Indian, hunting and following a buck, came across an animal that he believed to be a big bear. To his astonishment, upon taking aim at the animal, the creature looked up and **spoke** to him **in his own tongue**. "It was a man about seven feet tall, and his body was very hairy."

If we decide that our informants are exaggerating, we're back to the conspiracy theory. All right, the Natives may be exaggerating, in line with their mythology and beliefs, but what motive or interest could Ostman and Janice have had in so grossly exaggerating or conspiring with the Natives? I see none at all. Since they faced disbelief and suspicion, one can expect them to belittle things, not exaggerate.

Some critics maintain that if what we have in Tennessee is true, then those creatures are not bigfoot/sasquatch, but something else. This doesn't make our problem easier: instead of one mystery we're getting two. Yes, in theory it is possible that some superanimals have crossed the "Rubicon of mind," while others have not. Let us note that Ostman's adventure, with its "accepted" bigfoots, took place in British Columbia, far away from Tennessee. Thus there is no ground at all to exclude the Tennessee adventure and its furry fellows from the bigfoot problem.

In my book, *In the Footsteps of the Russian Snowman*, 1996, I told about a local teacher in Eastern Kazakhstan named Mstislav Kushnikov, who:

> ... has heard stories about ksy-gyik (wild man) from the locals and personally saw huge footprints at the place of one sighting. He opened a regional museum and one of its paintings, done by a local artist, is an enlarged copy of a picture in a text-book of anthropology showing a Neanderthal in front of his cave. Once a young tractor-driver from a nearby village scrutinized that picture for a long time and said: "I've seen a type of this kind, only female. She had dangling breasts and her skin was not as naked as in this one in the picture ... her body was covered with long, shaggy, dark-brown hair" (p. 104).

I recalled that episode when reading the following lines in Jan's story:

> At that time in school we were studying the subject of prehistoric man. I made the crucial mistake of pointing out that we had a family of bigfoot on our place that looked a lot like Neanderthal man except much hairier. I will never forget the consequences for this slip of the tongue. My teacher told me I was a liar and my classmates made me out an outcast. ... I ended up changing schools and attending another high school for two years thereafter (p. 171).

Thom Powell, in his book *The Locals*, 2003, tells of the capture of a seven-foot hairy male during a forest fire in Nevada in August 1999, as reported by an anonymous witness (why anonymous is well explained there). The captive had "multiple burns to hands, feet, legs, and body." He was given medical care, "tranquilized and moved to unknown location." The witness mentions "human-like arms and legs, face not like man or ape but mixed between." Witness "felt he was in the presence of a very human creature," said he "will always be a believer of their existence beyond any shadow of a doubt as seen with my own eyes, smelled with my own nose and heard with my own ears. His image is still as visible as it was then. No monster, no animal but a lineage of native man. ...

Specifics, features, anatomy? Well, stand in front of the mirror and think of man's evolution" (pp. 219, 224, 228).

So it's time to think of man's evolution and bigfoot's place in it. The pet theory of the "ape model" proponents is that bigfoot is the product of so-called parallel evolution. According to genetic findings, man and chimpanzees are more closely related than chimpanzees and orangutans. It is argued that bigfoots descend from the orangutan line of primates, and therefore genetically more different from man than chimpanzees. Why then are bigfoots bipedal? Just as a result of parallel evolution—independent development of upright locomotion in a separate from man line of primates.

That is conceivable. Insects, birds, and bats, for example, had mastered flying quite independently and each kind in its own way. Birds and our primate ancestors became bipedal as a result of independent causes. What about intelligence? It also develops independently and parallel in different kinds of animals. Crows and parrots are very intelligent, pigeons are not. Dogs, cats, and rats are intelligent, rabbits are not. So it is conceivable that primates of a separate evolutionary line from ours could reach a very high level of intelligence. If they happen to be bipedal and show high intelligence, how can we know they represent a parallel line of primate evolution? By means of a biochemical analysis of their proteins, for example.

One such analysis is mentioned in Grover Krantz's book. On page 127 he tells that Jerold Lowenstein, a biochemist in San Francisco, analyzed a few hairs collected by Bob Titmus and supposed to be bigfoot.

> Lowenstein was able to compare the protein structure and found it similar to human and African apes; it was less similar to orangs, thus eliminating them and all other animals from consideration. Differences in protein are better indicators of relationships than are visible structures because these are nonadaptive.... Lowenstein's test was not fine enough to say "yes" or "no" to the closest matches (human, chimp, and gorilla), or whether it was a new type within this group.

As to visible structures of supposed bigfoot hairs, they have been analyzed under microscope by Dr. Fahrenbach. In this work human hair was for him the only "confounding variable." This means that in its structure bigfoot hair and human hair are variables of one type of hair, different from the hair of apes and other animals. DNA analyses of supposed bigfoot hair and scat also show them to be "human," even without any variables. The analysts tend to interpret these results as contamination of the material by human DNA from the people who collected those samples. Will Duncan, who initiated DNA analyses of such material from the Carter farm, thinks otherwise, and refers to the opinion of an expert: "A scientist in Michigan began to independently suspect that the human DNA he was getting from various purported BF hair samples was, in fact, not contamination but from BF." It is thought, writes Duncan, that our close relatives, and perhaps other closely related hominids of unknown types, "would have nuclear DNA matching the human pattern. Without having knowledge of what difference would be there, and at which point of the genome to look for them, we don't presently know how their DNA would differ from ours."

Thus there is no indication that bigfoots are the result of parallel evolution and only distant relatives of humans. On the contrary, there are signs of a very close relationship. If so, crossbreeding can be banked on. In Europe, Asia, and Australia there are legends, as well as old and not so old reports, of crossbreeding between "wild men" and normal humans. The Zana case in the Caucasus is one of the best known of this kind. Some years ago I learned that a similar case is on record in America, though never mentioned in the books of leading investigators. I came across it in "S'cwene'yti And The Stick Indians Of The Colvilles: The Interaction of Large Bipedal Hominids with American Indians," as reported by Dr. Ed Fusch, anthropologist, 1992. The paper was sent to me by the late Don Davis, and I understand this material is posted on Bobbie Short's website.

The following are quotations from Dr. Ed Fusch's report.

> "The "Old Timers" among the Spokane Indians had some very strong and unshakable beliefs about him: Belief number 1—and strongest of these beliefs—was that S'cwene'y'ti was NOT an animal". "S'cwene'y'ti was accepted by the Indians as part of their environment, like the deer and bear. He was not considered an animal, but people." "Sasquatch was referred to by several different names but a common conceptual thread permeates all their beliefs. He was always considered a human being, members of their own species. Prior to the arrival of the white man, the only people known to the Indians were other Indians and Sasquatch. The Lake Band of Indians called him "Skanicum," which translates to "Stick Indian." "S'cwene'y'ti was known by the Spokane Indians to interact also with human females." "Laura states that there are areas just south of Nespelem and about two miles north of the Columbia River where she can call Skanicum (in his language, she knows how) and he will answer. She believes that they

live in the area. At one time she encountered a large male Skanicum on the highway near Nespelem. It tried to converse with her, making organized sounds, leading her to believe that they have a language. She left quickly." "One girl was kidnapped by the S'cwene'y'ti people and not returned until she was grown woman." "While cooking dinner one of the women, a recent bride through bride-purchase, took a kettle and went off after water. Minutes later she was heard screaming. The men rushed to the scene but could only stand and watch as Skanicum carried her off. They knew that Skanicum was very vengeful and if harmed the captive may be injured and the mountains would not be safe for any Indian. ... She was with Skanicum all summer, or at least a couple of months, when the men searching for her on horseback saw her gathering wild potato roots. Skanicum was asleep nearby. Upon seeing the men she emptied her lap of the potatoes, crept quietly to them, leaped on one of the horses behind its rider, and thus escaped. Upon return to camp all of the Indians immediately broke camp and hastily departed the area. During her stay with Skanicum the woman had gathered roots, etc., which they shared. Skanicum eats anything that other people eat but lives primarily on roots such as that of the thule (tooly) or cattail plant, which they gather, dry, and store in caves. They build fires with flint stone and steal hides from Indians, which they use for bedding and to cover the entrance to their cave. During her stay with Skanicum the woman became pregnant and bore a son named Patrick, who grew up on the reservation. Patrick's body structure was very different from that of other Indians as his arms were very long, reaching about to his knees. He was very short, about 5'4" tall (his mother was described as "tiny"), possessed a sloping forehead, very large lower jaw, a very large wide mouth with straight upper and lower lips, and straight protruding teeth. He was kind of stooped, or hump-backed. His ears were elongated upwards (peaked) and bent outward at the top. He had very large hands and long fingers, is described as very ugly although extremely intelligent. He attended school on the reservation, was "very smart," operated a ranch in the area, died at about the age of 30, and is buried on the reservation. ... From his marriage to Laura's cousin were born three daughters and two sons." The oldest daughter, Mary Louise, "now about 65 years old," "Mary Louise lives near Omak."

Believe it or not, according to Janice, bigfoots even have a word of their own for bigfoot-human crossbreeds. In the vocabulary she compiled, under number 112, we read: "Hanke-Wasichun = Half blooded (as in if a bigfoot and a human have a baby together)."

Legends? Well, I wish our North American colleagues would devote as much time and effort to verifying the Patrick legend as Igor Bourtsev has devoted to exploring the Zana legend.

Of all the questions raised in this discussion the question of bigfoot language is the crucial one. No wonder then that Henner Fahrenbach hastened to send me a warning already: "Dear Dmitri, with all due respect to Jan's involvement and efforts with Fox and her other sasquatches, I would preach extreme caution about accepting any of her assessments of their language abilities."

He further explains his caution by her unacceptable assessment and interpretation of the sounds on the Sierra tapes. I am grateful to Henner for his response, and his advice is well taken. Yes, Jan's account regarding the bigfoot language sounds quite fantastic at our stage of knowledge and ignorance. I am impressed, though, by the way she describes how they speak. In theory it is expectable that pre-*sapiens* hominids should speak in the way Janice describes. How could she know that? And the notebooks in which she wrote down the alleged bigfoot words are real and still existing.

And this is what Will Duncan hastened to put in:

> Hello Dmitri and Henner, from the perspective of three years' involvement in the Carter Farm situation, I presently am telling people this: We have absolutely no hard evidence of the language abilities of her BF. By this I mean we have no sound recordings, no videos of them talking (or otherwise) and so on. We do have anecdotal evidence that they may be able to speak or to mimic speech. Specifically, Jan says they can. Lila says they can, Gene and Michelle McCauley told me they heard Fox say "Hello," Paul Coy told me he heard Jan talking with Fox but couldn't understand him, and I personally heard murmuring vocalizations sounding like moderately deep human voices coming from the area of the main barn while I was in front of the house and there were no people around. ... I do not contend that any of this information is conclusive. But it is very suggestive and points to avenues for future investigation. It would be foolish to ignore it. If BF has human DNA it would make its purported language abilities easier to understand. Certainly Jan's story, as difficult as much of it is to swallow whole, indicates they are people of some sort rather than apes, just as Albert Ostman and many of the Native reports suggest.

It is said that the significance of a scientific theory can be measured by the time it impeded scientific progress. Let's hope the "ape model" theory won't go down in history as very significant in this respect. Still it plays a major part in causing the bigfoot research community to turn an unseeing eye to the Carter Farm case and the book *50 Years with Bigfoot*.

In my opinion, after this book, business-as-usual is not on the cards for hominology. The idea that the North American homins may be people is coming full circle, from the reports of J. W. Burns and Albert Ostman of sasquatches in British Columbia to Janice Carter Coy's story of bigfoots in Tennessee. Should the idea be confirmed, all our books will turn into short introductions to the subject, while *50 Years with Bigfoot* will become the first textbook in hominology. Admittedly, its drawback and limitation are in the fact that the authors are lay-persons, not scientists. Let's hope that a second or a third textbook will be authored by diploma'd hominologists. In the meantime many thanks should go to John Green for publishing Albert Ostman's story and to Mary Green for publishing the story of Janice Carter Coy.

While mainstream science is turning its back on hominology, primatologists lost no time in altering the meaning and taxonomic level of such useful terms as "hominoid" and "hominid."

> When scientists use the word hominin today, they mean pretty much the same thing as when they used the word hominid twenty years ago. When these scientists use the word hominid, they mean pretty much the same thing as when they used the word hominoid twenty years ago. ... If you're more confused now than you were before, you are just about where you should be. We scientists really need to clean up shop in this area." (Thomas M. Greiner, Associate Professor of Anatomy / Physical Anthropology, "What's the difference between hominin and hominid?")

But this muddle of terminology doesn't concern the problem we're discussing here. And the banter about "naked apes" and "hairy apes," mentioned by Loren Coleman in his book, is good only for fiction, not science. There are two notions and terms in science which have not changed their meaning so far: "human primate" and "nonhuman primate." Russians and Americans are human primates, chimps and gorillas are nonhuman primates. The clear question, in need of a clear answer, is this: What kind of primate are such homins as bigfoot—human or nonhuman? My answer is this: If they have a language as mentioned by Albert Ostman and described by Janice Carter Coy, they are definitely **human**. (Let us recall that back in the 18th century Linnaeus proclaimed two kinds of man: *Homo sapiens* and *Homo troglodytes*.) I would hold this true even if the words of their language are largely borrowed from *Homo sapiens*. How this could have happened is another question and mystery.

If they don't have what can be called human language, then they must be nonhuman primates on the threshold of humanness. This judgment is based on the independent evidence of those who claim to have seen or even interacted with sasquatches and dared voice their unpopular accounts and opinions, even if they are at loggerheads with the prevailing opinions and theories of those who have never seen these hairy bipeds.

Hominology came into being in a no-man's land of science between zoology and anthropology. It has been shifting ever since from the zoological side of the area to the anthropological side. Accordingly, there is reason for hominologists to be shifting from cryptozoology to what could be called crypto-anthropology. Frankly speaking, I've always felt that the partnership between hominology and cryptozoology is a marriage of convenience rather than of love and mutual understanding. It has been good for cryptozoology and, under the circumstances, good for hominology, but for the latter NOT good enough. And this because the partnership relegated hominology to pure zoology, concealing its paramount anthropological and philosophic aspects. The International Society of Cryptozoology and its good journal, *Cryptozoology*, let hominology down by completely ignoring hominology's major asset, the Patterson/Gimlin film, and one of hominology's major problems, the Iceman. This was so because the Society and its journal were fully focused on "mere animals" and zoology, while the majority of academic cryptozoologists found it too risky for their reputations to plunge into hominology. What world science and humankind itself badly need, without realizing it, is The International Society of Hominology and its journal, *Current Hominology*.

Finally, let me remind you of these words by Grover Krantz: "It might be argued that we don't really know enough about sasquatch behavior to be absolutely certain about this judgment as to its animal status. But if we are in error, isn't it imperative that we find out as soon as possible?" (*Big Footprints*, p. 12). Find out **how**? By killing one of them? No way! To find out the truth as soon as possible we would need a repeat of the Ostman adventure, but with an anthropologist, say Dr. Jeff Meldrum, in the shoes of Albert Ostman.

*July 2005*

SECTION 11

# Updating the Iceman

## Old and New on the Frozen Mystery

I was recently requested to provide my comments on the Iceman, and this led me to take the issue off the back burner. It is remarkable that the Iceman started its "career" the same year as Patterson's Bigfoot movie. According to John Napier, "At the beginning of the summer of 1967" Frank Hansen "started touring the Iceman"(*Bigfoot*, 1973, p.109).

Both the Iceman and the movie are still officially believed to be fake. But if the film subject is taken now as real by the majority of bigfoot researchers, the Iceman is denied authenticity by most of them, in spite of the fact that the Iceman initially made a louder noise in science than the Bigfoot film. Here is a quote from the article by Magnus Linklater, "Neanderthal Man?", published by the *Sunday Times* of London on March 23, 1969:

> A strange ape-like creature frozen in a block of ice is providing American anthropologists with one of the most intriguing questions they have faced in recent years. Is it a fraud, a freak, or is it a form of human being believed to have been extinct since prehistoric times? One thing is certain: it has two large bullet-holes in it. Just as a precaution the FBI have been called in...

The concluding lines of the article ran as follows:

> Whatever the explanation, a capital crime may have been committed. Accordingly the FBI has been informed. However fanciful all these suggestions, the anthropological world may be on the verge of one of the most exciting discoveries in the study of man. Dr Heuvelmans's ape-man might just provide the evidence of a missing link in the evolution of man. Even if it doesn't it could become as great a cause célèbre as the Piltdown Man.

The Iceman owes its *cause célèbre* status to Bernard Heuvelmans and Ivan Sanderson, the founding fathers of cryptozoology, and in part to Boris Porshnev, the father of hominology. In February 1969, Heuvelmans published, in the *Bulletin of the Royal Institute of Natural Sciences* of Belgium, a paper entitled, "Notice on a specimen preserved in ice of an unknown form of living hominid: Homo pongoides." Later, in 1974, he devoted a voluminous book to the case, *L'Homme de Néanderthal est toujours vivant* (Neanderthal Man is still alive).

Sanderson committed his findings to paper in the report, "Preliminary Description of the External Morphology of What Appears to be the Fresh Corpse of a Hitherto Unknown Form of Living Hominid" (*Genus*, Vol. XXV, N.1–4, 1969). Porshnev's role lies in the fact that Heuvelmans referred to Porshnev's ideas in claiming the present-day survival of Neanderthal man, supposedly evidenced by the Iceman. Porshnev in turn dwelt at length on the case in the Russian popular press and asked a very pertinent question: If the Iceman is a model, WHAT is it a model of?

The relevance of this question became especially clear with the publication in *FATE* (March 1982) of the piece, "The Iceman Goeth," in which debunkers referred to one Howard Ball, "who died several years ago" and who "made models for Disneyland." "He made (the Iceman) here in his studio in Torrance (Calif.)," Ball's widow Helen told Emery. "The man who commissioned it said he was going to encase it in ice and pass it off, I think, as a prehistoric man." Ball's son Kenneth helped his father build the figure.

He says its "skin" is half-inch-thick rubber. "We modeled it after an artist's conception of Cro-Magnon man and gave it a broken arm and a bashed-in skull with one eye popped out" (p. 59).

That the Iceman is a model of Cro-Magnon man is sheer nonsense and the height of anthropological ignorance. There exists no artist's conception even of Neanderthal man as hairy as the Iceman. A "prehistoric man" of this kind was posited only by Boris Porshnev's anthropological theory, which was not widely known at the time and is not recognized

In May 1969 Argosy magazine featured extensive coverage of the Iceman in an article by Ivan Sanderson. It was this article that brought the issue to the attention of the general public. Remarkable photographs and artwork by both Heuvelmans and Alika Lindbergh, are provided. In many circles it was indeed thought that the "missing link" had been found.

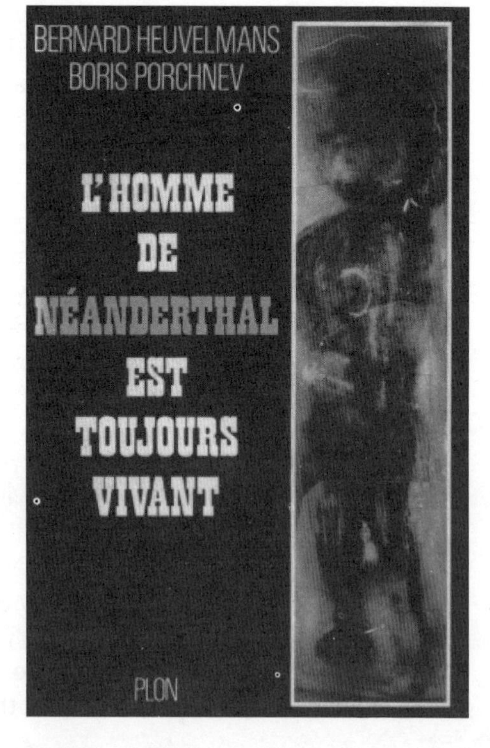


Heuvelmans/Porshnev book, 1974, showing the '"Iceman" on the front cover. It consists of two parts: "The Struggle for Troglodytes" by Porshnev, translated into French, and the Iceman investigation by Heuvelmans. It is a monumental work of 500 pages with extensive coverage and numerous photographs.

even today. So the question persists: If the original Iceman is a model, WHAT is it a model of?

But the most crucial question concerns the exhibit's authenticity. There are two episodes in the story which seem to indicate more than anything else that what originally lay in the ice was not a fabrication. In July1989, Minnesota sasquatch researcher Mike Quast visited and interviewed Frank Hansen at his ranch. In his good book, *The Sasquatch in Minnesota* (1996), Mike has this to say on the matter:

> The reports published by Sanderson and Heuvelmans brought an incredible amount of attention Hansen's way, much to his anger because he had insisted on no publicity when he allowed them to examine the Iceman. He was particularly upset with Heuvelmans, whose report appeared first.
>
> According to Hansen, what does not appear in either scientist's report is just how they became convinced the Iceman was real. To get the best possible view of it they had hung bright lights over the glass under which it lay, and while Hansen was away from them for a moment one of them placed one of the hot lights directly on the ice cold glass. It shattered, and a pungent odor like that of rotting flesh rose from the ice. This convinced them that an actual corpse, freshly killed, lay before them. Hansen will never forget what the distinguished scientists said when he reminded them of their promise not to publicize the story at that point. "We are scientists first," they told him, "and gentlemen second." (He doesn't say exactly which one of them said this.) (p. 144).

Ivan Sanderson, in his report, refers to this important incident in this way: "The corpse, or whatever it is, is rotting. This could be detected by a strong stench—typical of rotting mammalian flesh—exuding from one of the corners of the insulation of the coffin. Whatever this corpse may be, it would seem to include flesh of some kind" (*Genus*, p. 253).

Why did Hansen insist on no publicity when he allowed the two scientists to examine his exhibit? How could publicity from such examination harm his carnival sideshow business? And why did publication of the scientists' conclusion that the corpse was real cause the showman's anger? The answer is in Magnus Linklater's words cited above: "a capital crime may have been committed." This must have been the reason for Hansen's subsequent actions, maneuvers, and conflicting stories. Let us also note one of his recurring statements that was as little believed as all his other declarations, namely, that the Iceman did not belong to him, but to a millionaire in California.

A second episode indicating the Iceman's reality happened in July 1969 when, after a tour of Canada with his exhibit, Hansen was held up by US customs officials at a border post in North Dakota. The episode was related by Sanderson to Heuvelmans and is mentioned in the latter's book (pp.283–84). Customs demanded from Hansen special permission by the US Surgeon General for carrying the corpse of a "humanoid creature." Hansen argued that it was not a real corpse but a "fabricated illusion" made of latex rubber, and offered documents of its fabrication. That did not impress the officials, who demanded that a piece of the Iceman be taken for examination. Hansen protested, saying this would damage the exhibit.

In desperation he even phoned Sanderson and asked for advice. The latter, thinking that this time Hansen toured the model, advised that the customs x-ray the exhibit, to which suggestion Hansen cried out: "Impossible! The owner will never allow this!" (My translation from the French—D.B.).(Heuvelmans remarks in brackets that there was no need to inform and ask the owner because x-rays leave no traces). Hansen then sought by phone the help of the Iceman's owner in California, as well as that of his own senator in Washington, Walter F. Mondale, subsequently Vice-President in Carter's Administration. Twenty-four hours later Hansen was released with the Iceman unchecked.

To quote Mike Quast again:

> Some call the Iceman by the name "Bozo," a carnival clown, nothing more. To most serious investigators now that's all he was—a phony, no more real than a mannequin. He has, for the most part, been written off as a big joke. But the joke is on them, because the Iceman was real.
>
> At the 1967 Arizona State Fair he (Hansen) met a man who to this day he will not name, but he says "It was a name I recognized immediately," and that it was someone connected to the entertainment industry. The man said he had a very interesting specimen in storage in California and asked Hansen to consider taking it on a carnival tour. Shortly thereafter, in Long Beach, Hansen first laid eyes on the Iceman.
>
> The man explained that an agent of his had discovered the creature in its frozen state in a refrigeration plant in Hong Kong and that it had originally been found floating in the sea by Chinese fishermen in a 6,000-pound ice block. He was a deeply religious man, Hansen explained, and he thought this creature seemed to go against the theory of creation as told in the Bible, thus he wanted no connection to it.

Hansen agreed to display it, but first the ice was temporarily shaved down for his benefit and he saw that it was indeed a real corpse, not a fake.

Hansen was given permission to use whatever phony advertising he wished in order to draw crowds. Stories about "Bigfoot" in the news at that time helped as well, and the display was very popular. ... After some time, however, Hansen began to worry that he might get into serious legal trouble if what he had turned out to be a human corpse. So, returning to California, he had a replica manufactured from latex rubber and hair, intending to switch it with the original if he ever had to." (pp. 143–44).

In 1994, Quast got a surprise when Hansen himself gave him a call.

But the biggest surprise came when he said that he had recently heard from the real owner of the Iceman, who he had not talked to in a long time and didn't even know if the man was still living. He still wouldn't name him, of course, but he said the owner claimed to still be in possession of the original Iceman and that it was still frozen and in good condition. Also, he might (just might) consider presenting it to the public once again in the near future. Well, that was a couple of years ago. No word yet. ... The last word, however, belongs to that anonymous owner, who once stated to Hansen that if he was ever identified he would dump the Iceman in the Pacific Ocean. (p. 146).

Having read all that, I contacted Mike Quast in 2002 and in April received a letter from him, with the last paragraph reading as follows:

I have had one theory—and that is all it is—about who the anonymous owner of the Iceman might have been. I am not saying that I necessarily believe it as fact, but the only name that comes to mind is the late actor Jimmy Stewart. Hansen said it was someone in the entertainment industry and that when he met him it was a name he "recognized immediately," and that the man did not want to be publicly identified with the Iceman because of his strong religious beliefs. I believe Stewart was known as being rather religious, was a world traveler, and he did have some interest in such subjects as he was involved with Peter Byrne and Tom Slick in getting a yeti hand smuggled out of Nepal (according to Loren Coleman's book on Tom Slick). Stewart was still alive when Hansen told me he had just been in touch with the owner, but died a couple of years later. That is the only idea I have come up with ... If investigators visit Hansen today, they might try mentioning this theory to him and just studying what his reaction is to the name.

I heartily thanked Mike Quast for the information in his book and the theory in his letter, and proposed to Alan Berry to try and verify that theory with a visit to Hansen and one more interview. Alan was too busy to go to Minnesota, but interviewed Hansen by phone on April 7, 2002. Here are some excerpts from that interview:

Berry: What do you think the Iceman represented?

Hansen: I can tell you I don't associate it with Bigfoot.

B: You mean if it was real?

H: Yeah ... well, I mean if it was real, I would think of it like might be some kind of early man, but I don't know.

B: What did the owner tell you about where it came from?

H: He was in the business of producing movies, and he (?) was in the Tokyo (?) Bay area, and saw a block of ice with this thing in it. He asked the fishermen, "What do you want for it?" They dickered and he ended up trading a case of whisky for it. He said he didn't know what it was, just that it was interesting and something his people might use, you know, as a prop. The owner leased space with refrigeration on a ship and the block of ice with the Iceman was shipped to the U.S.

B: What was the owner's interest in exhibiting the Iceman?

H: Just to see what the public would think of it ... what kind of furor or controversy it might create. He wasn't looking for anything out of it himself. He was a religious man. He just wanted to see how people would react if they thought there was really a primitive form of man that came before us in time, you know, evolution and such...

B: What kind of person was the owner?

H: He was very, very religious. He didn't want the Iceman exhibited as anything real, only wanted the public's reaction. Like could it be something almost human from prehistoric times?

B: Who was the owner?

H: I can't tell you, I am under oath. I can just tell you that he was a big name... Anybody would recognize his name right away today even, but he's dead. He passed away.

B: What had become of the body?

H: I tried to take it (the exhibit) into Canada for a show, was stopped at the border. It was the Bureau of Customs, and they stopped me because they thought I was transporting cadaver across the border. It was seized at the border. I explained to them it was just an exhibit, neither man nor beast, but they didn't believe me until a US Senator bailed me out. Because of, who was he, Irene? Well, I was a good friend of him, and had given him a lucrative donation, yes, it was a senator in Washington. It was through Walter Mondale, the senator, that they got an order from Agriculture and Forest Products to "let them go." After the border incident and with "all the people" that were after me, I got tired of the whole thing and phoned the owner to take back the Iceman.

It is most important that Hansen confirmed the border incident of which we learned first from Heuvelmans, even though there are certain differences with Heuvelmans's words in Hansen's description. Why did he mention Tokyo instead of Hong Kong as the place where the Iceman came from? Was his memory failing?

In September, 2002, Dr. Peter Rubec talked to Hansen on the phone. Here's a quote from Rubec's email to me: "I did ask Hansen about Jimmy Stewart. There was a fairly long pause, but all he would say is that the owner of the real Iceman (he was fairly emphatic there was a real one) was in the movie industry and had died. But he would not reveal who it was."

I then discussed the matter with Loren Coleman, who, when writing his book about Tom Slick, had contact with Stewart. Loren confirmed to me that the latter was very religious and referred to the opinion of Mark Hall, who had two separate interviews with Frank Hansen in the 1990s: "It appeared the owner did not value it (the Iceman) in the way many of us would... The true owner of the Iceman did not want to be the one who presented the 'missing link' that would undercut the truth of Biblical creation. The owner was interested in seeing people's reaction to the 'missing link' and so allowed the Iceman to be displayed" (*Living Fossils*, 1999, p. 85)."Mark Hall senses," wrote Loren to me, "that the mysterious owner was a pro-creationist."

I then addressed Peter Byrne, saying that I've been trying to crack the Iceman riddle in recent years, urging Krantz, Greenwell and others to do so while Hansen was alive, and continued, in part: "When I read *The Sasquatch in Minnesota* by your friend and follower Mike Quast I asked him to help. And he did by supplying information about Hansen which is not in the book. In his letter he shared with me his opinion and hypothesis regarding Jimmy Stewart. So the credit for it goes to him."

Peter Byrne wrote back on August 30, 2003: "Your hypothesis [I had told him it was not mine.—D.B.] concerning an Iceman connection with Jimmy Stewart is very interesting and indeed is one that has surfaced previously, mainly because of my connection with him going back to Yeti and Himalayan days. So, let me talk with some family members and what they have to contribute to it and then I will get back to you."

His email of September 4 added this: "In the matter of the Iceman these leads definitely need to be examined and followed up; as you say, anything is possible and actually there is a faint but persistent rumour in entertainment circles that Jimmy Stewart did have an association of some kind with some large and mysterious animal." The email of October 30 said the following: "As of now I do not have a lot to report. There are, as I said previously, grounds to believe that Jimmy Stewart was definitely connected/associated to/with a large animal of some kind; however there are conflicting reports on exactly what it was. This (confliction) of course could be part of a cover up; it is my finding that in many cases families of people who have these associations like to have them brushed under the carpet, so to speak, after they (the finders or investigators) pass away. This was indeed the case with Tom Slick whose family, after he died, seem to have destroyed all of the evidence that he gathered on the Bigfoot mystery and whose foundation, the Southwest Research Institute of San Antonio, Texas, now state that they had nothing to do with his BF research when in fact some of the expense and salary checks that I received when I ran the first northern California Bigfoot project were on the institute's bank account. So it may be the same thing with the Stewart family."

As of today, I have not heard from Peter Byrne anything more on the Jimmy Stewart connection.

The latest information on the Iceman that reached me comes from Curt Nelson, who on April 11, 2005 emailed me the following:

> I live in Minnesota just north of Minneapolis/St. Paul. Hansen, I'd heard (from Mike Quast), was last known to live near the small town of Rollingstone, about 100 miles south of me. I drove down there in February just to see if I might find him (the phone number for Hansen Mike Quast provided me was no longer in service). In the town of Altura (a few miles from Rollingstone), I stopped at a bank and went in and inquired about Hansen, about his whereabouts. The woman (a bank teller) I spoke to said she knew where Hansen's home was but that he was gone, that he died two years prior. She said that his wife and son still lived in the area, though, and she looked them up in the phone directory for me. ...
>
> I went out into the parking lot of the bank and called Mrs. Hansen

(Irene) using my cell phone, and I reached her. She was not enthusiastic about talking about The Iceman but she did speak to me for about 5 minutes. ... She said her husband died with the secret as to the true story on the Iceman, that even she didn't know it. She seemed to think that was quite appropriate and she seemed sincere about it. And at least twice she said, in reference to the secrecy surrounding the Iceman, that it was "to protect the innocent ones."

The son is an attorney and I reached him at work just after speaking to his mother. He was in a bad mood in the first place, I would say, and was just barely polite to me on the subject of his father's iceman. (I'm sure the Hansen family has been bothered plenty about it over the years.) He told me the second body, the one widely thought to have been an obvious fake, was gone. That it had been cleared out long ago.

I called Mrs. Hansen back again while driving home to ask for a clarification on something (can't recall what just now), and she asked me if I knew Roger Patterson. She said she and her husband visited Patterson in California. She just volunteered that, seemingly just to make conversation. She didn't remember anything about the meeting, but it tells me Frank Hansen had an interest in bigfoot. I find that interesting—that the carnival man showing off The Iceman would look up a man who claimed to have filmed a bigfoot, a man thousands of miles away in California. It suggests to me that Frank Hansen believed what he had might be a bigfoot. (If Hansen's body was a fake why would he be interested in bigfoot?)

In summary here is what I took away from my conversations with Irene Hansen and her son: Nothing is final, it is still all a mystery. ... The son is a lawyer and if there is a concern about legal issues (the creature might be considered human) he has certainly counseled his mother on how to answer questions—with no real answers. It seems to me that the simplest truth behind this story would be that it was all a hoax perpetrated solely by Frank Hansen. If that were true why, decades later, wouldn't he and now his family just say, forget it, it was just a carnival trick!(?) ... Please feel free to use what I've told you in any way you like.

Thank you very much, Curt Nelson, for your most important information. It is news to hominologists that Frank Hansen has died. Regrettably, the event passed unnoticed two years ago. I agree with your inferences and conclusions, especially the one concerning Hansen's visit to Patterson in California, which is a big surprise. The news should be verified and discussed with Patricia Patterson. You are right, if the Ice-man was a fake why would Hansen be interested in Bigfoot? He must have been interested in Bigfoot because he was keenly interested in the exact nature of the carnival exhibit he displayed. Was it a human or non-human primate? The very legal status of the exhibit depended on the answer. The leading Bigfoot researchers, such as John Green and Grover Krantz, called Bigfoot an ape, a giant non-human primate. Was it not for this reason that Hansen for a time presented the Iceman as a bigfoot he himself killed during a hunt in Minnesota? The different signs he used for the exhibit in sideshows are also indicative in this connection: "What is it?", "Siberskoye Creature," "Found in the Woods of Minnesota," "Is it Prehistoric?"

The question "What is it?" must have been heavy on his mind when he allowed Sanderson and Heuvelmans to examine his exhibit and asked them not to publicize their findings. Heuvelmans's published conclusion that it was the corpse of a killed Neanderthal Man must have alarmed Hansen a lot. From his words to Alan Berry, "I don't associate it with Bigfoot" and "it might be some kind of early man," we can conclude that Heuvelmans's verdict stuck in Hansen's mind and determined his words and actions to the end.

Of special interest are Irene Hansen's words that the secrecy surrounding the Iceman serves "to protect the innocent ones". This brings up the question: And who are "the guilty ones"? They can well be inferred from Hansen's own words. First, the Iceman owner who smuggled a corpse into the U.S. and kept it illegally; second, Frank Hansen who displayed a smuggled dead body without permission; third, ex-Vice-President Walter Mondale through whom Customs got an order to let a cadaver across the US-Canada border.

And who are "the innocent ones"? Apparently the families of the guilty ones. They know the truth and for obvious reasons are determined to keep it secret, no matter what the detriment to science.

Mike Quast again: "It is certainly a case that seems to deserve any researcher's undivided attention, for in it we supposedly have what Bigfoot people have sought for so many years: the actual corpse of a hair-covered humanoid" (p.137). I am convinced now that the words "we supposedly have" could be changed to "we do have" if not for the fact that the actual corpse is still out of our reach. As I wrote not so long ago, "The negative impact of indifference on one side, and hidden or open hostility on the other, leaves the tiny number of hominologists little chance to quickly obtain traditionally acceptable biological proof." The Iceman case illustrates this point with utmost clarity.

Let us note that after Sanderson and Heuvelmans the case was followed up and bits of truth gleaned and collected not by scientific institutions, such as the Smithsonian or the International Society of Cryptozoology, whose express task was to investigate such cases, but by private researchers, such as Mark Hall, Mike Quast, Alan Berry, and Curt Nelson. Well, long live private enterprise!

I wish the Stewart connection would finally be established. I wish its confirmation for two reasons. First, to the usual question "Where is hard evidence?" we'd have a ready answer: "Ask the Stewart family." Second, I'd offer Hollywood a scenario of a film, based on facts stranger than fiction, from the scene of getting the body of an ape-man for a case of whisky to the final shots of dumping it in the Pacific Ocean. The story would be the opposite of the Piltdown Man. In the latter a fake was used to fool scientists. In my scenario a Hollywood pro-creationist film star makes the anthropological world (on the verge of one of the most exciting discoveries in the study of man) take a real "missing link" for a "fabricated illusion." The film would be titled, "The Carnival Cover-up."

SECTION 12

# Substantive Interviews

## The Heinselman Interview

The interview took place via email during February 2001 between Craig Heinselman of Francestown, New Hampshire (USA) and Dmitri Bayanov of Moscow, Russia.

**QUESTION 1: Regarding the terms Hominology and Homin, can you explain the reason why such terminology is needed in this area of research, and how these terms originated? Why the separation from the general term, Cryptozoology?**

**Answer from Dmitri Bayanov**: The terms *hominology* and *cryptozoology* came into use a generation ago independently of each other. Boris Porshnev is the founder of hominology, Bernard Heuvelmans of cryptozoology. But being still in a cryptozoological phase of development, hominology is an integral and most important part of cryptozoology.

Why should it retain a separate name, despite its inclusion in cryptozoology? Partly because of its importance, but even more so, because "homin," i.e., "non-human hominid," is both "a new word" and concept in science.

Anthropology has never dealt with a primate having a human form but leading a non-human way of life. Zoology has never dealt with a beast having a human form. As soon as a cryptid is recognized by zoology, it stops being a cryptid and comes under the auspices of one or another branch of zoology. Such was the case of the okapi, the giant squid, the gorilla, and a number of other cryptids. But the homin has no discipline to come under, except hominology.

Zoology, primatology, and anthropology, as we know them today, have no place for living wild humanlike bipeds—the creatures' specifics are not dealt with by any of the established sciences. Hominology is necessary and inevitable because between zoology and anthropology there is a big gap in knowledge; a no-man's land of science. The hominologist is a hunter for knowledge in this no-man's land.

When the "snowman" affair touched my country in the 1950s, terminology was quite a problem. Science had come across an object that it not only had no knowledge of, but not even a name. In the West they called the Yeti "a bipedal anthropoid," putting it in the family of apes, despite the fact that apes are quadrupedal. As for Porshnev, he viewed the "snowman" as a relic of ancient hominids, and yet regarded it as non-human, in accordance with his own unorthodox views.

A way out of the apparent contradiction was proposed by Pyotr Smolin, chief curator of the Darwin Museum, a dedicated supporter and partner of Porshnev. Smolin proposed the term "relict hominoid," i.e., a humanlike relic, without putting into it any strict taxonomic category. The term was accepted by Porshnev, and his monograph (1963) was titled "The Present State of the Question of Relict Hominoids."

He writes in it of "the emerging science of relict hominoids." In the 1970s, after Porshnev's passing, I began to call this science "**hominology.**" The drawback of the term "hominoid" for us is in the fact that in the systematics of primates, it embraces both apes (pongids) and hominids (to which our subject belongs). And the term "hominid" is not convenient either because it includes modern man, *Homo sapiens*, and his fossil ancestors. To avoid confusion, the term "**homin**" is introduced to indicate extant non-human hominids.

Actually, **homin-ology** can be nothing but the science of homins.

**QUESTION 2: In hominology, there appears a separate debate over "lumping" and "splitting" of homins. Part of the debate is if creatures/beings such as the Yeti, Alma, Yeren, Orang-Pendek, and so forth exist, are they the same kind of creature/being or different kinds? Various people from Mark Hall, Loren Coleman, Patrick Huyghe, and Ivan Sanderson (to name a few) have attempted to classify or separate the geographical or morphological differences in these homins by names such as Marked Hominid, Proto-Pigmies,**

**Sub-Humans, and Neo-Giants. What do you think of this debate and the classification/separation scenarios?**

**Answer:** I wrote in 1992:

> In cryptozoology, we know of certain fossil, or living, or supposedly extinct forms, and find that certain cryptids more or less match these known forms.... Hence, all the talk of 'living dinosaurs,' 'living gigantopithecines,' etc.... Of course, a cryptozoologist can go further, and propose a name for the species or subspecies of the cryptid he is after. But that is part of cryptozoology's fun, not science, because the proof of the 'pudding' is beyond the realm of cryptozoology. The proof only comes when a cryptid turns into an ex-cryptid by becoming a zoological accepted taxon.
>
> (*Cryptozoology*, Vol. 11, 1992, pp. 129-30)

This is also true for hominology, which is still in a cryptozoological phase of existence. In this realm we have first to distinguish between fact and hypothesis. As a dedicated hominoiogist, I take it for a scientific FACT that most homins in Eurasia, Australia and the Americas are HOMINIDS. But what kind(s) of hominid is still HYPOTHETICAL.

> Relic hominoids being flesh and blood, their existence does not depend on any classification, but the existence of any hominid classification is bound to depend on the nature of relic hominoids.... In short, to size up the creature we seek with the existing taxonomy is like measuring an object with a measure which is bound to be changed when the measurement is finished. Yet engage in this strange procedure we must, if only to show that our 'wards' are no freak sapiens or visitors from outer space.
>
> (*Current Anthropology*, June 1976, p. 313)

As hominologists depend in their classification hypotheses on the taxonomy and nomenclature situation in paleoanthropology, let us take a closer look at it.

> A glance at paleoanthropological nomenclature shows that the names given to different hominid forms are not more than appellations to distinguish them from each other, even if these names have been clothed in the scientific cloak of Linnaeus' principles. They express no classificatory meaning, even if the donors of the names or those

who use them believe that they do. (Franz Weidenreich, *The Skull of Sinanthropus Pekinensis*, 1943.)

The nomenclature of fossil forms in anthropological literature has been in a state of confusion — almost of anarchy— ever since the beginning. In the last 20 years or so, there has been a more conscious effort to make it conform to the standard of zoological nomenclature, but most anthropologists have had little or no training in taxonomy, and do not really know what these standards are. Not only fossil men, but also fossil primates in general, have a very confused nomenclature.... Many early writers had the habit to create a new species, or even a new genus, for almost every newly discovered specimen.... Hominoid classification has been particularly difficult (because of the limited amount of material available for study and comparison), and this has contributed to confusing the nomenclature.

(Anna K. Winner, *Current Anthropology*, Vol.5, No.2, April 1964)

Thus, in the beginning, "splitting" in paleoanthropology prevailed in quite an unjustified and unscientific manner. As "lumping" took over and developed, I want to note one particular move in this direction which is of great significance for hominology. I mean the "lumping" of Neandertal with *Homo erectus*, performed by Grover Krantz in his article "Sapienization and Speech" (*Current Anthropology*, December 1980). He found that Neandertal had more common morphological traits with Homo erectus than with Homo sapiens, and concluded that Neandertals "could all be classed with *erectus*" (p. 714).

He also wrote:

*Homo erectus* existed for over a million years with relatively little change—a kind of evolutionary plateau— and then was transformed rather quickly into *Homo sapiens*. (p.773).

So it is logical to surmise that today's wild bipedal primates in Eurasia, Australia, and the Americas are relics of that evolutionary "standstill," which lasted long enough for homins to penetrate and settle whole continents before the advent of *Homo sapiens*. Adapting to local environments and changing ways of life, they must have more or less departed in their physique and habits from the fossil *Homo erectus* (Neandertal included), forms presently known to science. It would take much discussion to go into a detailed substantiation of this view here. Still,

... the hominologist's dream is that all the hominid forms known from the fossil record, and even those not known from it, will turn out to be alive. The dream, however, has to be checked against reality. (*Current Anthropology*, June 1976, p.312).

**QUESTION 3: In a reprinting of some of your letters to Roderick Sprague that appeared as "A Hominologist's View from Moscow, USSR" in *The Scientist Looks at the Sasquatch* (1977, Anthropology Monographs of the University of Idaho No. 3,), and previously in Northwest Anthropological Research Notes 11 in 1977, you wrote the following passage: "Since hominoids can not be expected to fall from the blue on their own, as is the case with meteorites, I wish some pranky UFOnauts would dump a load of bigfeet on the heads of skeptics among modern academics."**

**This appears at first look as a clever jab at conventional rejection of hominoids such as the Yeren, Alma, Sasquatch, and so forth. However, the context mentions UFOnauts. There has been, and continues to be, a perhaps growing sub-section of cryptozoology and hominology that feels UFO's and paranormal answers are the solution to these mysteries. How do you feel about this growing subsection in this area? How have its implications affected the biological study of these Homins?**

**Answer:** In the beginning of my hominological career a noted UFO investigator offered us cooperation, a sort of marriage between UFO studies and hominology. With all respect for serious UFO watchers I declined the offer, thinking it would be a "misalliance." I take hominology to be a young science, a branch of primatology, whereas UFO studies are still at a stage loosely comparable to a time when philosophers and naturalists were arguing whether stars and planets are gods or natural objects.

Once, on a TV program, a UFO enthusiast was pitted against me, saying the "snowmen" were biorobots controlled by UFOnauts. I said I agreed if he accepted my idea that gorillas were also such biorobots. He did not know what to answer. I also said that if UFOnauts use homins, or humans, or cows, or sea serpents for their own needs, it's their own business. As for hominology, it needs the UFO connection in its present pursuits no more than zoology needs it in studies of chimps and gorillas, or ethnology in studies of the Tasadays.

Interestingly, a group of hominologists, including M.J. Koffmann, happened to see a UFO in the Caucasus, but never an almasty. So my words, mentioned by you, regarding "pranky UFOnauts" dumping "a load of bigfeet" on the heads of academics, were nothing but a jab at the latter.

**QUESTION 4: You mention cryptozoology and hominology in a recent article entitled "Could Bipedalism be Initial" from BIPEDIA (number 19, January 2001). In it you say: "Naturally, anyone is free to proclaim and support any hypothesis, including that of Initial Bipedalism, but the latter has no place in cryptozoology and hominology, which are based on evolutionary theory, with Darwinism in its core."**

**Now the debate over Initial Bipedalism isn't the topic here; that can be done better within the confines of other publications. However, it does bring up a question. As we are all free to support a particular theory, some researchers favor a creationist basis for hominology and cryptozoology, while others an evolutionary basis (be it Darwinian or other). Is this contradiction between the two detrimental to research acceptance in your view? Is this debate over hominology and cryptozoology as argued in your country?**

**Answer:** One is science, the other religion. As there are religious physicists, chemists, zoologists, etc., no wonder religious cryptozoologists and hominologists are also on record. When they collect evidence for cryptids, no problem.

But problems may arise if they drift into the domain of evolution. Homins and humans have, for example, dermal ridges on their fingers, palms, toes, and soles. The evolutionary explanation is that this feature is a leftover from the arboreal stage of primate evolution. What is the creationist explanation? Did God envisage and intend fingerprinting? If man was created in the image of God, does He also have dermal ridges? What for? Or take the navel. Mammals have it, reptiles don't. Did Adam have it? Scholars and artists in the Middle Ages racked their brains over the riddle. One problem the less for hominologists/evolutionists.

**QUESTION 5: In Cryptozology Volume 1, you wrote, in part, that "...realists and folklorists in hominology should sit down together and sort out the mountain of folklore on hominoids." Have you seen this sorting out take place in your country in the last few decades or from your correspondence with other researchers over the years?**

**Answer:** During my first expedition to the Caucasus as a member of M.J. Koffmann's team, in 1964, I was struck by the fact that the locals often referred to the almasty quite matter-of-factly by such names as "devil," "satan," "goblin".... Back in Moscow, I plunged into reading literature on folklore, demonology, and the history of religion. I was fascinated by what opened to my eyes, my mind already opened by the Porshnev theory, and what I learned in the expedition.

It became clear to me that folklore and demonology, or what John Napier [following F.W. Holiday] called the Goblin Universe, is the richest source of hominology, and is very realistic, but is totally misunderstood and misinterpreted by academic specialists on folklore and mythology. Soon I came up with a work whose title could be translated into English as "In Defense of Devilry." The work was never published in the Soviet years and no folklorist ever agreed to cooperate with me, despite my friendly approaches.

When the country's political situation began to change I enlarged my original work, changed the title to "Wood Goblin Dubbed Monkey: A Comparative Study in Demonology," and after fruitlessly approaching many publishers, at last succeeded in finding one who published it in 1991. I sorted out in it volumes of published folklore of the many peoples of the Soviet Union, focusing on the most realistic descriptions of the appearance, behavior, and habits of their "demons."

Folklore not only supports what we learn from eyewitnesses, but also provides details and particulars that I never heard from them because it contains observations and memories amassed and compressed over hundreds of years.

I wish the book would be translated and published in English. I sent a copy to Nancy Logan, who together with Kira Gindina did an excellent translation of Oleg Ivanov's "Avdoshki," published by Ray Crowe in *Special Newsletter* # 13, December 1999.

In a recent message to me, Keith Foster, a keen bigfooter of Colorado, refers to Theodore Roosevelt's book *Wilderness Hunter* (1893) to the effect that "Roosevelt's native companion did not want to go into a certain area for fear of the devils there. Roosevelt called them forest hobgoblins." I am glad to see such similarities in the "demonology" of Russia and America. This is yet another feather in the hat of hominology.

**QUESTION 6: As the 30th Anniversary of the 1967 Patterson-Gimlin film came and went, with skepticism and new waves of hoax claimants, what was the feeling in your country from hominologists? In regards to the film, and the continual debate over its authenticity, do you have any comments beyond what was recorded in your 1997**

**English-language book *America's Bigfoot: Fact, Not Fiction* (Crypto Logos. Moscow, 1997)?**

**Answer:** The first Bigfoot documentary film will always retain a special place, not only in Bigfoot research but also in the history of science in general. Why was it rejected in America in the first place? Why turned down in England? Why proven authentic in Russia? Why is it still being rejected in America?

The shortest answer is in the word *hominology*. The analyst had to take a scientific approach in examining the film before accepting or debunking it. A scientific approach of what science? The answer is hominology. Back in the 1970s, Russia had already a tiny team of hominologists who tested and verified the film in a scientific manner. The world outside Russia had no such new specialists at the time and the documentary was inevitably rejected.

Even today hominology is cold-shouldered by U.S. cryptozoologists, and that is why the film is still largely believed to be a hoax in America. The truth of that is aptly demonstrated by the history of the International Society of Cryptozoology. When it was founded in 1982, hominology had already been for some years the most advanced field of cryptozoological research, with its profound theory and its richest collection of evidence, both modern and historic, such as sightings, tracks, and a documentary film, plus mythology, folklore and ancient and medieval artistic images. Four years had already passed since our conclusion was announced at the Vancouver conference:

> The Patterson-Gimlin film is an authentic documentary of a genuine female hominoid, popularly known as Sasquatch or Bigfoot..... Thanks to the progress of this research, we know today that manlike bipedal primates, thought long extinct, are still walking the earth in the second half of the 20th century. We also know how such a biped looks and how it walks, this knowledge being available now to anyone who wants to use their eyes.... The success of this research is a triumph of broadmindedness over narrowmindedness and serves as an example to the world at large, which seems to be in dire need of such a lesson.

But nobody took the least notice of all that in the newly founded ISC. Hominology was admitted into print, but just as a word, not a science. Instead of seeking official verification of the Patterson/Gimlin film in America and the solution of the Bigfoot problem in its very backyard, the ISC first of all devoted its efforts and resources to the problem of Mokele-Mbembe in far-off Africa. There is no denying it was a romantic and gripping idea, trumpeted by the media around the world, but today, with hindsight and sobering up, it is clear that Bigfoot had enormous priority over Mokele-Mbembe for the ISC's love and attention. The ISC's second love was darling Nessie. On record is my letter of June 30, 1983 to Richard Greenwell, saying:

> I see that Nessie-related photo materials are given minutest consideration in the ISC publications, while the best Bigfoot film to date, with a wealth of surprising details, is being kept mum about.

You can say that, aside from the film, the ISC devoted also space and attention to the Bigfoot problem. Yes, that is true, but was it sufficient? Was the supply meeting the demand? If yes, why then did the ISC President "debunk" the Bigfoot documentary in *Cryptozology*, Vol.7, 1988 without giving any explanation? Grover Krantz gives an "explanation" today:

> The ISC cannot take an "official" stand on any cryptid, but only report the opinions of various people, including Heuvelmans' and my own.

I don't know what Grover's own opinion was at the time, I only know that he did not then defend the Patterson-Gimlin film, while my protest was not admitted into print. Let us also note that Krantz, Bayanov, and Heuvelmans were not "various people" in the ISC: the former two were board members and the latter the Society's president. If he found "the Patterson film showing an alleged Bigfoot with some grievous anatomical inconsistencies," why should the scientific community, outside of the ISC, think otherwise?

Philip Lieberman is Professor of Cognitive and Linguistic Science at Brown University. In the June 1976 issue of *Current Anthropology*, I referred to his work because it supports certain aspects of Porshnev's ideas. Last year I contacted Professor Lieberman by e-mail and was dumbfounded by what he said in response about the Patterson/Gimlin film.

> The supposed Bigfoot film appeared to primate specialists to be that of a human walking, wearing a crudely modified ape costume.

That is 33 years after the film was taken and 28 years after we in Moscow verified its authenticity. To paraphrase Keith Foster, U.S. cryptozoologists have hidden the truth of the Bigfoot documentary not only from themselves but also from the whole of America.

What's to be done? First let me say what could have been done. Article VI of the ISC Constitution states:

> The Board of Directors has the power to appoint and dissolve committees and to reassign duties thereto as may be necessary to effect policies and programs instituted by the Board.

What's the use of this article if it's never used? When I was on the ISC Board of Directors I proposed to use it in order to set up a Hominology Committee for dealing properly and efficiently with various aspects of hominology, including the Bigfoot documentary. The board voted down my proposal on the ground that it:

> Would weaken the Society. Such a precedent could lead to moves for a Lake Monster Committee, a Giant Octopus Committee, a Mokele-Mbembe Committee, etc., since recognizing relict hominoids as worthy of such special attention could be seen as downgrading the importance of other areas of cryptozoological interest.

Being a hominologist I may be biased, of course, but in my opinion, nothing has weakened the Society so much as pussyfooting in the matter of Bigfoot. And what could be wrong with moves for a Lake Monster Committee, etc., etc.? Zoology is appropriately subdivided into many subdisciplines and bodies. Why shouldn't cryptozoology, if corresponding cryptozoologists so desire? And if they don't, if they feel no need for a special committee at present, why deny such a chance to hominologists? Finally, is my belief that Bigfoot research is more important than, say, Onza research, unjustly downgrading the importance of other areas of cryptozoological interest?

I am glad that in recent years Richard Greenwell has turned his attention to the Bigfoot problem and mounted expeditions to Bigfoot country in northern California. In my view, hominology could and should be the locomotive of cryptozoology. If it truly succeeds, the status of cryptozoology will be radically changed and bolstered accordingly. That is why I have again addressed Richard Greenwell and Grover Krantz with the idea of a Hominology Committee within the ISC. Let the ISC Board of Directors appoint such a committee to effect a program of the objective examination of all evidence relating to the Bigfoot issue. Under the program, a panel of scientists will be set up, including Dr. Grover Krantz, Dr. Jeff Meldrum, Dr. Henner Fahrenbach, Dr. John Bindernagel, as well as disinterested experts such as J.H. Chilcutt, latent fingerprint examiner, etc., etc. Work done, the committee will publish the panel's conclusions for the whole world (including Professor Lieberman) to learn and remember.

As for the Bigfoot movie, it calls for action in two aspects: scientific and humanitarian. I wrote to Grover Krantz in 1982: "It's a crime against science and common sense that the film has not been studied by science in Patterson's own country."

I am satisfied that he agrees with me on this point. The film is no less important to anthropology than all of the hominid fossil record taken together. Regarding the humanitarian aspect, it's a shame and disgrace for America to have let Roger Patterson die without his exploit having been recognized and awarded. It's a shame and disgrace for America to have let the names of Roger Patterson and Robert Gimlin be mocked and slandered for decades by unscrupulous *glory vultures*. I wonder how long such shameless violation of human rights in America can be tolerated.

**QUESTION 7: Aside from hominology, have you been involved with other Cryptozoological research? Your country has reports of living mammoths and lake creatures (Dracontology), have these been an interest to you as well?**

**Answer:** *Professionally*, no. But I follow cryptozoological events in other areas and we discuss them at our seminar. In general, I approve, encourage, and assist efforts of my colleagues in this regard.

(Originally published in *Hominology Special No.1*, April 7, 2001)

# Linda Coil Suchy Interviews Dmitri Bayanov

The following is an interview of Dmitri Bayanov that took place in June 2008 and was originally published in *Who's Watching You?* by Linda Coil Suchy, Hancock House Publishers, 2009, p. 283.

**LCS: Hello, Dmitri, it's an honor to interview you. Could you tell us about yourself and how you became interested in the bigfoot phenomenon?**

DB: I am a student of a new discipline that I named hominology, and I am a "freelance philosopher." I received a college education, but have no scholarly degrees because my scientific interests and works have been in the subjects ignored or rejected by mainstream science. I was born in Moscow, Russia, in 1932. I am married and have a son and two granddaughters.

As a boy, I was very interested in animals, and visits to the zoo excited me much more than visits to the cinema (not any longer, for I hate seeing animals in cages).

On June 22, 1941, Nazi Germany attacked the Soviet Union, and history's most terrible war started. Moscow was repeatedly bombed, and my father took the family (mother, sister, and me) to Tajikistan (then part of the Soviet Union), far away from the front. He was soon called-up into the army, and we stayed in Tajikistan until the end of the war in 1945, when we returned to Moscow.

While in Tajikistan, we lived in a small town called Shakhrinau, and it was there and then that I heard for the first time of "hairy wild men" living in the mountains, but could hardly believe the rumor. I recalled it decades later, when I revisited Tajikistan in 1982 on a hominological expedition, as described in one of my books.

At the time, I continued to entertain great interest in animals and dreamed to become a famous zoologist, like the eminent German naturalist Alfred Brehm (1829–84), whose big, well-illustrated volume, translated into Russian under the title The Life of Animals, I read and re-read all the time. One trait of my character was high curiosity, and later, in Moscow, schoolmates used to make fun of it, teasing me with the phrase of a fictional character who liked to ask improbable questions, like this one: "What if from an egg an elephant is hatched?"

I guess I was eleven, and still in Tajikistan, when a strange and hardly believable thing happened to me. One late evening, in the fall, during one of my "nature study" outings, I noticed a flight of goldfinches descend in a tree for a night rest. Suddenly, a daring thought struck my mind: Why not catch one of these lovely birdies? It was an "if" thought: if I succeed, if I catch one, my dreams will come true. I'll become like Alfred Brehm, or get notable in some other way.

Very cautiously, very slowly and quietly, mimicking a hunting feline, I started climbing up the tree. It was semi-darkness, and I discerned one bird, closest to where I was in the tree. I well remember that my hand trembled as I stretched it out toward the bird. I snatched it; it let out a piercing call of distress, and the whole flock instantly and noisily took off from the tree.

Carefully holding the bird in my hand, I managed to climb down, returned home, showed my trophy to mother and told her why I got it. Then I released the captive.

Well, I am not like Alfred Brehm, but some success, in a field beyond my imagination at the time, has been achieved.

That episode was fine and marvelous, and it happened during the daily manslaughter of war in the west of the country, on a thousands-mile front from north to south. Tens of millions perished in it, and there was hardly a family in the country that did not have a member killed or maimed. One of my uncles, the best-loved one, was killed in the Stalingrad battle. I always feel and remember that I, and my dear ones, are alive because he, and millions like him, gave their lives. There would have also been little chance for me to survive if I had been born a few years earlier and took part in the battles.

Some people today imagine that Nazism in Germany and "communism" in Russia (it was not at all what Marx had termed communism) were equally bad. That is wrong. The Soviet system was an inadequate and in some ways evil and criminal execution of fine and humane ideals ("the road to hell is paved with good intentions"), while German Nazism was a highly efficient execution of utterly evil and criminal ideas. Whatever good was in the Soviet Union, especially in culture, medicine, and education, it was due to the ideals of justice and humanism, proclaimed in its ideology. Whatever evil was in Nazi Germany, it was due to its evil ideology. This shows the supreme importance of ideas that people follow, and consequently, of people's humanistic education and enlightenment. On the whole, the Second World War, in which Russians and Americans were allies, was a virtual Armageddon between the forces of relative humanity, on one side, and those of absolute inhumanity, on the other. That is, if the events are taken and viewed on a historic and global scale. As to individuals involved in that global beastly disaster, there were, as usual, upright and humane people on both sides.

I mention here these historic events because their impact had much to do with my pursuit of science, including hominology in later years. But to finish with my experiences in Tajikistan, I should add that I learned there full-well the feel of starvation caused by war deprivation. It's a most humiliating condition, turning a human being into an ever-hungry animal, unable to yearn for anything but grub. One winter, our food situation was so bad that I missed one school year, unable to do homework for loss of memory caused by starvation.

On the positive side is the memory of the delicious taste of canned margarine we began to get in 1943 or 44 under the American Lend-Lease program of aid to the Soviet Union, as U.S. ally in the war. To this day I can't help feeling that margarine tasted better than any butter I ever ate in my life.

Another marvel was U.S. Studebaker trucks and Willys jeeps that I saw for the first time. Once in the mountains, I gaped with admiration at a Willys jeep that with unusual ease dashed up a very steep road. That was in Tajikistan, a remote corner of the vast Soviet Union, which covered one sixth of the whole world land area. Think therefore, what a stupendous amount of technological and food aid that was being sent by America to its ally to reach and cover the country's still huge unoccupied territory. While America was donating so generously food and technology, the Soviet people were sacrificing for the sake of victory a sea of blood.

So as a youth and young man back in Moscow, I gave much thought to what happened to mankind as a result of a second and much nastier world war; what happened to the German people under Hitler, and the Soviet people under Stalin. From zoology my interests shifted to philosophy, sociology, and anthropology. It was due to these interests that in 1964 I met Professor Boris Porshnev, who acquainted me with the problem of so-called relict hominoids, and that was the start of my hominological investigation, including the bigfoot phenomenon.

**LCS: What are the Russian names for what we refer to as bigfoot or sasquatch?**

DB: These names are divided into old and new. Old names are many and retained in folklore, in everyday speech (in sayings and proverbs), and dictionaries of the Russian language. The most common among them is the name Leshy, derived from the word les (wood, forest). Russian–English dictionaries translate Leshy as wood goblin. According to folklorists and ethnologists, Leshy is a figment of people's superstition and has no basis in reality. Hominologists, on the contrary, have enough evidence and arguments that Leshy is an old and original Russian name for a "forest wild man."

The mention of wood or forest is present in numerous other ethnic names of our hero. As to superstitions, mythology and magical beliefs connected with this subject, they are based on the real extraordinary physical and psychological powers of these hominids.

The new name that appeared without any connection with Leshy, and before the birth of hominology, is snezhny chelovek (snowman). It's a shortened translation of "abominable snowman" that entered the mass media vocabulary in the middle of the 20th century, especially as a result of the British 1954 yeti expedition in the Himalayas.

There is still no generally accepted scientific term for Leshy and bigfoot in the flesh. My handy working term is homin. Hominology is the science of homins.

**LCS: Strange "wild men" are depicted in Russian drawings, paintings, sculptures, and engravings. How far back can you trace the history of these wild men?**

DB: Images of "wild men" are to be found all over the world. We see them in prehistoric petroglyphs, in the art of the ancient world and the Middle Ages, and in recent and modern folk art. Their tell-tale signs are depiction of hirsute bodies, usual lack of clothes, and certain features of the face and figure that distinguish these images from those of modern humans. This is an important category of evidence and source of information for hominology.

**LCS: What do you speculate is the nature of the Russian forestman and bigfoot? Are they descendants of Gigantopithecus, or entirely something else?**

DB: The exact biological, i.e., genetic and taxonomic, nature of these bipedal primates is unknown. I strongly believe they are not descendants of Gigantopithecus. I believe they are as close to modern man, Homo sapiens, as are Neanderthals and/or Homo erectus. It is noteworthy that in three cases known to me (one in Russia and two in America), witnesses among common people said that these "wild men" look like Neanderthals (whose pictures were known to them from books).

**LCS: Can you tell us how the Russian forestman differs from the North American sasquatch in relation to size, footprints, vocalization, and appearance? And do you believe they are of the same or a different species?**

DB: Judging by available data, I don't think our forestman, Leshy, is noticeably different from your forestman, bigfoot. I think that if they are different, then at the level of races, i.e., subspecies.

**LCS: How does the Russian forestman differ from the North American sasquatch in regard to its habits and interactions with humans, dogs, or other animals?**

DB: Here too, I see no difference. Even in their habit of braiding the manes of horses they are alike.

**LCS: Tell us about some of the techniques that you have employed during research expeditions.**

DB: The main technique during our research expeditions is the correct way of approaching and interviewing witnesses. As a rule, people are not eager to share their experiences of encountering or interacting with forestmen. And when they agree to talk, it is essential to ask the right questions and record the answers correctly. For this purpose we have special instructions and list of questions. Coming to a village in a homin habitat, we introduce ourselves not as hominologists but as zoologists and ecologists researching local wildlife. Thus, from questions about elks, wolves, bears, etc., it's easier to pass on to the question about a creature that is like a bear, but walks erect on two feet.

As to fieldwork, searching for homin footprints and their photographs and casting are the usual practice. Knowing that many witness accounts have been received from shepherds and farmers, I dreamed of creating and using "mobile animal farms" of our own, placed in promising areas, hoping that our farm animals (horses, goats, rabbits, etc.) would attract the homins. But to be realized, such plans require lots of funding, and we have always had none.

Our late Leningrad (now St.Petersburg) colleague, zoologist Rostislav Danov, who said he had tasted the flesh of nearly all animals that died in the zoo, intended to adopt the exact lifestyle of a "wild man" in case he happened to see fresh tracks of the creature. In this way he hoped to make friends with the forest people and begin directly observing them. Rostislav made a living by catching and selling poisonous snakes (for medicinal purposes) in the deserts of Central Asia, and before he had a chance to test his technique, he died of multiple complications following a snake bite.

The best technique that is the most fruitful so far, has appeared spontaneously in America among people, women in particular, who befriend and observe bigfoot on their property. Your grandmother was a pioneer in this regard. In truth, you don't find bigfoot; bigfoot finds you.

**LCS: Are there some areas or regions of Russia where the forestman has been spotted more than other areas or regions? And what is their preferred habitat?**

DB: In the beginning, following the yeti expeditions in the Himalayas, we believed that the forestmen could only be met in the Pamir Mountains of Tajikistan, closest to the Himalayas. Then sighting reports began coming from the Caucasus, much closer to us in Moscow. Now we have sighting accounts from all over Russia, including the Moscow Region. I conclude that in some areas, where the pressure of civilization on wildlife is the least, extant hominids enjoy permanent habitats; and in areas where pressure is high, they probably lead a Gypsy-like vagabond life, appearing here and there from time to time.

**LCS: What is the one thing you want to learn and know about these creatures (both the Russian forestman and/or bigfoot)?**

DB: The one thing I dearly want to learn is their linguistic ability. This will show how closely they are related to us, and if we have a means of mutual understanding. From almost all historic and modern evidence it is inferred that they have no speech ability. On this account, and for certain theoretical reasons, my teacher of hominology, Boris Porshnev, maintained that only modern man, Homo sapiens, developed the gift of speech, which is his exclusive possession and the cause of origin. I believed this theory and spread it. True, there were a few exceptions in the evidence, but they were either ignored or explained away as inaccuracy in witness accounts. Now, after the 2002 book, 50 Years with Bigfoot, by Mary Green and Janice Carter, I accept the possibility of verbal language, at least in some non-sapiens hominids. This means that the Porshnev theory may have to be seriously revised. Igor Bourtsev holds strongly that some homins, thanks to their high mimicking talents, “picked up” language from humans. And I’ve always supposed that, even if Porshnev was right and speech is modern man’s invention, a young homin could be taught to speak. Anyhow, the question of their linguistic ability is the most pressing and educational.

**LCS: If you could have the attention of the whole world, what is the one thing or message you would want to convey about these intriguing creatures (both the Russian forestman and/or bigfoot)?**

DB: I would want to convey three wishes: May these nature people teach us to live in peace and harmony with Mother Nature. Seeing in the dark, may they help us see the light of reason. May they help us learn to be Homo sapiens.

**LCS: Thank you, Dmitri, for this interview. I greatly appreciate your thoughts and insights.**

DB: You are welcome, Linda. Good Luck with your book.

# Intervention Theory and Darwin's Mistake

Creationism is intellectually so weak that Pope John Paul II had to endorse Darwinism. Intervention Theory, being a hybrid of Creationism and Darwinism, looks stronger. Is it conceivable at our present stage of knowledge and ignorance? Why not? Maybe only because all anti-naturalistic theories of man's origin seem to reveal not his great intelligence but his great vanity. They are a kind of intellectual vanity fair.

In very many respects the difference between the amoeba and the chimpanzee is greater than between the chimpanzee and man. Do we accept that the chimpanzee evolved to its level of intelligence without Cosmic Intervention? If yes, why not accept that man evolved to his level from the chimpanzee level also without such intervention?

By the way, the theory of the origin of species by natural selection has two authors: Charles Darwin and Alfred Wallace. One thing they disagreed on was the origin of man. Wallace held a kind of Intervention Theory, far ahead of any gene manipulation practices.

The two great scientific revolutions of our civilization are the Copernican and the Darwinian. The former is fully victorious, the latter not yet. There is no anti-Copernican movement today, but much anti-Darwinian bustle and funding. If only all that money could be siphoned to promote hominology. Maybe then the Darwinian Revolution could be fully victorious, too.

So far the only viable argument against Darwin is this:

**DARWIN'S MISTAKE**
**(author unknown)**

Three monkeys sat on a coconut tree
Discussing things as they're said to be.
Said one to the others, "Now listen, you two,
There is a certain rumor that can't be true
That man descended from our noble race.
That very idea is a disgrace.
No monkey ever deserted his wife,
Starved her babies or ruined her life.
And another thing you will never see:
A monkey build a fence around a coconut tree
And let the coconuts go to waste
Forbidding all the other monkeys to taste.
If I put a fence around this tree,
Starvation would force you to steal from me.
Here's another thing a monkey won't do:
Go out at night and get on a stew,
And use a gun, or club, or knife
To take some other monkey's life.
Yes, man descended, the ornery cuss—
But, brother, he didn't descend from us.

Looks like man's origin is a Cosmic Mistake.

# A Popular Lecture in Hominology

A Popular (i.e., with Rhyme & Reason) Lecture in Hominology, delivered on April 1, 2004, at The International Center of Hominology

**PRIMATES OF HOMINOLOGY**

Primates of Hominology
Don't care for technology

Raw food they quite admire
And feel no need of fire

Nor need they electricity
But least of all publicity

Great connoisseurs of honey
They have no use for money

They are naturalistic
More than egocentristic

And more animalistic
Than we are globalistic

Full of primeval dynamism
They've long adopted globalism

Full of parapsychology
They never harm ecology

Full of anomalistics
They still escape statistics

Sometimes they favor human bids
Especially from little kids

And they are fond of women
But not of macho seamen

Devoid of subtle gimmicks
They are superb as mimics

You needn't take assumption
For seeing their gumption

The ancients called them Troglodytes
For the enormous appetites

Believed to be divine
They drank a lot of wine

Swift took at them a nasty look
Describing Yahoos in his book

Their standing in athletics
Defies apologetics

Though champs of kinematics
They flee from cinematics

But more than any Nazi
They fear paparazzi

In lakes and ponds they rival frogs
When on land they scare dogs

Their smell is not delicious
But attitude not vicious

Their voice is strong and booming
And they don't practice grooming

Despite abundant hair
They melt into thin air

Some buffs with no apology
Connect them with ufology

And with no hesitation
Ascribe them levitation

So homins being cute
Do things that all dispute

No doubt they are cool
And make of me a fool

Excuse my terminology
In teaching hominology

When people grin and smile
The talk is worthwhile

In closing let me mention
Best thanks for your attention

# A Biggie Epilogue

Two things stand out sharply in my 43-year-long involvement in hominology. One is the 1967 Bigfoot documentary film whose 40th anniversary will be marked in October 2007. Why this 715-cm-long strip of cellulose was destined to be so important in the short history of our young science must be clear from the pages devoted to it in this book. One of the film's assets is its overall agreement with sighting accounts by numerous eyewitnesses. I mean agreement on bigfoot's appearance, as noted by John Green right after he viewed the subject on the screen for the first time. As the film has been found genuine by all thorough and accurate investigators, most sighting reports have to be genuine, too. Film and sightings are mutually supportive. But given their great scientific value, both film and sightings are limited to the evidence of bigfoot's appearance and locomotion. They tell us nothing about the giant's intelligence and other mental abilities. As remarked by Will Duncan, "the prevailing ape-like model preferred by many long-time investigators is based on very limited observations of sasquatches in remote settings." The striking exception whose contrary message had been ignored for decades was the account of Albert Ostman.

The other crucial thing for me was e-mail contact in the spring of 2002 with Mary Green and Janice Carter, followed by publication of their book *50 Years with Bigfoot*. The significance of that event is dealt with, perhaps not sufficiently, in "Is a Manimal more Man than Animal?". Perhaps it's no overstatement to say that Janice Carter's story augurs revolution within revolution. Bigfoots of North America were believed by all investigators, myself included, to be less humanlike than their counterparts in Asia and Europe. If they turn to be a kind of human, as it follows from what we learn from Albert Ostman and Janice, then bigfoot counterparts in Eurasia are also human. The universal historical term "wild man" may happen to be literally correct. If recognition of the existence today of non-*Homo sapiens* hominids means a revolution in scientific knowledge, then recognition of these hominids as humans will mean a revolution in our knowledge of the subject. A new chapter, nay, a new stage of hominology will have to begin, demanding a radical change of our thinking and our approach to the matter.

Although the situation is "iffy", current input tends to be more and more in favor of Janice's views. It's most unfortunate that my appeals to some leading North American hominologists to support Janice Carter's investigation at her site were ignored. Igor Bourtsev then flew from

Moscow to Tennessee, spent five weeks at the Carter Farm as a guest of Janice, and became convinced that bigfoots visit the farm. But our financial situation did not allow him to stay longer or aid Janice. Due to material constraints she had to sell her property and relocate. What has become of Fox and his family is not known to me. But I know that some brave American investigators have offered support to Janice at last, formed an investigative group with her participation and are busily gathering evidence in favor of the humanlike model. They are doing this secretly on a non-disclosure agreement until enough proofs are hopefully in hand to sway the prevailing opinion of the bigfoot researchers.

I am not in a position therefore to share the important tidbits that reach us in Moscow thanks to our friendship with Janice and some members of her group. But what I'm glad and free to report is that the Russian "No Kill" and "No Harm" approach to the problem has won us the friendship and trust of four more bigfoot contactees in the U.S. besides Janice. (I guess there are hundreds or thousands of them in North America). Characteristically, the four are women, but unlike Janice they refuse to go public, revealing their names and locations (which are in different parts of the country). Like Janice, they observe bigfoots on their respective properties, but not as closely or for such a long time as in the Carter case. I correspond with them by e-mail. Forbidden to reveal their names, I'll refer here to these contactees as Lady Number One, Lady Number Two, Lady Number Three, and Lady Number Four.

Naturally, I was eager to ask them what they think of Janice's experiences described in *50 Years with Bigfoot*. Lady Number One wrote back:

> So far what I've read is so stunningly familiar that I am at a loss for words. There are things that Jan has witnessed that I have not, such as killing animals. (...) I know there are countless things to compare, which gives a fuller picture of the whole.

Here are a few examples of her own experiences:

> I love the interaction we share and am rarely frightened even at night alone. This is how I have spent most of my time in close proximity to them. A large male will come to within approx. 12 inches from my face if I take the screen off my upstairs window. I estimate his height at above 9 1/2 ft., probably closer to 10 ft. tall.

No wonder, her name for them is the Big Guys. She continues:

Although my first experiences with the Sasquatch terrified me, I have actually learned to have much affection for them, particularly the children and adults at my second habituation area. The huge male who lived there and 'adopted' me as part of his family actually protected me from a mentally deranged neighbor (who poisoned my dog and continued to harass me. He cut the fence and let my horses out along with many other things). He would often prowl around my house until the day I heard him outside in the bushes calling "kitty, "kitty". The huge male Sasquatch, who I named Galahad was in the area that day and headed straight for him, tearing brush and stomping loudly. I heard the neighbor scream and run away. Galahad did NOT chase him, but the neighbor never bothered me or my animals again.

Why she does not go public:

The thought of having my habituation sites "invaded" has been a deterrent for me.

She's been taking notes of her observations and to my suggestion of writing a story like Janice, replied:

As long as my "friends" could be protected I would not rule out writing about them.

Some more of her experiences:

I have left gifts (to sasquatches - D.B.) many times over the years and been given gifts in return. First was a pair of kittens, then a long haired domestic rabbit, a turtle, numerous 'food' items and last a baby goat which was taken from the neighbor. Actually 3 goats were taken that particular morning before dawn, one doe and two kids. One of the kids was placed on my deck as the Big Guy passed my house. It woke me up circling the house on the deck bleating.

What is remarkable here is the fact that it is only the human animal, as far as I know, that has invented the custom of "giving gifts in return." What happens if you stop giving gifts or begin giving irregularly?

I had a little boy who showed up every night without fail. If I didn't have his peanut butter and honey sandwitches ready he would actually have a tantrum and throw little rocks and sticks at my house and windows. If I was on the phone I'd actually have to hang up to feed him! Others I know who have fed the B.F. have had things tossed at their house, too, also tapping on wood and disturbing sleep! I had horses part of the time and I know the Sasquatch were fascinated with them and regarded me kindly becasue of my care of them. I remember caring for a crippled puppy, and the Sasquatch children coming close to watch me play with it. I also have grandchildren, the first is 13 years old now and played a major part in earning my trust with the Sasquatch. I recall one evening at dusk holding him up in the air as an infant, singing to him and kissing him. I noticed a large female Sasquatch looking intently at me at the edge of the woods. She bent down and picked something up and raised it so I could see it. "It" was her own infant, probably about the same size as my grandson. Fascinating and wonderful, a beautiful gift of trust and understanding passed between us.

Mary Green is acquainted and in contact with this most remarkable bigfoot befriender, and this is what she wrote me about her and Janice:

> The little ones (sasquatches - D.B.) have been allowed to play in the yard of (Lady Number One) by their mothers. The mothers often clap their hands at her grandchildren and their own young playing together. I have never heard of them clapping their hands in pleasure, much like we applaud a good performance. I never mentioned to (Lady Number One) before talking to her the other day about Janice's bigfoot clapping their hands on the farm. So when she also told me this, I was absolutely amazed. This just is not a coincidence, these ladies really know their bigfoot.

To my question what she thinks of the Patterson film and how its subject compares with the Sasquatch she observed, Lady Number One replied:

> I think the Patterson film shows a true Sasquatch. As to the general appearance, it is the same as what I have seen here. Each Sasquatch, as with people has individual characteristics. I have noticed that in my area there appear to be two distinctive types of Sasquatch (a conclusion I heard from other investigators as well - D.B.). The first looks like the Patterson Sasquatch, perhaps with even shorter hair on the body and head, but the second has much longer thicker hair, and the largest male I have seen approx. 10 ft. tall was this type. The

> shorter-haired type tend to have a more human look, although most wild. (?) The heavy-hair type have a more apelike appearance. I know the two types are interacting sexually as I have witnessed this, so there is also potentially a third type which is a combination of the two distinct types.

A hypothesis comes to mind that the shorter-haired type that tend to have a more human look are actually hybrids of humans and homins or may be descendants of such hybrids. Also note that Janice claims to have seen naked, hairless bigfoots.

The same contactee-lady sent me what she calls "A list of emotions as one gets to know the Sasquatch":

Terror
Panic
Horror
Illness
Vulnerability
Shock
Amazement
Curiosity
Wonder
Trust
Privilege
Friendship
Responsibility
Protectiveness
Initiation
Blessing
Humor
Kindness
Expanded view of "Reality"

Note how the contactee's emotions change from negative to positive with increasing and deepening contact. Philosophically, the last one, Expanded view of "Reality", is most significant.

Epilogue is not the right place to tell in detail of my correspondence also with Ladies Two, Three, and Four, but some highlights are in order. As a bigfoot contactee, Lady Number Four was a novice, for the family had bought a property at which no bigfoot activity was noticed until they got a horse. Then things started to happen which gladdened her friend,

Lady Number Three, but became stressful and unnerving for Lady Number Four. This is how she described the situation:

> I can't ignore something that's constantly here, not knowing how it affects my safety. It unnerves me to find butt- and hand-prints around the horse trailer in the afternoon that weren't there in the morning. And me being home all day.

She also mentioned that the house was tapped for about twenty minutes at night. Lady Number Three, who hoped for conducting a habituation effort at that site, reported at the time:

> This whole thing is extremely upsetting, not only because my new friend is upset and scared, but because this priceless opportunity has come to a screeching halt. I feel like crying. They have been considering selling the house and moving. I would hate to see that happen. I feel that her family and the Ss are suited for each other for co-existing peacefully.

And this is what Lady Number Four wrote me some time later:

> It seems everyone else can go to a particular area, observe the BFs (ha!ha! - D.B.), and then *leave and go home*. I wasn't able to do that, and as they were coming up to the house at night, along with some frightening incidents, I had become extremely afraid of them.

At my request Ladies Number One and Number Two shared their experiences with the scared novice and that calmed and reassured her. Lady Number One's advice was this:

> I know EXACTLY how you felt at first discovering you had BF at your property. My first sighting would have given me a heart attack I am certain if I'd not had a strong healthy heart! I was totally terrified and there was NO ONE to talk with about what I was experiencing. I am glad you did not have to continue for as I did for 2 years, learning what I was dealing with by myself! (...)
>
> Regarding criticism from others, unfortunately there are lots of skeptical people in the world and you will find some of these to be insensitive as well. Just remember YOU are the one who has been trusted with remarkable experiences. This makes you wonderfully special. This is not something that is "transferable". Others cannot possibly

imagine what you are experiencing unless they have had some sort of similar experience.

After they talked on the phone, Lady Number One wrote to Number Four:

> I so much enjoyed talking with you, and learning of your interaction with the "Big Guyes". I know sometimes it can seem like a "curse", but I believe God only allows those who are very special to have the rare insight into this mysterious species. I have the distinct feeling that "they" like you and want to be friends. I hope in some way I have helped calm some of your fears regarding them. I know I feel privileged to be allowed a relationship with them for so many years and find each new discovery fascinating.

She wished the best "to all of us who take on this difficult but wonderful research". This helped a lot as shown by Lady Number Four's message to me, saying:

> I'm feeling so much better just after reading the emails from the two ladies, and I'm so grateful. I particularly enjoyed reading how (Lady Number Two) sits out on her deck and enjoys having them around. I was doing that for a while, listening to them feed in the ravin at night, until I became afraid of them. Now I sit out there in the afternoon and enjoy listening to two young (?) ones who like to come spy on me.

She also wrote this:

> We seem to have a playful young BF that's been kicking up the last few nights. We have an above-ground pool in our deck, and he's taken to hitting the side of it. It makes a wonderful racket much like thunder, and puts deep dents in the side (recall a similar diversion by almasty in the Caucasus, p. 342 in this volume, - D.B.). Fortunately it's not a pool we ever plan to use, or he'd be destroying it. The other night he was apparently trying to play with our cat. He was underneath our deck and rapping on the deck right under where ever our poor cat happened to be. I think the BF was enjoying the game more than the cat. The next morning there were tracks approximately 8" long around our deck. I glimpsed a second BF last week. It was a little black one, maybe 4 or 5 feet tall. While I'm still too unnerved to stay

here myself at night, when my husband is out of town, I'm thoroughly enjoying my daytime encounters, as well as the night time ones as long as my hubby is here.

But the best discovery and contribution to hominology by Lady Number Four in the short history of her interaction with the Big (or mostly not so Big) Guys was her discovery of intricate braids in the mane of her mare. This is what she wrote as the phenomenon started:

> I'm stunned by the braiding a horse's mane thing. (The mare, name given - D.B.) almost always has a least one section in her mane that looks badly braided. I've never been able to figure out how it gets there. It takes baby oil, a large toothed comb and a lot of time to pick it out. I've actually wondered if a neighbor came during the night to mess with her mane. I just couldn't figure it out. I just went out to check and what's braided in her mane right now is actually fairly intricate, in a way. I'll take pics of it (followed detailed description of braids - D.B.). I'm absolutely dumb-founded.

Lady Number Three, who had our book *In the Footsteps of the Russian Snowman* where several pages are devoted to this phenomenon in Russia and other parts of the world, put Lady Number Four wise in this regard. She also corresponded with Janice, who appraised that horse- mane-braiding had also been observed at her farm but she missed mentioning it in her book. Hearing that I asked Janice to fill me in and this is what she said:

> The bigfoot do braid horses manes and tails some also. (See pp. 39, 247 in this volume. – D.B.) We called them "witches braids". The bigfoot usually wait until after dark to approach a horse that is not so frightened of it and they do an old trick of pinching over the withers of a horse to quieten them and it temporally placed them in a trance-like state where they do not move on you. Native Americans used and still use this pinching trick. Anyway, if the horse isn't frightened of the bigfoot and the bigfoot isn't in a hunting mood the horse can sense they will stand still after the pinch of a few seconds and then the bigfoot start stroking and petting them like we would stroke and pet a horse and they talk to them or whisper to the horse in soothing voice. Then they will braid the mane and the tail also if the horse stands still long enough for them to do so.

The mane-braiding thing, although reported by numerous sources (including Shakespeare) throughout the world down centuries, remains a mystery regarding its purpose. Let me note that Lady Number Four's mare was not in lactation. The important thing we learned in this case is that sasquatches are on par and in solidarity with the rest of the world's homin community regarding this strange activity. Their only difference is that they like to decorate with intricate braids not only horses but llamas as well. A U.S. contact has sent us corresponding photos.

As Lady Number Four continued to observe ever new braids in the mane of her horse and share photos of them with us, I proposed that she offer colored ribbons to the maker of braids and see what would happen. She wrote back:

> That's a great idea to tie or braid colored ribbons into her mane and see what they do. I'll do that.

She did and some time later came up with:

> CHECK THIS OUT!!! The Ss braided in the middle ribbon that I'd left unbraided!!! The ribbon looks like it was slid down the piece of mane I'd tied it around, and then it and that piece of mane were worked into the existing second braid. I'm thrilled out of my mind, this is so cool! As the next part of the experiment, I unbraided the other two ribbons and simply tied them to the mane, too. Pics #009 & 011 re of those experiments.

As a result, Lady Number Four, as well as Russian hominologists thanks to her, have now scores of excellent photographs of horse mane braids, with colored ribbons and without, being to my knowledge the best collection in the world to document the phenomenon. Seeing these braids you can't believe they were made by a human, nor can you believe sasquatch fingers are all thumbs. Lady Number Three, who closely monitored these developments, wrote that she wasn't trying to outdo "these talented beings" but tried to see if she "could at least duplicate the braids." She found she couldn't. "I could not incorporate the strands into the other side like they had though I have had experience with many kinds of braiding before, herringbone braids, French braids, and regular braiding, also knotting."

With permission of both ladies and without revealing their real names, Igor Bourtsev, who studied the phenomenon in the Caucasus in the 1970s, published an article in a Moscow popular science magazine (academic press is still out of reach for hominologists) about horse mane braiding in America. For some time I hoped that sasquatch fingerprints could be taken resulting from the practice of mane braiding combined with ribbon manipulation. But the experiment did not work because hair and ribbons are not suitable material for taking fingerprints.

Buoyed up by my relationship with these fine sasquatch contactees, I encouraged them, somewhat with a touch of humor, to create the nucleus of a future Friends of Bigfoot Labor Union. I told them, "you four are such marvelous Heroines of Science!!! And I am so lucky and honored to know you and be in touch. I was delighted with (Lady Number One's) words that 'God only allows those who are very special to have the rare insight into this mysterious species'. How wonderful! On my part, as a scientist, can't guarantee that God is on our side or interested in hominology, but can guarantee that we, Russian hominologists, do see you as very special folks, and that we are happy and proud to be in contact with you, that we envy you and wish you the best of luck."

Then quite in earnest, I wrote them about the priorities of our research, pointing out the need for good photographic images of bigfoots as priority number one. I quoted Boris Porshnev in this book who wrote in 1963 that "If proceeding most cautiously we succeed in conditioning the creature to come and take food in a definite place, that would be a real scientific victory. (...) Scientific work could be launched in such a case even without direct contact of researchers with the specimen, for modern zoology boasts of an excellent means of taking color films with a telephoto lens at a great distance. A relict hominoid would then appear on the screen showing its usual movements and habits against a background of its natural environment."

In theory the idea is fine, but as is known the mother of problems is practice. As a rule, homins visit feeding spots at night, and night photography is not so simple, especially for non-professionals who lack appropriate training and equipment. And when homins are glimpsed in broad daylight, another problem arises. "Ss don't like to be photographed", quipped Lady Number Three. Lady Number One added:

> The Sasquatch take advantage of everything which allows them to remain unseen, this is their nature (just like that of the Russian domovoy, see p. 41 . – D.B.). I notice they will stand in the thickest brush, and in shadows. I have even seen them lie on their belly and cover their body in leaves. Night vision equipment which I tried just this year caused an almost complete disappearance of the Sasquatch until I took the camera down. Now they are coming to the cabin again.

Lady Number Four:

> I was curious to read that they've been observed lying on their belly and covering themselves with leaves. About 8 weeks ago I thought I observed the very phenomenon but had not heard of it at that point. I don't know how they respond to a regular camera, but I have observed that they neatly avoid our infrared surveillance camera that is mounted under the back eaves of our house. It's stationary but can be turned manually to point any direction. When I have it pointing to the left, they will appraoch from the right, even boldly coming right up onto the deck underneath but out of view of the camera. When it's pointing to the right, they will come right up to the house from the left. They seem to be quite savvy avout it.

I strongly doubt they understand and avoid photography per se. They probably fear cameras (and binoculars) pointing at them, like they fear firearms pointing at them. Lady Number Four: "I've also observed that binoculars pointed in their direction cause a flurry of anxious activity." So it's not a problem of photography per se – no wonder under the circumstances there have been taken lots of snapshots, films and videos but of a very poor and therefore scientifically useless quality. The problem is to get visual images not inferior at least to those obtained by Roger Patterson.

Incidentally, Roger Patterson's victory is due to two separate factors: 1) Patty, the subject of the film, did not hear the men approach because of the water murmur in the creek, and 2) Patterson evinced exceptional courage and adroitness, plus full readiness for an encounter. After that lucky event such a coincidence of positive factors has simply been lacking.

Sharing my advice with the four contactees about how to try to take quality pictures, I wrote:

> Re " Sasquatches don't like to be photographed," imagine Roger Patterson asking the Furry Lady at Bluff Creek, "Dear Madam, don't you want to be photographed?", and getting no film without her saying, "Thank you, dear. Yes, I do." Hominology would have suffered immensely, being deprived of its best visual proof. The film is still distrusted by scientists because it is the only one of acceptable quality. One swallow does not make a summer. We need many photo proofs of acceptable quality. Had Patterson been in the shoes of Robert Carter Sr or Janice, we'd have miles of BF film, and "a burden of proof" would be no burden.

I proposed what I called Feeding-Time-Shifting Technique, which to my knowledge has not been tried in earnest so far. Luckily, a reliable rumor has reached us in Moscow that videos of excellent quality have already been taken in the U.S. on a farm with a long history of sasquatch habituation. The event is kept secret and will be made public "at the right time." The right time should certainly be this year, for the Patterson-Gimlin documentary, 40 years old in October 2007, badly needs backing up with additional visual evidence. If our information is correct and lots of good visual evidence is truly obtained, this must be a result of enough funds applied at last to the task executed most patiently in the right way. We send congratulations then to our overseas colleagues.

Now to the crucial question of Bigfoot linguistic ability. Lady Number One said this:

> As to language, I know that the Sasquatch do indeed have their own language and have heard it on more than one occasion. I am not however able to understand it. I have also heard them speak a few words in English which they no doubt learned from hearing them repeated several times. My grandchildren call me "Grammie" and I have heard the Sasquatch call to me by this name. There are a few other words, such as "kitty" and "Mama" which come to mind.

She also wrote that "They make various type of sounds back and forth to each other."

From Lady Number Two I have in part the following:

> Sasquatches do seem to *use* and employ symbols, including speech, to communicate. (...) The adult male seems quick to correct misbehavior with a quick, firm "one-word" verbal command. I've heard them converse...

Lady Number Three put in this:

> I've heard them talking from a small distance and they have walked to the house window and verbalized my first name.

Like Janice, she claims that the sasquatches she is dealing with "do utilize telepathy as well as verbalization." She claims to be able to interact with them telepathically.

Is the question of the reality of speaking bigfoots answered now and closed? By no means! It's just a few more steps in a new direction, but steps of great importance since they show the relevance of the problem and the pressing need to explore it in depth.

I hope these bits and pieces of evidence from the four witnesses quoted by me show the kinship of bigfoots/sasquatches with their counterparts around the world and through history, as described in this book. The picture will become much clearer should these witnesses publish in full their observations. Three of them have been taking notes and I urged them to share their stories with the public at the earliest opportunity. So far so good. Now I have to mention problems and complications of a different and most serious order. In a November 2006 e-mail Mary Green informed me that (Lady Number One) was gravely ill and that she "doesn't want the bigfoot there to come to any harm once she's gone." The message continued:

> She has quit feeding them because of this situation. We discussed that the bigfoot become very dependent on feeding and will become quite pushy and insistent that you feed them after a while, a true danger to habituation methods. Also this makes the sasquatch vulnerable to others. Once they lose their fear of a few kind humans, they are in extreme danger here. Even Janice worries about this. Janice's Toby sometimes goes to others' doorsteps in his effort to beg for food in the harsh winter months here in TN where much food is not available. Even now, where Janice has moved to, Fox has already been bothering her neighbor friend for food. He frightened her very much the first time he knocked on her trailer door for food. It's a good thing that Janice had spoken with the woman and her husband because they might have shot at him. However, the lady had heard Janice say that Fox was old and lost much of his hair and therefore she recognized Fox when she saw him. (...) So you can see, Dmitri, that habituation methods have drawbacks as far as the safety later on of the bigfoot. Maybe you need to study this phenomenon and write something more about it. Is it truly wise for us to befriend them to the point they become dependent upon us for food? There are dangers involved for the giant hominids themselves when we do. So there has to be a responsibility toward the bigfoot/sasquatch if we choose to use habituation methods.

Writing in response, I recognized the possible adverse effects of our sasquatch habituation efforts, saying, "I agree such risks are involved, and our approach should be judicious and depend on the circumstances and possible outcome in each individual case. But as a general rule and principle I stick to the idea that the existence of sasquatches should be recognized on the basis of evidence obtained without any violence and harm done to them, i.e., such evidence as photography, filming, footprints, etc. If that could be achieved without feeding them - great! But could it? That is the question. There is no question though that lots of violence and harm are done to sasquatches and planned against them under the conditions of their non-recognition by science. (...) Exactly on these two points — their existence, their reality and their near-human nature — your work and that of Janice (and I mentioned the names of the other contactees) are of such great importance. The question is how to make the results of this work mutually supportive and known to the scientific community without any adverse effects to sasquatches or any of you. Let us ponder this question together."

I can add now that one thing of the habituation efforts has been practiced successfully, that of feeding, while the other and the very purpose of the endeavors, that of getting good photographic and other proofs, has been disappointing. Conclusion can be made that feeding should be practiced only in cases of high probability of getting photographic and other objective evidence.

Even so I face most serious objections from two of my lady friends that go to the heart of our research. Lady Number Three tells me this:

> I do not wish to prove that the Ss exist to any person and only wish to observe and interact. I have concluded, based on my observation of other "researchers" beliefs on message boards that exposing the Ss to public identification will endanger them. People fear what they do not understand and they destroy what they fear. Knowledge that there is another hominid/human out here that is bigger, faster, stronger, and more adept at survival will bring out the base nature of our kind. It's bad enough that members of our species cannot get along based on skin color or religion alone. It's a nice thought that we would instantly as a whole afford the Ss the same rights as we have. No, if found to be human, they will have to adhere to our rules. Land they live on and may have for generations will be off limits to them and they will become "trespassers". Hunting deer out of seasons will make them "poachers". Having more than one mate will make them criminals in our monogamous societies, they may be corralled and made to stay in reserves, and the list goes on and on. I cannot and will not be a part of their ultimate destruction.

Recall John Green saying (p. 166 in this book), "To take the silly side of the argument (as I see it) if it were decided these things were human they would have to be counted in the census (...). If they were able to understand all that and express an opinion, I expect they would prefer to be apes and take their chances about getting shot." That was said in jest thirty years ago but what a sasquatch contactee says today is damned serious. The arguments of Lady Number Two are even more alarming and telling:

> In our scientific "get-whatever-grants-you-can" and "use-the-most-expedient-methods-possible", as well as our focus on fame and gain, concerns, at least for me, are very real. As you know, too often "acquire, confine, and experiment" are the scientific methods used for efficiency, expediency, and control. I just can't ever see such methods changing for sasquatches, or exceptions being made.
>
> If proven to exist, I so fear a race would be on to obtain live subjects and dead specimens. The scientific community would never leave subjects as rare as sasquatches alone — particularly if they felt something could be gained personally, professionally, or scientifically. Even if American scientists were prohibited from obtaining subjects here, there would be nothing to prevent them from obtaining subjects elsewhere, such as what has occurred within the scientific community, timber industry, and other such industries. If logging restrictions are imposed here, the timber industry simply goes abroad to harvest timber. Logging doesn't end, and other countries often willingly comply, often for economic gain. It wouldn't stop at our borders. Industries, including scientific ones, cleverly find ways to skirt obstacles to get what they need. And heaven forbid the scientific community here would establish their own insulated breeding colonies as they've done with chimpanzees and other primates. It's a horrible, horrible business for primates. And in science primates have no or few protections. They aren't protected by animal welfare laws. In science, institutional animal care and use committees, which are "in-house" committees made up of in-house peers, decide, using a checklist, when and how animals, including primates, are used. If a study justifies that animals (including primates) be used, the in-house committee makes that decision, and "rubber-stamping" and *quid pro quo* among colleagues is known to be prevalent and rampant. If, for example, the committee determines that using anesthesia might affect the results of a study, anesthesia can be withheld regardless of the pain inflicted on the subject. As long as the use of subjects, procedures, and methods can be justified, anything goes. If it's determined a primate needs to be kept

in isolation, a primate can be kept in isolation for years — as long as it's justified. Farm animals and animals used in science and scientific research are not protected by animal welfare laws. There are no laws in science. Only institutional animal care and use committees made up of in-house peers, and a checklist of guidelines.

If sasquatches are not *H. sapiens*, their fate will potentially be in the hands of institutional animal use and care committees — just like chimpanzees and all other primates. Scientists will be able to do whatever they want, as long as it can be justified. There won't be *anything* we can do to protect them.

Well, recall Margaret Atwood's *Oratorio for Sasquatch, Man and Two Androids* (see p. 211). That sounded like satire and exaggeration. Not quite so after this most realistic outcry of a sasquatch befriender. So Lady Number Two makes this conclusion: "At this point in time, where most people think they're myths, perhaps ignorance is bliss for the species."

My reply was this:

I have to admit now that your eloquent pessimism beats my wishful optimism. The gloomy perspective described by you is very disheartening. So it remains for me to say this. Discovery and cognition make the destiny of mankind. That is our main difference from animals. Discovery and cognition make the meaning of a researcher's life. Neither we nor anyone can stop the progress of knowledge. What we can and must do is defend and promote humane methods of obtaining knowledge in the spheres of our competence, and even beyond them whenever possible. As is known, you are second to none in this regard.

She wrote back:

Thanks so much for your nice message, especially your understanding and support conveyed. As you can hopefully gauge, I have no problem with our species' quest for knowledge, only in how we acquire and gain such knowledge. The methods we use. And, in some cases, what we do with it. By nature I'm generally very optimistic and positive, but I'm also a realist. I also see and realize we can be a very exploitive, cruel, and self-serving species, sometimes at a very huge personal cost to others (especially animals) and the environment. Clearly some among our species simply lack personal "checks and balances" — or simply don't care. When it comes to sasquatches and their well being, every aspect and every possibility have to be taken into account and considered. Heavily weighed.

Regarding "a very exploitive, cruel, and self-serving species", the words of a monkey-philosopher on a coconut tree come to mind: "Yes, man descended, the ornery cuss, but, brother, he didn't descend from us."

And what a relief to know that the four ladies I quote are part of our species. While I am preoccupied with the question how recognition of the discovered hominids will affect mankind, its science, philosophy and religion, my lady friends are painfully concerned as to how it will affect the lives of our hairy relatives themselves. This brings to mind the following words of Benjamin Radford (who failed to acknowledge our letter of protest, see p. 148):

> Ultimately, the biggest problem with the argument for the existence of Bigfoot is that no bones or bodies have been discovered. (Bigfoot at 50 – Evaluating a Half-Century of Bigfoot Evidence. *Skeptical Inquirer – The Magazine For Science and Reason*, March/ April 2002, p.34)

It's wonderful how these admirable ladies, acting at the grass-roots level and having no foothold in the science establishment, are making a scientific revolution without discovering a single bone or body. Just by telling honestly some of their life experiences. In this regard how futile and irrelevant are the voices of highbrow *scoftics* (Roger Knights' fitting neologism) who never stop calling for bones and bodies. The scoftic species are simply impervious to calls of reason.

The situation is getting more and more reminiscent of the novel *Les animaux dénaturés*, 1953, by the French writer Vercors (Jean Bruller, 1902 – 1991). Its first English translation was titled *The Murder of the Missing Link*, 1955, the title today is *You Shall Know Them*. To cite the advertisements, "A group of primitive hominids is found in New Guinea, and the question arises as to what rights, if any, they have. For example, is there any reason why they should not be used as slave labor, as an Australian businessman plans to do?" A female specimen is brought to England and "Not science nor philosophy, not Parliament nor clergy, can decide if she is manlike ape or apelike man." They cannot decide because another question has never been posed and answered in law: What is man? The very first words of the novel are thought-provoking: "All man's troubles arise from the fact that we do not know what we are and do not agree on what we want to be."

The novel is mentioned in Boris Porshnev's documentary story *The Struggle for Troglodytes* (1968, in Russian) with the opinion expressed that Vercors' fiction has nothing to do with the subject of relict hominoids because the latter are definitely non-human. *50 Years with Bigfoot* and subsequent developments tend to indicate that my teacher of hominology may have been mistaken. Credit here goes to the French hominologist Jean Roche who in the book *Sauvages et Velus* (Wild and Hairy), 2000, devotes many pages to the question *Hommes ou bêtes?* (Humans or animals?) in the light of Vercors' novel and states his disagreement with Porshnev. He also numbers differences between Vercors' fictional hominids and the real ones studied by us.

The greatest difference, of course, that never occurred to Vercors even in a night dream is the fact that the philosophic and scientific questions he raised in the novel are being revived by the study of *real* hominids observed NOT in the jungles of New Guinea or some other remote wilderness but in the woods and mountains of Oregon, California, Tennessee, and many other states of America! Such a turn of the Snowman Saga was beyond the dreams of even the most accomplished dreamers. Going in 1964 on my first expedition to the Caucasus I could never foresee that the solution would be coming from the other side of the globe and the world's most developed country.

How to figure out this paradox? How come a live "missing link", or something reminding of it, is living secretly and officially ignored on the territory of a country with the most advanced science and technology? Well, as to its living there nothing unusual as long as the country retains unspoiled woods, lakes, rivers, mountains, and other features of the natural environment. Wilderness is not necessary, what is necessary is sufficient wildlife. In this regard America is on a par with or even ahead of some regions of Asia, Europe, and Australia that still provide homins with a livelihood. The native people of these regions have always been well aware of the existence of "wild men". The European settlers in America also encountered these beings, as shown by numerous reports in the American press of the 18th and 19th centuries. Like in Europe and Asia, scientific discovery was not made then because world science (actually West European in concepts, paradigms and methods) was not ready at the time for such a discovery. In the 20th century it became ready, and the discovery *was* made almost simultaneously in Europe and North America by such investigators as Bernard Heuvelmans, Ivan Sanderson, Boris Porshnev, M.-J. Koffmann, George Haas, John Green, Peter Byrne, Roger Patterson, and others.

But scientific discovery and its recognition by the scientific establishment are things apart. Opposition to this particular discovery is extremely strong, and in America perhaps the strongest of all. And yet it seems that America is going to be the first country where recognition of the discovered live hominids will be made officially. Just because it's a country of highly developed science and technology. As anthropogenic pressure on the environment continues, as land development proceeds and more people get properties on the borders and territories of homin habitat, more and more human-homin encounters and interactions take place. Under these conditions, thanks to high technology, the Internet in particular, the existence of bigfoots is rapidly becoming an open secret. And an open secret is bound to become an established fact because of another factor of American life: its two-party political system and freedom of the press. Under such a system dissenting scientists cannot be totally silenced. The difference in the scientific careers of the Russian professor Boris Porshnev and American Grover Krantz is a telling example.

So American hominologists are bound soon to prove the truth of their discovery, and that without recourse to bones or bodies, the boneheaded critics notwithstanding. Then the truth of discoveries by Russian, Chinese, and Australian researchers will become obvious as well.

This optimistic scenario is prompted also by an important upcoming event: the launching of the **Relict Hominoid Inquiry**, a scholarly peer-reviewed on-line journal and a newsletter, edited by Dr. Jeff Meldrum, associate professor of anatomy and anthropology at Idaho State University, with the assistance of an international editorial board. Jeff Meldrum is the author of the sturdy volume *Sasquatch: Legend Meets Science*, 2006. He boldly took up the torch from the pioneering researcher Dr. Grover Krantz (1931-2002). The project will be funded by John Green, the distinguished veteran of sasquatch investigation and author of books on the subject.

It will be a step in the right direction, for our research is marking time because researchers are not united by a scientifically-specific organization and journal. As a result they endeavor under the thumb, so to speak, of zoologists and anthropologists whose knowledge on the subject is usually very inadequate or simply irrelevant. This situation will end as soon as hominology stands on its feet and plays in science a special role of its own.

Bringing closer these days, let us remember and never forget the behests of our selfless sasquatch contactees: to do all we can to safeguard our wild hominid brethren.

*September 2007*

# An Appeal to the US Government

Concerned with the fate of bigfoot/sasquatch and conditions of their study, Bayanov appealed for help to the then US President Bill Clinton and sent his books to the White House.

To: Mr Bill Clinton
President of the United States

From: Dmitri Bayanov  February 14, 1999
. . . [address]
Moscow, Russia

Subject: Bigfoot

Mr President,

I am a Russian author and researcher. First of all, may I congratulate you on the happy ending of the "bury Clinton" campaign which won no glory to its organizers. (...) Secondly, may I use this occasion to let you in on a "secret" of both American and Russian life. It's revealed in two of my books, "In the Footsteps of the Russian Snowman"(1996) and "America's Bigfoot: Fact, Not Fiction -- U.S. Evidence Verified in Russia"(1997). I am sending you those books, published thanks to political changes in my country, but as you may have no time to read them I'd better tell in brief what it's all about.

You can see it's not a joke from the fact that Theodore Roosevelt gives a dramatic account of an encounter with a creature, later dubbed Bigfoot, in his book "Wilderness Hunter", published in 1892. Bigfoot, or Sasquatch (or Snowman) is part of wildlife both in Russia and America.

The New York Times said on September 23, 1974: "It stands 8 feet, weighs 500 lbs. and walks like a man. Call it Bigfoot or hoax, it's been leaving huge footprints all over the Pacific Northwest for 160 years. ... If Sasquatch does turn out to exist, zoologists and anthropologists will have a great deal to explain."

A Russian scientist, Dr. Boris Porshnev, did provide an explanation 35 years ago, saying that Bigfoot heralds a scientific revolution in anthropology and primatology, plus reforms and revisions in a number of other disciplines. Porshnev was ostracized by anthropologists and viciously attacked personally by some zoologists. When, in 1972, his book, connected with the subject, was banned for publication, he died of a heart attack. He called his endeavors "the struggle for troglodytes."

The fate of his U.S. counterpart, Dr. Grover Krantz, of Washington State University, is fortunately not so tragic, but he also somewhat suffered in his career for telling the truth of this matter.

In 1967, two bold and enterprising Americans, Roger Patterson and Robert Gimlin, managed to catch a Bigfoot on film. The film was declared a hoax by U.S. scientists who had no knowledge on the subject. In 1972, in Moscow, the footage was studied in depth by a team of researchers, myself included, and found to be authentic. Two decades later, Dr. Grover Krantz has confirmed our findings.

Roger Patterson, alas, is long dead, but Bob Gimlin is still with us, and is still being mocked and humiliated in the U.S. for having allegedly participated in a Bigfoot hoax.

So I wonder, Mr President, if you could do something to defend the good name of the late Roger Patterson and the honor of Robert Gimlin. They are both real heroes of science.

Bob Gimlin's address: . . . . [address]

With all good wishes,
Dmitri Bayanov

In August 1999, he received the following answer:

> Thank you for your kind gift and for sharing your thoughts and concerns. It's important for me to know your views. I'm glad you took the time to write.
>
> Bill Clinton

No action followed, but at least the US Government is aware of the situation.

# Image Credits

| Page |  |  |
|---|---|---|
| 4 | Boris Porshnev | I.Bourtsev/D.Bayanov |
| 15 | Koffmann and Bayanov | D.Bayanov |
| 17 | Zhanna/Car | D.Bayanov |
| 18 | Group | D.Bayanov |
| 19 | Expedition members | D.Bayanov |
| 20 | Lizard | D.Bayanov |
| 25 | Jewish Demon | Author's collection |
| 29 | Devil's Footprints (all) | Author's collection |
| 31 | Bald Demon | Author's collection |
| 32 | Reconstruction | Author's collection |
| 34 | Assyro-Bab. Demon | Author's collection |
| 39 | Domovoy | Ivan Bilibin |
| 51 | Jonathan Swift | Public Domain |
| 53 | Map | Author's collection |
| 61 | Yury Krashnikov with donkey | D.Bayanov |
| 61 | Yury Krashnikov | D.Bayanov |
| 63 | Tapestry | V. Kinov |
| 67 | Oleg Ivanov | D.Bayanov |
| 72 | Wildman | Public Domain |
| 73 | Wild woman | Public Domain |
| 74 | Feifei | Public Domain |
| 75 | Sinsin | Public Domain |
| 77 | Gong Yulan | Zou Guoxing |
| 136 | Recreation | BBC |
| 136 | P/G film creature | R.Patterson (Public Domain) |
| 140 | Book Cover | Hancock House/E&M Dahinden |
| 147 | Film Site - C. Murphy | C. L.. Murphy |
| 151 | Logo | Int. Center of Hominology |
| 155 | Book cover | Lydia Bourtseva (artwork) |
| 176 | Haas, Dahinden, Buckley | Author's collection |
| 176 | Jane Goodall | Jane Goodall |
| 209 | Cartoon | Daniel Perez (by Tim Curry) |
| 250 | Gilgamesh epic | Author's collection |
| 250 | Bowl image | Author's collection |
| 251 | Kelermes mirror | Author's collection |
| 252 | Pan | Author's collection |
| 252 | Silenus | Author's collection |
| 253 | Painting – Icon | Public Domain |
| 254 | Persian Illustration | Public Domain |
| 255 | Drollery | Public Domain |
| 255 | Wildman family | Public Domain |
| 256 | Wildman | Author's collection |
| 257 | Wildman | Author's collection |

COLOR SECTION

# Index

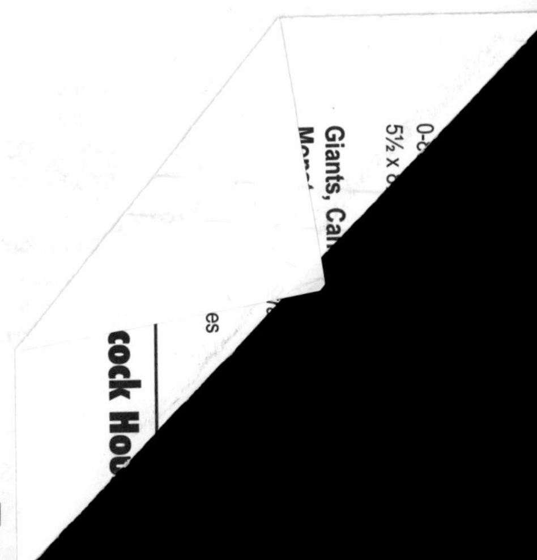

# Other **Hancock House** cryptozoology titles

**Best of Sasquatch Bigfoot**
*John Green*
0-88839-546-9
8½ x 11, sc, 144 pages

**Bigfoot Discovery Coloring Book**
*Michael Rugg*
0-88839-592-2
8½ x 11, sc, 24pages

**Bigfoot Encounters in Ohio**
*C. Murphy, J. Cook, G. Clappison*
0-88839-607-4
5½ x 8½, sc, 152 pages

**Bigfoot Encounters in New York & New England**
*Robert Bartholomew*
*Paul Bartholomew*
978-0-88839-652-5
5½ x 8½, sc, 176 pages

**Bigfoot Film Controversy**
*Roger Patterson, Christopher Murphy*
0-88839-581-7
5½ x 8½, sc, 240 pages

**Bigfoot Film Journal**
*Christopher Murphy*
0-88839-658-7
5½ x 8½, sc, 106 pages

**Bigfoot Sasquatch Evidence**
*Dr. Grover S. Krantz*
88839-447-0
½, sc, 348 pages

**nibals &**

*ain*

**Hoopa Project**
*David Paulides*
0-88839-653-2
5½ x 8½, sc, 336 pages

**In Search of Giants**
*Thomas Steenburg*
0-88839-446-2
5½ x 8½, sc, 256 pages

**In Search of Ogopogo**
*Arlene Gaal*
0-88839-482-9
5½ x 8½, sc, 208 pages

**Know the Sasquatch**
*Christopher Murphy*
978-0-88839-657-0
8½ x 11, sc, 320 pages

**The Locals**
*Thom Powell*
0-88839-552-3
5½ x 8½, sc, 272 pages

**Meet the Sasquatch**
*Christopher Murphy, John Green, Thomas Steenburg*
0-88839-574-4
8½ x 11, hc, 240 pages

**Raincoast Sasquatch**
*J. Robert Alley*
978-0-88839-508-5
5½ x 8½, sc, 360 pages

**Rumours of Existence**
*Matthew A. Bille*
0-88839-335-0
5½ x 8½, sc, 192 pages

**Sasquatch: The Apes Among Us**
*John Green*
0-88839-123-4
5½ x 8½, sc, 492 pages

**Sasquatch Bigfoot**
*Thomas Steenburg*
0-88839-312-1
5½ x 8½, sc, 128 pages

**Sasquatch/Bigfoot and the Mystery of the Wild Man**
*Jean-Paul Debenat*
978-0-88839-685-3
5½ x 8½, sc, 428 pages

**Shadows of Existence**
*Matthew A. Bille*
0-88839-612-0
5½ x 8½, sc, 320 pages

**Strange Northwest**
*Chris Bader*
0-88839-359-8
5½ x 8½, sc, 144 pages

**Tribal Bigfoot**
*David Paulides*
978-0-88839-687-7
5½ x 8½, sc, 336 pages

**UFO Defense Tactics**
*A.K. Johnstone*
0-88839-501-9
5½ x 8½, sc, 152 pages

**Who's Watching You?**
*Linda Coil Suchy*
0-88839-664-8
5½ x 8½, sc, 408 pages

**Yale & the Strange Story of Jacko the Ape-boy**
*Christopher Murphy*
978-0-88839-712-6
5½ x 8½, sc, 48 pages

**use** *titles at* **www.hancockhouse.com**